THE INORGANIC CHEMISTRY
OF
BIOLOGICAL PROCESSES

THE
INORGANIC CHEMISTRY
OF
BIOLOGICAL PROCESSES

M. N. Hughes

Lecturer in Chemistry, Queen Elizabeth College,
University of London

A Wiley–Interscience Publication

JOHN WILEY & SONS
London · New York · Sydney · Toronto

Library of Congress Catalog Card No. 72–5717

Reprinted 1974

Printed in Belgium by Ceuterick, Printers since 1804 B. 3000 Louvain

PREFACE

The overlap region between inorganic chemistry and the biological sciences is one where exciting and significant developments are taking place. It is now appreciated that metal ions control a vast range of processes in biology; that life is really as dependent upon inorganic chemistry as organic chemistry. New developments in instrumental techniques have further accelerated the growth of 'inorganic biochemistry' so that it is now probably true to say that this subject involves one of the most rapidly expanding areas in the chemical and biochemical sciences.

This book is intended to present an introduction to this most important field. It has its origins in third year undergraduate courses at Queen Elizabeth College and is written primarily for chemists, particularly inorganic chemists. The material presented includes a survey of the occurrence and role of the metal ions of biological importance and shows how the function of these ions may be studied experimentally. While most topics of current interest are discussed, the coverage is not intended to be exhaustive. The book does not depend upon a prior knowledge of biological subjects, some relevant material is summarized in Chapter I. It is also hoped that this book may be of interest to workers in the biological sciences, and so, primarily for this purpose, a brief survey of the relevant properties of transition metal complexes is presented in Chapter 2, together with an account of the mechanisms of their reactions in solution.

I am happy to acknowledge the assistance of a number of colleagues and friends; in particular Dr. K. J. Rutt for his helpful comments on the early chapters, Dr. C. W. Bird for his encouragement throughout all stages of writing this book and Miss Jane Cooper for her excellent typing of the manuscript. I am grateful to Professors J. Brachet, J. Coleman and S. Lindskog for permission to reproduce Figures 1.1, 4.3, 4.4 and 4.5 respectively and also to the appropriate Editors as specified in the text.

CONTENTS

CHAPTER ONE

INTRODUCTION

Metal ions play a vital role in a vast number of widely differing biological processes. Increasing knowledge will almost certainly serve to demonstrate this fact more effectively. Some of these processes are quite specific in their metal ion requirements in that only certain metal ions, in specified oxidation states, can fulfil the necessary catalytic or structural requirement, while other processes are much less specific, and it is possible to replace one metal ion by another, although the activity may be reduced.

Metal ion dependent processes are found throughout the Life Sciences and vary tremendously in their function and complexity. Three examples, from biochemistry, physiology and cytochemistry respectively, are given to illustrate this point. Thus the metal ions potassium, magnesium, manganese, iron, cobalt, copper, molybdenum and zinc are all important catalysts of a variety of enzyme reactions such as, for example, group transfer, redox or hydrolytic processes. Not only, however, are these metal ions involved in such processes, but, in certain cases, there are other protein systems involved in storing and controlling the concentration of the metal ion, and then in transporting it to the appropriate site for incorporation into the necessary enzyme system. Sodium, potassium and calcium, on the other hand, are heavily involved in certain physiological control and trigger mechanisms, while potassium, calcium and magnesium ions are all important in maintaining the structure and controlling the function of cell walls. The metals cited in these examples are not the only ones involved in biological processes, others although quantitatively less important, also have biological functions. Of the cited metal ions, Na^+, K^+, Mg^{2+} and Ca^{2+} are present in much greater amounts than the heavy metal ions. Thus, in the human body these four cations constitute some 99% of the total metal ion content.

The physiological and biochemical function of the metal ion in all these processes is obviously a matter of fundamental importance, but the difficulties involved in attempting to clarify their role should not be minimized. Such a study presents many difficult problems, the solutions of which often require an overlap of disciplines. For the Inorganic chemist, without doubt, the field which holds out most hope of effective exploitation is that of the role of metal ions in enzyme and similar systems. It should be stressed that here there are often additional advantages associated with the presence of the metal ion which contribute very markedly to an under-standing of the system. This is particularly true of transition metal ions. Thus, as a result of the electronic properties of the metal, a variety of powerful instrumental techniques may be brought into play. Again the presence of the metal ion provides an extra guide in elucidating the mechanism of the enzyme action, if only in providing an extra check on its correctness in terms of correlating the specificity of the system for that ion in the light of modern awareness of the preferred environment, stereo-chemistry and electronic properties of the ion.

Recent developments in inorganic and organometallic chemistry have resulted in a very significantly increased understanding of the bonding, structure and reactivity of coordination compounds. This has been practically reflected in certain areas of inorganic chemistry, for example that of the activity of small molecules such as CO, H_2 and olefins on coordination to a metal, and this is leading to an increased understanding of certain important catalytic processes. Equally, however, the border area between inorganic chemistry and the Life Sciences should present a challenge to the inorganic chemist to apply his increased understanding in the design of model systems that throw light on the behaviour of metal ions in biological processes, and ultimately to look more closely at these processes themselves. These developments in inorganic chemistry have been matched by developments in biochemistry in that it is now possible from the biochemical point of view to consider processes at the molecular level. Certain progress in the field of metal ion activated enzymes has been made, particularly in the case of metallo-enzymes where the metal is firmly bound to protein. Most metal cations in living organisms will in fact be associated with proteins and so the subject of metal-protein binding is a most fundamental one in the overall context of this book.

BACKGROUND MATERIAL

It is necessary at this stage to introduce some background material in order to explain the terms and concepts used in later chapters. Much of this material is necessarily presented at an elementary level.

Amino acids, peptides and proteins

The proteins are macromolecules of great biological importance. They are made up of α-amino acids of L configuration, linked together *via* peptide bonds —CONH—. By suitable treatment they can be degraded to smaller peptides and finally to the constituent amino acids. Some twenty of these amino acids are found in nature, together with the α-imino acids

TABLE 1.1 Naturally occurring amino acids

$$\begin{array}{c} R-CH-COOH \\ | \\ NH_2 \end{array}$$

R	
$H-$	Glycine
CH_3-	Alanine
$(CH_3)_2CH-$	Valine
$(CH_3)_2CHCH_2-$	Leucine
$CH_3CH_2\underset{\underset{CH_3}{|}}{CH}-$	Isoleucine
$HOCH_2-$	Serine
$CH_3\underset{\underset{OH}{|}}{CH}-$	Threonine

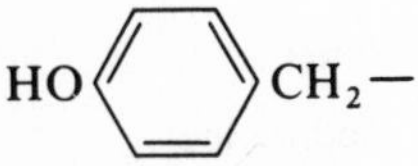

Tyrosine	
Phenylalanine	
Tryptophan	
$^-OOCCH_2-$	Aspartic acid
$^-OOCCH_2CH_2-$	Glutamic acid

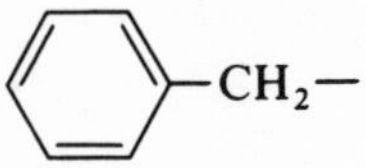

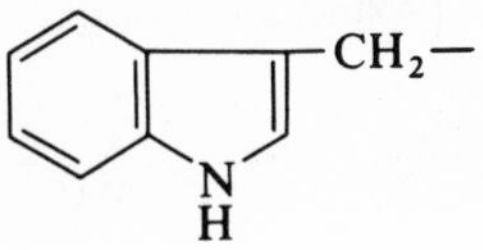

Glutamine

Asparagine

$\overset{+}{N}H_3(CH_2)_3CH_2-$	Lysine

TABLE 1.1 (cont.)

R	
$\underset{H_2N}{\overset{H_2N}{\diagdown}}\hspace{-0.3em}\underset{+}{C}\!=\!NH(CH_2)_2CH_2-$	Arginine
[imidazole ring]—CH_2-	Histidine
$\overset{+}{N}H_3CH(CH_2)_2CH_2-$ with OH	Hydroxylysine
$SHCH_2-$	Cysteine
$CH_3SCH_2CH_2-$	Methionine
$\overset{+}{N}H_3$ $\diagdown$ $CHCH_2SSCH_2-$ $\overline{O}OC$ $\diagup$	Cystine
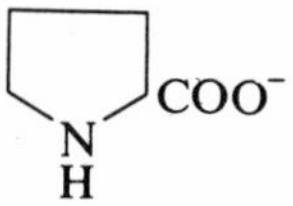 [pyrrolidine ring]COO^-	Proline (imino acid)
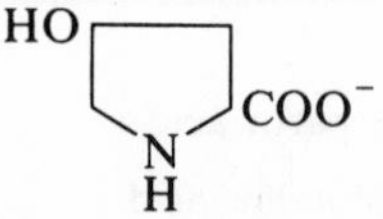 $HO-$[pyrrolidine ring]COO^-	Hydroxyproline (imino acid)

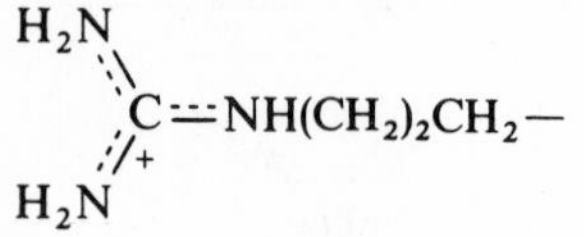

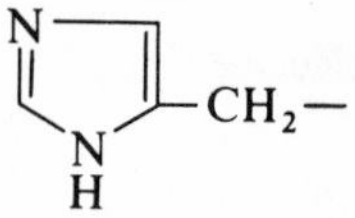

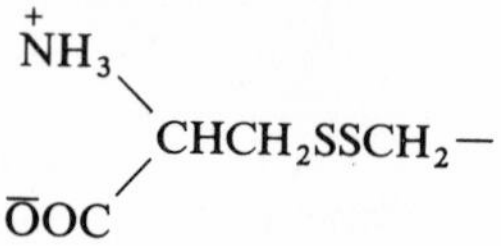

proline and hydroxyproline. These are listed in Table 1.1. Each naturally occurring polypeptide or protein involves a specific sequence of amino acid residues, which may be determined by chemical and biochemical methods. The sequence of amino acid residues will determine the physical and chemical properties of the protein in terms of the chemical and physical interactions occurring between the side chains R in $NH_2CH(R)COOH$. The important side chains in this connection are those involving aromatic groups, sulphur-containing groups, and $-NH_2$, $-OH$ and $-COOH$ groups. The nature of the residues may generate hydrophobic or hydrophilic environments in certain regions of the protein chain.

Each amino acid has at least two ionizable groups, the amino and carboxyl groups. The $-NH_3^+$ group is less acidic than the carboxyl group and so, between pH 4–9, the amino acid exists as a zwitterion, $H_3N^+CH(R)CO\bar{O}$. The α-carboxyl and amino functions are involved in the formation of the peptide link, and so each peptide has terminal $-NH_2$ and $-COOH$ groups, together with peptide links and side chains. These are all possible metal binding sites. Certain low molecular weight peptides are biologically important molecules.

Proteins

The molecular weights of proteins are in the range of 10^4–10^6 g. The determination of the chemical structure and the spatial configuration of proteins is a very important matter, as this is bound up with the biological function of the proteins. At the present time an increasing number of protein structures have been determined to a high degree of resolution by X-ray diffraction techniques. This is the only method for determining the complete structure.

The structure of proteins is discussed in terms of primary, secondary, tertiary and quaternary structure.

Primary structure

This is the sequence of amino acid residues in the chain. For a particular enzyme some residues are less important than others and may, in fact, in enzymes from different species, be replaced by other residues.

Secondary structure

This is concerned with the configuration of the protein chain that results largely from hydrogen bonding between peptide links. Pauling and Corey

$$\begin{array}{c} \diagdown \\ \diagup \end{array} C{=}O{\cdots}HN \begin{array}{c} \diagup \\ \diagdown \end{array}$$

began their classic study of this problem by determining the structures of a range of simpler compounds, including amides. This has demonstrated that for a stable secondary structure the peptide link is always planar; that the carbon atoms each side of the peptide link are *trans* to each other, so lowering repulsive forces; and finally that there is a maximum amount of hydrogen bonding between carbonyl oxygen and amide nitrogen atoms.

This hydrogen bonding may either be intramolecular or intermolecular. In the first case, it gives rise to an α-helical structure and in the latter case to a pleated sheet structure. X-ray studies have confirmed the α-helix structure and have shown that there are 3·7 residues per turn of the helix. Helical structures occur in globular and fibrous protein. Pleated sheet structures may involve peptide chains having all N-termini at one end or with every other N-terminus at one end. These are the parallel and anti-parallel forms respectively.

Secondary structure is also dependent upon the nature of the side chains in that interactions between these may, for example, lower the stability of the α-helix through repulsion.

Tertiary structure
This is concerned with the way in which the protein chain (with its secondary structure) folds upon itself, and the resulting shape of the molecule. Thus it may be globular or rod-like. This results from the interactions between side chains, and may reflect the formation of S—S bonds between cysteine residues, hydrogen bonding between side chains, so called 'salt' or ionic linkages between oppositely charged groups such as $-NH_3^+$ and $-COO^-$ and hydrophobic interactions between aromatic residues. It appears that the last named of these is most important.

The overall shape of the protein molecule may be studied in a number of ways. The most powerful technique is of course that of X-ray diffraction, but other techniques include the measurement of the intrinsic viscosity and the light scattering properties of the macromolecule. X-ray diffraction will also confirm or determine the primary structure.

A schematic representation of the structure of myoglobin is given in Figure 1.1. This globular protein is made up of eight helical segments

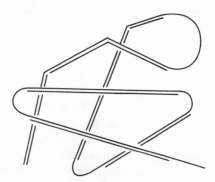

Figure 1.1   Schematic representation of the structure of myoglobin (the double lines indicate α-helical segments).

separated by regions of random coil. It has a high α-helix content of 77 per cent, and has no S–S bridges. It is rather atypical in this.

Quaternary structure

Many proteins are made up of linked subunits held together by other than covalent bonds. The quaternary structure reflects the way in which these subunits come together. Hemoglobin is made up of four such units, and substantial changes in interaction between these occur when oxygen is taken up or lost.

Proteins in solution, denaturation and other topics

One important question is the extent to which protein structures, determined in the solid state, change in solution. Only if there is good reason for assuming that little change has occurred can we extrapolate the results of X-ray structure determinations to the problems of enzyme mechanisms in solution. It appears, however, that solution and crystal structures are similar. Thus, the alkylation of side chains in myoglobin proceeds readily for those residues which the X-ray determinations show to be exposed and does not proceed for those deep inside the molecule.

It is clear from the preceding paragraphs that the secondary and tertiary structure of the protein is dependent upon a variety of interactions between various groups in the protein molecule. On heating or change of pH some of these interactions may be affected. The addition of solvents may break down some hydrogen bonding, while treatment with reducing agents may break S–S linkages. Clearly, therefore, proteins are readily subject to changes which convert an ordered structure to a disordered one. This is termed denaturation. Sometimes these changes are reversible and sometimes they are irreversible. Denaturation results in a number of changes in the chemical and physical properties of a protein. The sensitivity of proteins to denaturation is a major difficulty in experimental work.

Much work has been carried out on synthetic polypeptides such as poly-L-alanine and this has provided a basis for the study of naturally occurring proteins; for example, the value of certain physical techniques in measuring the α-helical content of proteins.

The determination of the helical content of proteins is an important measurement. Optical rotatory dispersion (O.R.D.) studies have been widely used, as the helix itself contributes to this in addition to the constituent amino acids. Other approaches have involved the use of infrared dichroism studies using plane polarized infrared light and hydrogen-deuterium exchange. Standard texts should be consulted for a full account of these techniques.

The use of models and the study of synthetic polypeptides with only one type of amino acid residue have also contributed to an understanding of the secondary structure of proteins. Thus it has been demonstrated that a mixture of D and L amino acid residues is not compatible with the α-helix. Similarly the requirements of the α-helix structure cannot be met by proline residues and so these must terminate the α-helix configuration. van der Waals' interactions between substituents such as in valine and isoleucine will also lower the stability of the helix.

One important property of proteins is the fact that they have a number of ionizable groups that may be involved in acid–base equilibria. These are the side chains of aspartic and glutamic acids, arginine, cysteine, histidine, lysine and tyrosine. As a result of the large number of titratable groups present in each protein the interpretation of protein titration curves is complex. In addition, the environment of the protein (Hydrogen bonding, medium effects) may well affect the pK_a of the amino acid residue so that it is greater or less than the value for the free acid by up to one pK unit. Certain residues may also be buried within the interior of the protein molecule and so not be accessible during the titration. By carrying out a back titration it is possible to discover if denaturation has occurred, resulting in inaccessible groups now being available. However, in favoured cases the analysis of titration curves has allowed the estimation of the number and type of the protonation sites available.

Enzymes

These are proteins which by reason of their particular three-dimensional structure are able to act as highly specific biochemical catalysts. The catalytic effect is considerable. Thus sometimes rate constants for enzyme catalysed reactions and model reactions differ by as much as a factor of 10^{12}. A number of explanations have been suggested for this, and we shall consider some later. Because the enzymes are proteins the general problems of protein stability hold here, and therefore pH and temperature must be carefully controlled.

In general the cell requires a different enzyme for each of its reactions, although a limited number will catalyse reactions of a general type. The esterases will thus catalyse ester hydrolysis. The need for such high specificity can readily be seen. Life processes usually consist of a series of interrelated complex reactions. Often the product of one reaction becomes the starting material for another. The reactions must be very specific in order to avoid complications from other simultaneously occurring systems.

The need for extra factors in an enzyme reaction can often be shown by the process of dialysis in which the enzyme solution, in a cellophane container, is suspended in distilled water. The cellophane pores allow the low molecular weight components to diffuse through into the surrounding water, leaving the protein molecules in the cellophane container. If the enzyme depended upon any of the low molecular weight substances it will fail to function until they are added back.

The enzyme activators can be metal ions or complex organic molecules such as nucleotides or certain B vitamins. These are termed co-enzymes and are bound to the enzyme protein, being removed on prolonged dialysis. Sometimes a co-enzyme is bound so firmly that it is not removed by dialysis, in which case it is termed a prosthetic group.

The mechanism of enzyme action

An old established analogy is that of the lock and key. This represents the complex three-dimensional relationship between an enzyme and the substrate on which it acts, a relationship which may require the incorporation of certain cofactors or activators before it is complete. The result of this is to activate the substrate so that it is able to react in the required way. This analogy is still useful in that X-ray studies on enzymes have shown, in all cases, the presence of a cleft at the active site into which the substrate must fit. The lock and key analogy is, however, inadequate in that it does not indicate the other effects which occur; for example, the substantial conformational changes that result when the substrate is bound to carboxypeptidase. The active site itself, that portion of the protein chain involved in the interaction with substrate, will only involve a few residues, although of course they may be from well separated parts of the protein chain as a result of protein folding. It is the configuration of the protein around this portion that provides the cleft into which the substrate must fit to become activated. This explains the specificity of the enzyme. Why, for example, an enzyme may only be able to attack certain optical isomers. Thus malate dehydrogenase oxidizes L-malate exclusively to oxaloacetate in the presence of co-enzyme 1 (NAD), while D-malate is unaffected, thus providing an effective method for obtaining the D-isomer from the DL form. The role of inhibitors can also be understood in a general sense, in terms of their modifying the overall molecular shape at the active site so that the substrate may not fit into it, or by their competing with the substrate for the active centre of the enzyme.

However, the portion of the enzyme known as the active site may not only act by generating a certain steric specificity. The nature of the side chains is also important in terms of their hydrophobic or hydrophilic properties. It appears that the active site is surrounded always by non-polar residues, thus providing an environment at the active site which is of lower dielectric constant than that of the aqueous solution in which the enzyme is found. This means that a number of interactions are different from what might have been expected; pK_a values are affected, ionic interactions will be stronger. All of these will be necessary for the correct functioning of the enzyme. Again, the active site provides the directional hydrogen bonding and van der Waals interactions to bind the substrate to the enzyme. In a number of cases X-ray studies have suggested that the initial interaction between substrate and enzyme induces further interactions that cause other parts of the enzyme to close in upon the substrate.

The actual source of catalytic power has been ascribed to a number of causes such as proximity and orientation effects and electron push-pull effects. It is, however, difficult to put these on a quantitative basis. One recent suggestion which has aroused some controversy is that of orbital steering, i.e. an orientation effect involving the orientation of orbitals in the reacting atoms of enzyme and substrate.

The cell

The complexity of the cell is illustrated by the fact that not only are there many celled organisms, but there are also one-celled species, such as bacteria, viruses and moulds whose behaviour must be entirely accounted for by the activities occurring within that single cell. The following comments apply, in general terms, to animal cells and unicellular plants (green plants will also contain chloroplasts, associated with photosynthesis) but not to bacterial cells which differ in many respects, e.g. they do not possess a nucleus and therefore belong to the prokaryotic class of cells.

Cells, in general, are encased by a membrane, the function of which is extremely important and which is dependent on metal ions. This membrane is selectively permeable to different metal ions (and other species) and this is readily associated with the function and distribution of, for example, the s block metal cations. Thus Mg^{2+} and K^+ are concentrated in the cell by the action of the membrane, while Na^+ and Ca^{2+} are rejected by it. This is associated with the utilization of Ca^{2+} as a structural factor in teeth, bones and shells, and as an activator of extracellular enzymes, while Mg^{2+} and K^+ are associated with intracellular processes, both as structural stabilizers and also as enzyme activators. The function of the membrane will be discussed at length in a later chapter. Plant cells are surrounded by an additional wall of cellulose giving extra rigidity.

The complex internal structure of the cell has been demonstrated by electron microscopy. Figure 1.2 shows a typical cell. In the cytoplasm of the cell are a number of structures, whose existence must be correlated with the various degradative and synthetic pathways involved in the

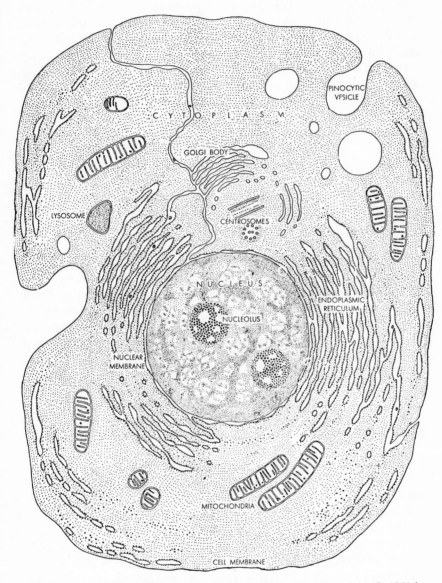

Figure 1.2   A typical cell (from *The Living Cell*, Brachet. Copyright © 1961 by Scientific American, Inc. All rights reserved).

working of the cell. All cells have a nucleus which contains the cell DNA, a little RNA and certain associated proteins, together with relevant enzyme systems such as RNA and DNA polymerases and those associated with membrane synthesis. The nucleus occupies an appreciable part of the cell volume and is surrounded by its own membrane. The remaining cytoplasm contains a large number of particles, the mitochondria, which are smaller than the nucleus by a factor of up to a hundred, while the general body of the cell contains the tubules of the endoplasmic reticulum and small groups of particles, the ribosomes, which contain most of the cell RNA.

Many types of cells also contain a large number of hydrolytic enzymes, such as DNA-ases, phosphatases and esterases, which are capable of destroying the cell components, but which are kept isolated by a membrane. Relatively large concentrations of the s block elements are associated with structural and functional aspects of the living cell, together with smaller quantities of the trace elements which are involved in enzyme activation. Other inorganic ions are also found in the cytoplasm, as are other enzyme systems.

The distribution of materials in the cell, the control of the influx of reactants and the efflux of products is dependent on the properties of the membranes of the cell and of the cell components. The properties of cell membranes have received much attention and are the subject of some controversy. However it appears that they are made up of a lipid layer, two molecules thick and surrounded above and below by protein.

The cell components

It is obviously of interest to separate all the components of the cell and to observe their specific function. The biochemist attempts to do this by rupturing the cell membrane and applying the technique of centrifugation, the various particles thus being separated in turn. The separated cell fractions can then be examined independently for their enzymatic activity.

Of great biochemical interest are the mitochondria. They contain several sets of enzymes and cofactors not found elsewhere in the cell. In this way, in the mitochondria, all the enzymes and cofactors for certain important cycles, such as the breakdown of pyruvic acid by the citric acid cycle, are kept together in an highly organized system. Each mito-chondrion is surrounded by an outer membrane. There is also an inner membrane with an irregular folded-type structure. Along this inner membrane is distributed a complex system involved in oxidation and phosphorylation. The complexity and interlinking of various biochemical processes certainly implies, in this case, a highly organized system of enzymes, co-enzymes and electron carriers. This is provided by attaching these systems to the mitochondrial membrane. The structure of the internal membrane increases the capacity of each mitochondrion for this ordered system. The result is a rigid system ideally arranged, for example, for hydrogen carrying, one that would not be achieved anywhere near so ideally if enzymes and cofactors were mixed in solution. The outer mitochondrial membrane is associated with such enzymes as succinic dehydrogenase and NAD dehydrogenase.

The control of the mammalian cell

The behaviour of the cell has to be regulated. As we have noted, the membrane plays a very important role in this by its selectivity towards outside species. However, there are external stimuli, such as nerve impulses and certain special chemicals (the hormones) secreted by the glands, which will affect the behaviour of membranes. Several of the hormones are important in controlling the concentrations of metal cations and other inorganic groups retained or disposed of through the urine *via* the kidneys. Thus the mineralo-corticoids, a group of adrenal cortical steroid hormones, are involved in increasing the loss of K^+ and decreasing the loss of Na^+. The parathyroid hormone stimulates the excretion of calcium and phosphate at the expense of bone calcium phosphate, while calcitonin inhibits the release of calcium.

The chemical reactions undergone in the cell are catalysed by the enzymes. The requirement for the appropriate coenzymes and cofactors is another controlling factor.

METALLOPROTEINS AND METAL-PROTEIN COMPLEXES

Most naturally occurring metal ions are bound by proteins. The question of the nature and extent of metal ion-protein interaction is therefore a most important one. It has been the subject of much work using amides, amino acids and small peptides as model compounds, as discussed in Chapter 3.

It is possible to classify metal protein systems under two headings. The classification is not an absolute one but is nevertheless a useful general guide to the nature of metal-protein interaction.

Metalloproteins (including metalloenzymes)

Here the metal ion and protein are firmly linked together so that the metal ion can be regarded as an integral part of the protein structure, from which it cannot be separated except by extreme chemical attack. The activity of the metalloprotein is usually lost if the metal is replaced by another.

Metal activated proteins or metal–protein complexes (including metal activated enzymes)

This group involves those examples where the metal ion is combined reversibly with the protein. Systems such as these are much less amenable to study than the metalloproteins for which it is usually possible to say something about the site symmetry of the metal and the nature of the binding groups that hold it to the protein. The metal ion-binding group interaction for this second class will of course be much less than that involved in the first class. This will be an important factor in the mechanisms of the reactions catalysed by these metal activated enzymes. A suitable example is the role of Mg^{2+} in the hydrolysis of ATP by the appropriate enzyme. The binding of metal ions to free ATP is through the phosphate groups. The hydrolysis of ATP does not change considerably from metal to metal in model systems. However, the state of affairs is very different in the enzymatic reaction. Cd(II), Ni(II) and Co(II) have an inhibitory effect while Mg(II) and Mn(II) have a catalytic effect. This phenomenon demonstrates, first of all, the importance of the role of the protein. Catalysis is a reciprocal effect, not just one due to the presence of the metal ion. Secondly, it is probably safe to say that the inhibition by certain metals is due to their blocking of a certain site on the enzyme which is necessary for catalysis, while the weaker binding of the metals magnesium and manganese has resulted in this site being free.

Specific tests can be applied to ascertain to which type any particular protein-metal system belongs. For metalloproteins (and metalloenzymes) the ratio of metal to protein tends to a constant value as the purity of the system is increased, finally becoming independent of further purification when in a highly pure state. This ratio will then be the stoicheimetric ratio of metal to protein molecule. The ratio may vary from one mole of metal ion per mole (such as zinc in carbonic anhydrase) up to, for example, eight moles of metal ion per mole, such as is the case for copper in ascorbic acid oxidase. Often for copper the metal ions occur in pairs, while enzymes such as aldehyde oxidase, and xanthine oxidase involve two different metal ions, in this case 8 moles of iron and 2 moles of molybdenum. Another example is that of ferredoxin where there are seven iron atoms per mole of protein (M. Wt. 6,000). Other examples are listed in the Tables of other chapters. Enzymes such as these, having a number of metal ions, have been termed 'many headed metalloenzymes'.

In the case of the metal activated proteins, metal ion analysis will vary as the metal can be removed very easily, e.g. by dialysis. This will result in reduced activity, but it is usually readily recovered on adding the metal ion again. The specificity in these cases, towards the metal ion, is usually much lower than that of the metalloproteins. Metal activated enzymes show a similar lower metal ion specificity and are usually easily inhibited by adding certain reagents. This in itself can provide useful mechanistic clues.

It is only possible to inhibit the action of a metalloenzyme by the addition of ligands of high activity for the metal ion and for its particular stereochemical environment. This is reversible when the metal still remains partly bonded to the protein molecule but it is often irreversible when the metal is removed completely from the protein. The resulting 'apoenzyme' (protein minus metal) can, however, in certain cases be reconstituted by adding the metal previously removed from the protein, but it can be very difficult to reconstitute with full activity.

The binding of metals in metalloproteins is a difficult question to resolve with certainty. There are many alternative donor sites, the side chains, peptide and terminal NH_2 and COOH groups. It is reasonable to suppose, however, that certain groups will have particularly enhanced basic properties and tend to dominate the competition among the various potential binding sites for the metal. Two common amino acids have such outstanding binding properties—these are histidine, through its imidazole ring, and cysteine through the thiol group. So the overall picture is

Histidine   Cysteine

probably less complicated than it might seem. There are many potential donor sites, but the cation must bind quite strongly to particular sites and only weakly to other sites to avoid having a number of alternative coordination compounds formed. Different metal ions would, of course, not necessarily be bound to the same ligand groups on the protein. For certain cases the full amino acid sequence of the protein molecule has been determined and the metal binding groups pinpointed with some certainty. Again, X-ray analysis is most important. In some cases it is possible through the techniques of transition metal chemistry to make reasonable predictions regarding the metal environment. These methods are discussed in Chapter 3.

THE ROLE OF THE METAL IN METAL-PROTEIN SYSTEMS. SOME EXAMPLES

The first point that must be stressed is that the relationship between metal and protein is a reciprocal one. The presence of the metal ion can influence the electronic and structural arrangement of the protein and so affect its reactivity. But the very fact that a complex protein rather than a simple structure is required indicates that the protein is equally important. So the protein can enforce unusual stereochemistries upon the metal ion in that the protein internal structural requirements do not allow it to provide a normal symmetrical binding site, or even normal metal-ligand distances, in some cases. This, in turn, can affect the reactivity of the metal. Thus the irregular stereochemistry forced on the iron atom in myoglobin makes the addition of oxygen to the iron atom a more energetically attractive process in that the iron is already in the transition state structure for a certain type of ligand substitution reaction.

Metal ions can act in a number of different ways,

(a) in trigger and control mechanisms,

(b) in a structural context, including 'template' reactions,

(c) as Lewis Acids,

(d) as Redox catalysts.

We shall briefly treat these in turn, prior to a full treatment in Chapter 3. It should be emphasized that one should not always attempt to limit the behaviour of any one metal ion in an enzyme reaction to one of these roles. It should also be noted that there are a number of metal proteins which appear at present to have no known biological function. It could well be that these are storage proteins, there being many such systems having this role. The storage, control and supply of certain factors in life processes is a critical one. Such proteins as conalbumen (in egg white) and transferrin (in blood plasma) reversibly bind and transport iron, while ceruloplasmin is a copper-carrying protein. The protein ferritin can store iron and then transport it to an appropriate site to aid in the biosynthesis of other molecules involving this particular metal. Other proteins control the concentrations of calcium, magnesium, zinc and possibly other metals.

Certain other metal-containing macromolecules are coenzymes. The best known examples are those of the cobalt-containing vitamin B_{12} series. Here cobalt is bound in the centre of a tetrapyrrole ring, this being the basic structure of the cobamide family of coenzymes.

(a) Trigger and control mechanisms

The cations Na^+, K^+, Mg^{2+} and Ca^{2+} are associated with a number of control and trigger mechanisms. Calcium is particularly important in controlling the permeability of semipermeable membranes which become porous in its absence. A specific example is its role in the excitability of nerve cell membranes.

Nerve impulses can be described as electrical impulses conducted along the membrane. This is associated with the rapid influx of Na^+ into the nerve cell resulting from the temporary reversal of the membrane's selectivity towards Na^+ and K^+. The sudden influx of calcium ions into muscle cells provides an example of a trigger effect. Normally calcium is excluded, but the presence of calcium ions results in the activation of certain enzymes and hence muscle contraction.

(b) Structural influences

Complex formation with metal ions could stabilize certain protein configurations and so affect the physical and biological properties of the protein. Metal ions also affect the structure and function of nucleic acids and nucleotides by structure stabilization and by their role in the un-winding and rewinding of the double helix.

The metal ions, calcium(II), magnesium(II) and manganese(II) can influence the equilibrium between native and reversibly denatured protein. Again, more specifically in enzyme behaviour, complex formation with a metal could be necessary to bring together certain parts of the protein chain which would normally be some distance apart in the free protein, and so constitute an active site. Some of the 'non specific' activations of enzymes by elements such as K^+ can probably be related to the ion organizing the protein in some way. Potassium appears to be particularly important as a stabilizing cation.

The simplest function for a metal ion is purely that of a template to bring the reacting groups into the correct relative orientation for reaction. In this case the metal will normally be bonded to both protein and substrate, other than either only to the protein or only to the substrate. The use of metal ions as templates in the synthesis of certain complex organic compounds has been demonstrated. Thus the two halves of corrin were synthesized separately and then cyclized by use of a nickel(II) template (Figure 1.3).

If metal ion cofactors in enzyme reactions have this type of role then it ought to be possible to understand the specificity of the metal ion in terms of the stereochemistry of non-enzymatic model compounds and the binding strength of different metals for certain groups. The similarity between Mg^{2+} and Mn^{2+} as enzyme activators can be understood on such a basis.

Metal ions are also important in maintaining the structure of cell walls. Calcium and magnesium are both metal ions whose concentrations are higher than that expected if their role was only that of enzyme activator.

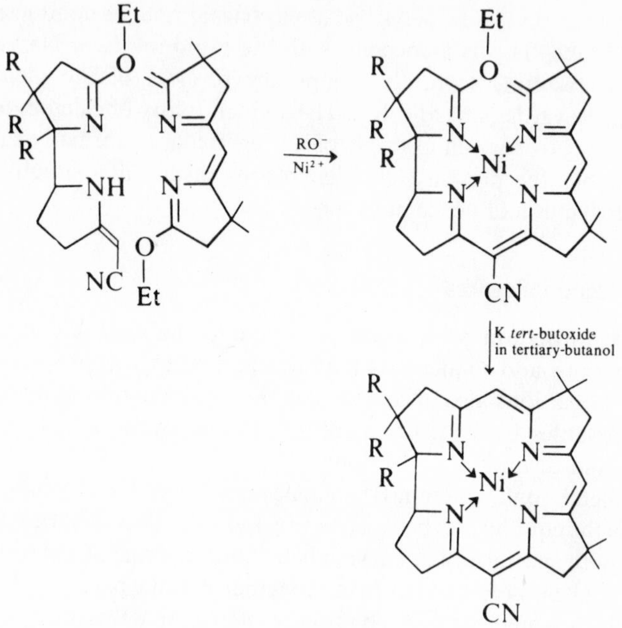

Figure 1.3   The use of nickel(II) as a template ion.*

These divalent metal ions could produce a stiffening mechanism for lipoprotein membranes by bridging neighbouring carboxylate groups.

(c) Lewis acid behaviour

Metal ions can accept electron pairs and so act as Lewis Acids. Different metal ions have different Lewis Acid strengths, which will clearly increase with the charge on the metal and with decreasing ionic radius. For the transition metals other factors are also important, and so for a series of divalent ions the following order generally holds:

$$Mn^{2+} < Fe^{2+} < Co^{2+} < Ni^{2+} < Cu^{2+} > Zn^{2+}$$

This order of activity is observed for the reactions of most model compounds involving Lewis Acid catalysis by metal ions, but does not hold completely for metal ion catalysis of certain enzyme reactions. The breakdown of this order is significant and will be discussed later.

The mechanistic role of the metal ion then is one of general acid catalysis, but differs from proton catalysis in that (a) the metal ion can coordinate to

* A. Eschenmoser, R. Scheffold, E. Bertelle, M. Pesaro and H. Gschwind, *Proc. Roy. Soc. A*, 1965, **288**, 306.

several ligands simultaneously, and (b) in that metal ion catalysis is possible in pH ranges where proton catalysis would be ineffective.

An example of an organic reaction catalysed by a metal ion in this way is the decarboxylation of dimethyl oxaloacetic acid. This is catalysed by ferric ion. The scheme is given below:

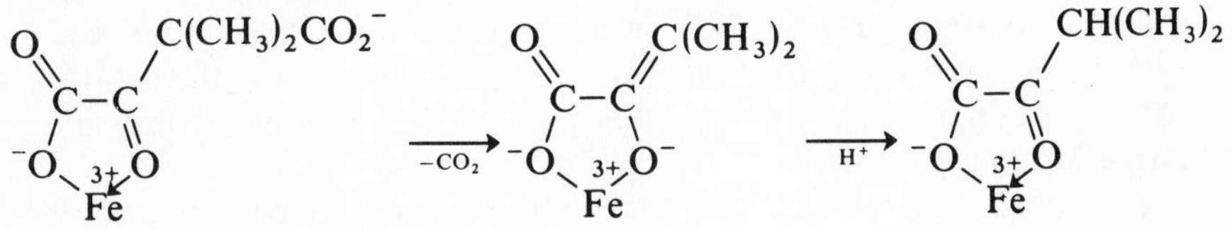

The enol chelation is confirmed by the appearance of a blue colour, due to a charge-transfer band, that is characteristic of ferric enolate complexes. The role of the metal ion can be easily understood, the Lewis Acid behaviour of Fe^{3+} allowing the CO_2 group to leave more easily.

Zinc and cobalt are good examples of strong Lewis Acid catalysts, for example, in the hydrolysis of phosphates (by phosphatases) and esters (by esterases), while Mg^{2+}, Ca^{2+}, Mn^{2+}, as has already been seen, are good catalysts for substrates involving weaker base centres, such as poly-phosphates.

Magnesium is particularly well known as an activator of those enzymes associated with phosphate systems. The formation of a phosphate ester is an important biochemical reaction and is often the first step in a complex synthetic sequence. Mono-, di- and triphosphate esters can all be formed, the biochemical activity of the triphosphate being greatest.

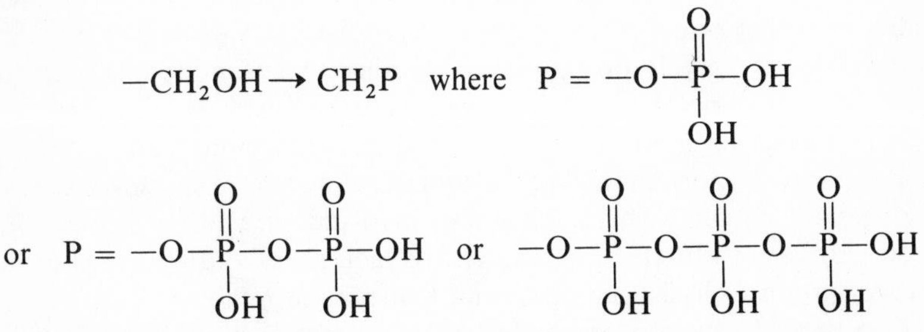

$$-CH_2OH \rightarrow CH_2P \quad \text{where} \quad P = -O-\overset{\overset{\displaystyle O}{\|}}{\underset{\underset{\displaystyle OH}{|}}{P}}-OH$$

$$\text{or} \quad P = -O-\overset{\overset{\displaystyle O}{\|}}{\underset{\underset{\displaystyle OH}{|}}{P}}-O-\overset{\overset{\displaystyle O}{\|}}{\underset{\underset{\displaystyle OH}{|}}{P}}-OH \quad \text{or} \quad -O-\overset{\overset{\displaystyle O}{\|}}{\underset{\underset{\displaystyle OH}{|}}{P}}-O-\overset{\overset{\displaystyle O}{\|}}{\underset{\underset{\displaystyle OH}{|}}{P}}-O-\overset{\overset{\displaystyle O}{\|}}{\underset{\underset{\displaystyle OH}{|}}{P}}-OH$$

Almost all enzymes that are involved in phosphate transfer require a metal, but magnesium is the most important. It is possible that the role of the metal ion may also be one of charge neutralization. Adenosine triphosphate is usually the substrate.

Examples of ATP-ases which require magnesium as a cofactor are myosin, which cleaves ATP to ADP (Adenosine diphosphate) and inorganic phosphate; and hexokinase which is involved in the phosphorylation of glucose.

$$\text{glucose} + \text{ATP} + \text{hexokinase} \xrightarrow{\text{Mg}^{2+}} \text{glucose-6-phosphate} + \text{ADP}.$$

This is a typical 'group transfer' reaction.

Again, Mg^{2+} is required in all stages of the biosynthesis of vitamin B_1, thiamine pyrophosphate (TPP, cocarboxylase) from pyrimidine and thiazole moieties and then in its enzymatic reactions, such as the decarboxylation of pyruvic acid. The first step in the biosynthesis involves the formation of a phosphate.

Zinc is also associated with hydrolytic enzymes, in particular with phosphatases, peptidases and esterases. Zinc-containing systems have been studied with some success because it is often possible to replace zinc by a range of transition metal ions and hence obtain useful comparative information. Detailed reference will be made in Chapter 4 to the role of zinc in certain systems such as carbonic anhydrase in order to illustrate the usefulness of this approach. Carbonic anhydrase is vital for respiration in animals as it catalyses the normally slow carbonic acid–carbon dioxide reaction.

(d) Redox behaviour

In these reactions redox changes in the metal ions catalyse valence changes in the substrates, such as in the nitrogen cycle. Here the specificity of the enzyme for the metal tends to be much higher. The transition metal ions are involved in a wide range of catalytic functions, while zinc also has a role as a catalyst of hydride transfer. A characteristic feature of these redox metalloenzymes is the irregular stereochemistry associated with the metal. This is very important in explaining their function. The redox potentials of the metal ions in these metalloenzymes are also of interest and may be related to the irregularity of the metal environment.

The more important metals in biological redox processes are iron, copper and cobalt with molybdenum involved to a lesser degree. The processes vary from electron transfer, oxygen atom and hydroxyl group incorporation to hydrogen atom and hydride ion removal.

Copper and iron are extremely important metals in biology. Both are involved in respiratory processes; iron in hemoglobin and myoglobin, and copper in blue hemocyanin are oxygen carriers, for example. Iron is a component of various cytochromes, peroxidases and catalase, which are all porphyrin enzymes. There are also iron transport enzymes (ferritin and transferrin) and concentration-controlling enzymes (transferrin). Ferredoxin is a non-heme iron–protein complex of considerable importance.

It is found, for example, in the chloroplasts of green plants and is the initial electron acceptor of the photoactivated chlorophyll molecule.

The copper proteins have been well studied. These include hemocyanin and ceruloplasmin. The latter is not an oxygen carrier but shows important oxidase activity. A number of substrates are reportedly oxidized by it, including p-phenylenediamine, ascorbate and ferrous iron. It has been suggested that a possible physiological role for ceruloplasmin is to incorporate iron into the protein constituent of transferrin (apotransferrin) and so facilitate the formation of hemoglobin and other respiratory pigments. Certainly there appears to be a link between the presence of ceruloplasmin and a requirement for iron incorporation. A major problem in the copper proteins is that of assigning a formal oxidation state to the metal; this will be discussed later in terms of the application of modern techniques.

Molybdenum is involved with several redox enzymes; it is well known for its association with nitrate reduction in the nitrogen cycle. Here it is thought that the metal changes from oxidation state V to oxidation state VI. It is also well known as being essential for nitrogen fixation in plant root nodules. These will be discussed in a later chapter.

Another transition element, manganese, does not in fact appear to be involved in redox reactions. As manganese(II) it is essential for the activity of many degradative enzymes. Manganese(II) is very similar to magnesium(II) and often can replace that metal. This has been used with some profit as manganese(II) can be studied by e.p.r. and other techniques. Manganese, and possibly copper, are involved in photosynthesis in that they appear to be essential for the biosynthesis of chlorophyll.

Even though it is not a transition metal, zinc is also found in certain dehydrogenases. Its role here is probably one of catalysis of hydride transfer. It is known in organic chemistry that pyridinium compounds react as if the *para*-carbon were positively charged; thus they can add hydride ions. In biological systems the most common pyridinium compound which carries out this reaction is the pyridine nucleotide coenzyme NAD^+ (nicotinamide adenine dinucleotide).

NAD⁺   NADH

A similar state of affairs holds for flavin co-enzymes. A metal cation will be able to stabilize the reduced form of the pyridinium or flavinium molecules by co-ordination.

Many NAD and flavin enzymes which are involved with hydride transfer are found associated with zinc and sometimes with transition elements. These dehydrogenases are often directly linked, in complex systems, to ferredoxins and cytochromes.

THE ADVANTAGES ASSOCIATED WITH THE PRESENCE OF A TRANSITION METAL ION

The presence of a transition metal ion in a metalloenzyme enables a wide range of physical techniques to be applied, giving information that cannot be obtained for metal-enzymes where the metal is magnesium, calcium, zinc, etc. Because of this it is very worthwhile to replace these metals by transition metals. Thus zinc can usually be replaced by cobalt and magnesium by manganese. In these cases the new metal ion probably occupies a similar site to the original metal. But the more metals that can be substituted, the more the comparative information that can be obtained. The presence of a transition metal, with its unfilled d orbitals, allows the standard techniques of transition metal chemistry to be applied to the problem of the site symmetry and the type of binding groups utilized. These are the techniques of magnetochemistry and electronic spectra, together with various other more specialized spectroscopic techniques. Certain non-transition metals may also be used as probes; for example, thallium(I) is an NMR probe for K^+.

The replacement of one metal by another has to be considered carefully, as, apart from the examples cited, there is always the possibility that the new metal may bind to different groups or tend to favour different stereo-chemistries. Again, the strength of the metal-ligand bond could be very important in determining the reaction pathway. This could vary with the metal.

It is very important that there are quantitative assessments of the interactions between differing metal ions and differing potential ligands.

This is measured by stability or formation constants and these are most important in the context of biological inorganic chemistry. There are now, however, many extensive compilations of stability constant data which provide some very useful information.

FURTHER READING

H. R. Mahler and E. H. Cordes, *Biological Chemistry*, Harper and Row, New York, 1966

S. Bernhard, *Structure and Function of Enzymes*, Benjamin, New York, 1968

M. F. Perutz, 'X-ray analysis, Structure and Function of Enzymes,' *Europ. J. Biochem.*, 1969, **8**, 455

Many excellent articles will be found in the following series:

Advances in Protein Chemistry, Academic Press, New York

The Proteins, (Ed. H. Neurath), Academic Press, New York (2nd Ed.)

The Enzymes, (Eds. P. D. Boyer, H. Lardy, K. Myrback), Academic Press, New York (2nd Ed.)

CHAPTER TWO

THE PROPERTIES OF TRANSITION METAL IONS

The important transition elements in biological processes are the redox catalysts iron, copper, cobalt and molybdenum. Manganese is also important, although its function is rather different. The study of metallo-proteins containing these ions is made easier by the presence of the transition metal ion, as its characteristic properties provide very useful information concerning the nature and symmetry of the metal-binding site of the protein. However, these advantages may also be made available in other systems containing non-transition metal ions as described in Chapter 1. Thus the metals magnesium, calcium and zinc, which are present in a number of hydrolytic enzymes, can be usefully replaced by certain transition metal ions. Indeed, a comparative study of a series of metalloproteins with the divalent metal ions $Mn(II) \rightarrow Cu(II) (d^5 \rightarrow d^9)$ will often afford additional information over and above that obtained from a study of any single one of them, subject of course to the transition metal ion occupying the same site as the naturally occurring metal ion. From the overall view point, therefore, it is most important that the electronic properties of these transition metal 'markers' be fully understood and utilized.

In this chapter it is hoped to present a summary of the aspects of transition metal chemistry that are particularly relevant to the study of metalloenzymes, that is those properties that throw light on the symmetry of the metal-binding site, the nature of the binding groups and the electronic state of the metal. Standard texts should be consulted for a further account of these properties.

THE TRANSITION ELEMENTS: COORDINATION NUMBERS AND STEREOCHEMISTRY

The transition elements may be defined as those whose ions have in-completely filled d orbitals, thus copper(II) is a transition metal ion, having a d^9 configuration, while copper(I) of d^{10} configuration is not. They have a number of characteristic properties, including the formation of a range of coloured complexes, with the metal in a variety of oxidation states. The majority of these complexes are paramagnetic, indicating the presence of unpaired electrons. The elements of the first transition series, those in which the $3d$ orbitals are being filled, tend to have a maximum coordination number of six, that is the metal ion can accommodate up to six donor atoms or groups. The most common coordination numbers are four and six, although five is much more common than was once supposed. The second and third transition series elements, in which the $4d$ and $5d$ orbitals are being successively filled respectively, can increase their coordination number beyond six but we will not be particularly concerned with these.

Transition metal complexes in which the coordination number is four may either have tetrahedral (T_d) or square planar stereochemistry, while the normal stereochemistry for six coordination is octahedral (O_h), although a rare alternative form involves a trigonal prismatic structure. In many cases, and often in biological systems, these structures will be distorted and will only approximate to tetrahedral or octahedral stereochemistry. There are two alternative structures for five coordination and it is difficult to predict which will be favoured for any particular system. These are the square pyramid and the trigonal bipyramid (Figure 2.1). In certain complexes it appears that the two structures undergo rapid interconversion, while in others it appears that the structure lies between these two extremes. These different stereochemistries are associated with certain characteristic properties (e.g. electronic spectra) and in many cases can be identified fairly readily. Table 2.1 lists the stereochemistries observed for certain of the elements of the first transition series in their various oxidation states.

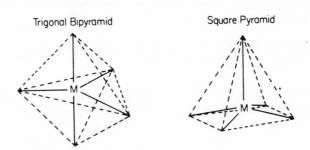

Figure 2.1   The stereochemistry of five coordination.

BONDING IN TRANSITION METAL COMPLEXES

The coordinate bond may be simply described in terms of electron-pair donation by the coordinating group (the ligand) to the metal ion. It is possible to account for certain features of the chemistry of the transition

TABLE 2.1 Stereochemistry and oxidation states of some $3d$ elements

Oxidation state	Coord. number	Stereochemistry	Oxidation state	Coord. number	Stereochemistry
$Cu^I(d^{10})$	2	linear	$Co^{III}(d^6)$	4	tetrahedral
	3	planar		5	square pyramidal
	4*	tetrahedral		6*	octahedral
$Cu^{II}(d^9)$	4*	square planar	$Co^{IV}(d^5)$	6	octahedral
	4	distorted tetrahedral	$Fe^{II}(d^6)$	4	tetrahedral
	5	square pyramidal		5	?
	5	trigonal bipyramidal		6*	octahedral
	6*	distorted octahedral	$Fe^{III}(d^5)$	4	tetrahedral
$Cu^{III}(d^8)$	6	octahedral		6*	octahedral
$Ni^{II}(d^8)$	4*	square planar		7	pentagonal bipyramidal
	4*	tetrahedral			
	5	trigonal bipyramidal	$Fe^{IV}(d^4)$	6	octahedral
	6*	octahedral	$Fe^V(d^3)$	4	tetrahedral
$Ni^{III}(d^7)$	5	trigonal bipyramidal	$Mn^I(d^6)$	6	octahedral
$Ni^{IV}(d^6)$	6	octahedral	$Mn^{II}(d^5)$	4	tetrahedral
$Co^I(d^8)$	4	square planar		4	square planar
	5	trigonal bipyramidal		6*	octahedral
	6	octahedral	$Mn^{III}(d^4)$	6*	octahedral
$Co^{II}(d^7)$	4*	tetrahedral	$Mn^{IV}(d^3)$	6	octahedral
	4	square planar	$Mn^V(d^2)$	4	tetrahedral
	5	trigonal bipyramidal	$Mn^{VI}(d^1)$	4	tetrahedral
	5	square pyramidal	$Mn^{VII}(d^0)$	3	planar
	6*	octahedral		4	tetrahedral

* Most common states.

metals ions by the use of Valence Bond Theory, but there are now a number of more satisfactory approaches which will account semi-quantitatively or quantitatively for the magnetic and spectroscopic properties of complexes, usually however by methods of increasing complexity, involving greater departure from a simple physical picture.

Crystal field theory

This approach completely neglects covalent bonding between metal and ligand and assumes that the interaction is only electrostatic, the ligands being treated as point charges. The basic assumption is clearly incorrect but nevertheless this theory is able to account semi-quantitatively for many aspects of transition metal chemistry and provides a basis for other theories.

Figure 2.2 schematically represents the orientation of the five d orbitals in space. It may be seen that these are not all spatially equivalent. Three orbitals, the d_{xy}, d_{yz} and the d_{zx} orbitals form one group, with their electron density distributed between the axes, while the other group, made up of the $d_{x^2-y^2}$ and d_{z^2} orbitals, has the electron density lying along the axes.

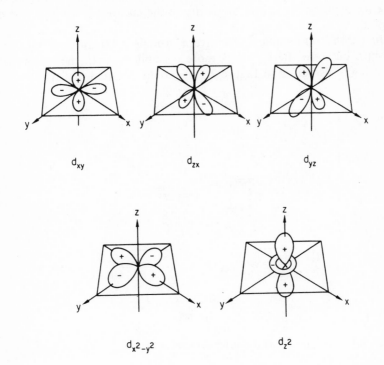

Figure 2.2   The d orbitals.

In the formation of complexes, as the negative (or polarized) ligand approaches the metal ion along the cartesian axes, so the orbitals of the metal ion will rise in energy. Not all the orbitals will be affected in the same way. If we consider an octahedral complex, then it may readily be seen that the more stable of these two sets of orbitals will be the one involving the d_{xy}, d_{yz} and d_{zx} orbitals (termed the t_{2g} orbitals) because the electrons in these orbitals will experience less repulsion from the electrons of the ligand. The other orbitals, the $d_{x^2-y^2}$ and d_{z^2} (e_g) lying along the axes, will be more destabilized. Thus the crystal field causes the splitting of the degenerate d orbitals of the free ion into two groups, one a set of triply degenerate orbitals, and the other a set of doubly degenerate orbitals. The splitting between these sets of orbitals may be obtained from electronic spectra and is given the symbol Δ_0 (sometimes 10 Dq). The crystal field splitting of d orbitals by an octahedral field is given in Figure 2.3. The actual energy difference between the split orbitals and the free ion is not known and so the splitting is usually represented with the energy of the free ion lying at the centre of gravity of the energies of the two sets of orbitals.

In tetrahedral complexes the five d orbitals are again split into two groups, but now the t_{2g} set is more destabilized than the e_g set. The splitting is also given in Figure 2.3; it should be noted that $\Delta_t = -\frac{4}{9}\Delta_0$.

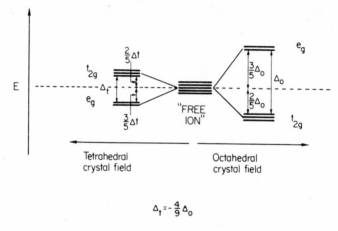

$$\Delta_t = -\frac{4}{9}\Delta_0$$

Figure 2.3   The splitting of the d orbitals by octahedral and tetrahedral crystal fields.

However, regular octahedral or tetrahedral stereochemistry is not always observed. Thus, two *trans*-ligands in octahedral complexes are often nearer or further away from the metal ion than the other four, so giving a tetragonal distortion. Let us consider the case when the two ligands on the z axis are further away than the other four. It may be seen that the degeneracy of the e_g orbitals is now removed, the d_{z^2} orbital becoming more stable than the $d_{x^2-y^2}$ orbital. Similarly the d_{yz} and d_{zx} orbitals will become more stable than the d_{xy} orbital. As the distortion becomes more severe a square planar complex will eventually be formed. The d-orbital splitting is shown in Figure 2.4. In certain cases the d_{z^2} level will cross the d_{xy} level and become almost as stable as the d_{yz} and d_{zx} pair.

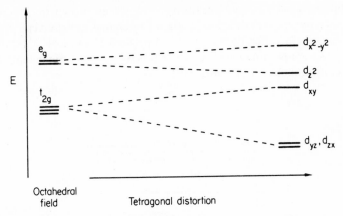

Figure 2.4   The effect of tetragonal distortion upon d-orbital splitting.

The implications of crystal field splitting

The splitting of the d orbitals into two sets is the key to the understanding of many of the properties of transition metal complexes. This is also the main feature of the other theories of bonding we shall mention and so much of the discussion that follows is of general application.

The number of unpaired electrons

The arrangement of electrons in an unfilled shell is governed by Hund's rules, which state that the stable configuration is the one with the maximum number of unpaired electrons, arranged with parallel spins. However, this state of affairs is complicated by the splitting of the d orbitals, as we now have to consider the possibility that it is energetically more favourable to pair electrons than to place them in the higher energy set of d orbitals. Two general cases exist:

(a) *the strong field case*, where Δ is high and hence electrons are paired (low spin complexes).

(b) *the weak field case*, where Δ is low, and it is more favourable to have a maximum number of unpaired electrons (high spin complexes).

The number of unpaired electrons may be determined by the use of a Gouy balance. If there are no unpaired electrons complexes will be diamagnetic, while the presence of unpaired electrons results in paramagnetic behaviour. The measured magnetic moment will indicate the number of unpaired electrons present.

For octahedral complexes only one possible arrangement of electrons exists for d^1, d^2, d^3, d^8, and d^9 configurations, but both high and low spin possibilities exist for configuration d^4 to d^7, as shown in Figure 2.5. d-Orbital splitting will therefore explain why change of ligand can produce complexes with different magnetic properties. Thus $[CoF_6]^{3-}$, a weak field species, is paramagnetic, while $[Co(NH_3)_6]^{3+}$ is diamagnetic, as are practically all cobalt(III) complexes.

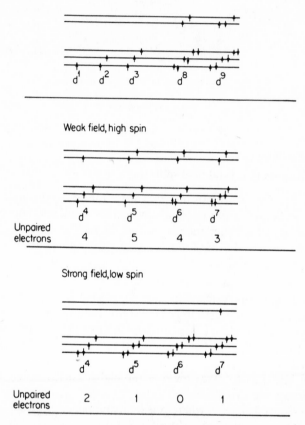

Figure 2.5 Weak and strong field arrangements for octahedral complexes.

A similar treatment may be applied to tetrahedral complexes. In theory $d^3 - d^6$ species should show both high and low spin complexes. However, in practice, as Δ_t values are low, only high spin complexes are formed. No low spin species are known at present.

Figure 2.4 shows why square planar d^8 complexes are diamagnetic, the separation between the d_{z^2} and $d_{x^2y^2}$ orbitals being greater than the pairing energy of the electrons.

A more detailed consideration of magnetochemistry will be given in a later section.

The spectrochemical series

The colour of transition metal complexes can be understood in terms of electronic transitions between the e_g and t_{2g} orbitals, the energy separations being of the same order of magnitude as that associated with the visible region of the spectrum. It follows therefore that the electronic spectra ($d - d$ spectra) of complexes will give information on the values of Δ for different ligands. It may be shown that for octahedral complexes of the first transition series, values of Δ_0 range from 7,500 to 12,500 cm^{-1} for divalent metal ions and between 14,000 and 25,000 cm^{-1} for trivalent ions. This explains why the species $[Co(NH_3)_6]^{3+}$ is a low spin complex while $[Co(NH_3)_6]^{2+}$ is high spin. Values of Δ increase as we move down to the second and third transition series. Phenomena such as these may be readily understood.

If all other factors are kept constant, the value of Δ will depend upon the ligand. If the ligands are written down in order of increasing Δ we have the *Spectrochemical Series*, so providing a relative assessment of the ability of ligands to cause d-orbital splitting.

Some common ligands lie as follows in this series; we have included a number of ligands of biological interest here for comparison. $I^- < Br^- < Cl^- < SCN^-$ (S bonded) $< F^- < OH^- <$ oxalate $\leqslant$ $H_2O < NCS^-$ (Ṅ bonded) $\leqslant NH_3 \simeq$ pyridine $<$ ethylenediamine $<$ dipyridyl $<$ O-phenanthroline $< NO_2^- < CN \simeq CO$. (SH$^- < -CO_2^- <$ amide $<$ imidazole).

A number of generalizations may be made at this point. The larger halide ions are poor d-orbital splitters reflecting the inability of the ion to approach closely to the metal ion. Ligands need not be formally negatively charged to split the d orbitals effectively, the metal ion will 'see' the ligand as a dipolar species, negative pole nearest. The position of the ligands nitrite, cyanide and carbon monoxide should be noted. This is due to ligand-metal π bonding, the existence of which is, of course, not allowed for in the basic crystal field postulates.

The positions of the bands in the electronic spectrum of a complex will reflect the positions of the ligands in the spectrochemical series. If a number of different ligands are present, then the overall Δ value will involve an averaging out of the individual Δ values. Thus the bands of ML_4Br_2 (L = unidentate ligand) will lie at lower frequencies than the corresponding bands of ML_4Cl_2. Again, if we have a complex $ML_4(SCN)_2$, the position of the bands, compared with those of the halogeno complexes, will indicate whether the thiocyanate is nitrogen or sulphur bonded.

It may be seen therefore that, if the electronic spectrum of a complex is known, it may be possible to make predictions concerning the nature of the binding groups by a consideration of the measured crystal field splitting.

Crystal field stabilization energies

The placing of electrons in the t_{2g} orbitals in octahedral complexes means that they are $\frac{2}{5}\Delta_0$ lower in energy than if they had been placed in the hypothetical degenerate orbitals. Hence a d^1 complex would be more stable than predicted by a simple electrostatic model. This energy is the Crystal Field Stabilization Energy (C.F.S.E.). This can be readily calculated for various d configurations from the orbital splitting diagram for the appropriate stereochemistry. Values are given in Table 2.2. Thus, for

TABLE 2.2 Crystal field stabilization energies for high spin complexes

No. of d electrons	Octahedral	Tetrahedral
1, 6	$\frac{2}{5}\Delta_0$	$\frac{3}{5}\Delta_t$
2, 7	$\frac{4}{5}\Delta_0$	$\frac{6}{5}\Delta_t$
3, 8	$\frac{6}{5}\Delta_0$	$\frac{4}{5}\Delta_t$
4, 9	$\frac{3}{5}\Delta_0$	$\frac{2}{5}\Delta_t$
(0), 5, (10)	0	0

an octahedral complex, $\frac{2}{5}\Delta_0$ is added for each electron in the t_{2g} level and $\frac{3}{5}\Delta_0$ is subtracted for each electron in the e_g level. Clearly for a d^5 high spin configuration there will be no C.F.S.E., and the values for d^n and d^{n+5} configurations will be the same.

The consequences of C.F.S.E.

The magnitude of crystal field stabilization energies can be significant and this is reflected in a number of ways as illustrated in the following examples.

(1) Hydration energies

Hydration energies (associated with Equation 1) have been calculated for a number of divalent ions of the first transition series by making use of thermodynamic cycles. These are shown in Figure 2.6.

$$M_g^{2+} + 6H_2O = [M(H_2O)_6]_{aq}^{2+} \tag{1}$$

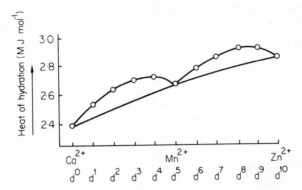

Figure 2.6 Variation in the heat of hydration with d-configuration for M^{2+} ions.

The results for the d^0, d^5 and d^{10} systems (no C.F.S.E.) fall on a smooth curve. The others lie above this curve by an amount equivalent to the C.F.S.E. This may be calculated and subtracted from the hydration energies, in which case the new values also fall on the curve. Alternatively, those results could have been used to provide independent checks on Δ_0 values obtained from electronic spectra. Results such as these are useful in that they confirm the correctness of the concept of d orbital splitting.

(2) Formation constants
The formation constants for a series of analogous compounds of the divalent metal ions of the first transition series with nitrogen donor ligands fall into the Irving Williams series; $Mn^{2+} < Fe^{2+} < Co^{2+} < Ni^{2+} < Cu^{2+} > Zn^{2+}$. (This will be discussed more fully in Chapter 3.) The formation of the complex involves the replacement of the coordinated water of the aquo species by the entering ligand. As this ligand will be higher than the water molecule in the spectrochemical series, there will be an increase in C.F.S.E., provided the ion is so stabilized. However, for Mn^{2+}, of d^5 configuration, there is no C.F.S.E., and so the formation of the complex does not result in any increased stabilization. For the other ions, the greater the C.F.S.E., then the greater the enhancement of stability on complex formation, so explaining why the Irving–Williams order in general follows the order of crystal field stabilization energies (with the exception of Ni^{2+} and Cu^{2+}).

(3) Coordination number
The preference of metal ions for tetrahedral or octahedral coordination is

also affected by C.F.S.E. A number of factors are obviously involved in this and play a much greater role in determining the overall enthalpy of a reaction such as:

$$[MCl_4]^{2-} + 6H_2O = [M(H_2O)_6]^{2+} + 4Cl^-$$

However, for a series of metal ions, such as those already considered, many of these factors will vary uniformly with change in metal ion, but C.F.S.E. contributions will not. Experimental studies have in fact confirmed that the variations in stability of tetrahedral and octahedral complexes do parallel the C.F.S.E. differences between the six coordinate and four coordinate species. Thus, tetrahedral cobalt(II) complexes are stable while nickel(II) tetrahedral complexes are much less stable with respect to an octahedral species.

(4) Redox potentials

Let us consider a transition metal ion existing in two oxidation states. The relative stabilities of these two species will be reflected in the redox potential for the couple. One factor that will affect the redox potential will be the C.F.S.E., for this will vary from one oxidation state to another, so enhancing the stability of one of them relative to the other. Therefore, variation in ligand (and hence Δ) will affect the C.F.S.E. contribution and so the redox potential.

Ligand field theory

There is good evidence for suggesting that metal orbital-ligand orbital overlap is important. We will not specifically discuss this but some aspects will be mentioned in the following sections. There are two general approaches to the problem of allowing for covalency in the metal-ligand bond. In *Ligand Field Theory* covalency is not formally introduced, but is allowed for by the adjustment of various parameters. For complexes in which the metal is in a normal oxidation state this usually works well as the amount of orbital overlap is small. The alternative approach, that of *molecular orbital theory*, begins with the assumption that overlap of orbitals does occur. By its very nature, however, this theory can allow for differing degrees of overlap and so can cover all possible cases, from the electrostatic picture to that of maximum overlapping of orbitals together with all intermediate possibilities. This theory will be briefly discussed in the following section.

There are three parameters of particular importance in ligand field theory. These relate to various interelectronic interactions and are the spin-orbit coupling constant, λ, and the Racah parameters, B and C, associated with electron repulsion. Fuller reference will be made to these later. The spin-orbit coupling constant is very important in determining the detailed magnetic properties of many transition metal complexes.

The presence of covalency means that the electrons of the ligand will be partly transferred to the metal ion so that the effective positive charge on the metal will be reduced. This will allow the d-electron clouds of the metal to 'expand' outwards from the metal ion. This has been termed the nephelauxetic effect and will vary from ligand to ligand. It is customary to arrange ligands in order of their tendency to cause cloud expansion—this is the nephelauxetic series, such as $F^- < H_2O < NH_3 <$ ethylenediamine $< -NCS^- < Cl^- \sim CN^- < Br < I$. Repulsion between electrons will be reduced as a consequence of cloud expansion and this is reflected in lower values of the Racah parameters in complexes than in the free ion. It is customary to express values for the complex by B' and C', usually $B'/B \sim 0.7$. Values of B' may be obtained from electronic spectra and provide the basis for the arrangement of ligands in the Nephelauxetic series.

Molecular orbital theory

The basic principles are those applied to simpler molecules. Molecular orbitals are constructed by linear combination of atomic orbitals, subject to the appropriate symmetry and energy considerations. Both σ and π molecular orbitals may be formed, but we shall first discuss the bonding in an octahedral complex considering σ interactions only.

The metal can provide nine orbitals, five $3d$, one $4s$ and three $4p$. Of these, the t_{2g} orbitals (d_{xy}, d_{yz}, d_{zx}) will not be suitable for σ bonding as they are not orientated towards the ligand orbitals. The six ligands will each provide an appropriate orbital, which will contain two electrons. We then allow each of the six metal orbitals to overlap with the appropriate symmetry σ orbital, two combinations being possible leading to bonding and antibonding (signified by *) molecular orbitals. This is represented schematically in Figure 2.7. The orbitals are given their appropriate symmetry designations. These are group theoretical in origin and will be used here only as convenient labels to identify the symmetry class to which the metal and ligand orbitals belong. It may be noted, however, the symbols a_{1g}, e_g and t_{1u} represent sets of singly, doubly and triply degenerate orbitals.

These molecular orbitals, as they are not constructed from atomic orbitals of equal energy, will resemble one atomic orbital more than the

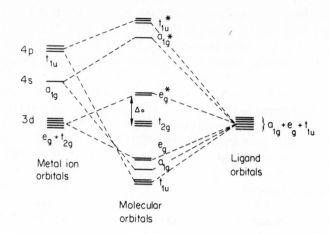

Figure 2.7   Qualitative molecular orbital scheme for an octahedral complex.

other. The six bonding M.O's will have more ligand character than metal character while the converse is true of the antibonding M.O's, while obviously the t_{2g} orbitals remain metal orbitals.

We now feed in the available electrons in accord with Hund's rules. Twelve electrons will occupy the six bonding M.O's, these may be regarded as essentially ligand electrons. The remaining electrons, equal to the number of d electrons of the metal ion, will then occupy the t_{2g} and e_g^* orbitals. In effect then the situation closely resembles that obtained by crystal field theory, provided we compare the e_g orbitals of the metal ion in crystal field theory with the e_g^* M.O's of molecular orbital theory. It is then possible qualitatively to account for magnetic and electronic spectral properties in a similar fashion to that of crystal field theory.

π bonding

This possibility must be considered whenever the ligand has suitable $p\pi$ or $d\pi$ atomic orbitals or π molecular orbitals available for overlap with the t_{2g} orbitals of the metal ion.

A number of situations are possible, depending upon whether the ligand π orbitals are filled or empty and whether or not they are higher or lower in energy than the metal t_{2g} orbitals. When ligands contain donor atoms oxygen and fluorine then the only π orbitals available are filled p orbitals, of lower energy than the metal t_{2g} orbitals. Other ligands have higher energy empty π orbitals. The two cases are depicted in Figure 2.8. It may be seen that the effect of (a) is to increase Δ, while in (b) Δ is decreased. Thus π

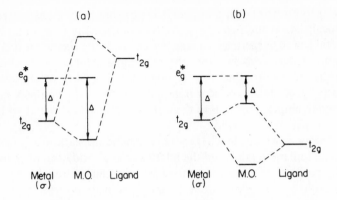

Figure 2.8 Interaction of ligand π orbitals and metal t_{2g} orbitals.

bonding is important in understanding the positions of ligands in the spectrochemical series.

It is also possible to have ligands, such as the other halide ions, where there are empty and filled π orbitals, the empty ones being d orbitals. Or again, in systems where there are π M.O's, there may be bonding and antibonding $p\pi$ orbitals. Well known examples are carbon monoxide and the cyanide ion. The overall result here reflects the relative importance of the two contributions. In practice, the π antibonding orbitals are closer in energy to the metal orbitals and have a larger effect, thus accounting for the position of those ligands in the strong field end of the spectrochemical series.

ELECTRONIC SPECTRA

The spectra of transition metal complexes in the range 25,000 to 5,000 cm^{-1} contain absorption bands of one or more of the following types. Absorptions in the near infrared region may be overtones of lower frequency infrared active ligand modes; these are not, however, usually of importance. Of much greater significance are charge-transfer and $d \rightarrow d$ bands. The former are usually distinguishable by their much greater intensity.

Charge-transfer bands

These sometimes appear in the visible region of the spectrum but are usually found in the U.V. region, often tailing off into the visible, however. They are associated with the transfer of an electron from an orbital of one atom to an orbital of another. It is possible for the charge-transfer to be localized on the ligand (e.g. SCN^-) but usually it involves ligand and metal ion. Charge may be transferred from the ligand to the metal in which case electrons in either the σ-bonding orbitals or π orbitals of the ligand are excited to the empty t_{2g} or antibonding M.O's of the complex. Alternatively, electrons of the σ-bonding orbitals of the complex may be promoted to empty π orbitals of the ligand, in which case metal $\rightarrow$ ligand charge-transfer has occurred.

The energy at which the charge-transfer band appears will depend on both metal ion and ligand, and the relative ease of oxidation or reduction of these species. For a constant metal ion, the charge-transfer band will be diagnostic of the ligand. In the $4d$ and $5d$ transition series, as Δ values are larger, it is sometimes more difficult to sort out charge-transfer and $d \rightarrow d$ bands. In all cases, comparison with the spectrum of a non-transition metal complex may help in the assignment.

$d \rightarrow d$ spectra

The energy difference between the e_g and t_{2g} orbitals of the metal ion is such that the excitation of an electron from a lower to a higher level can be achieved by the absorption of visible light, so accounting for the colour of transition metal complexes. The bands in the visible and near infrared regions of the spectrum, with the exception of those just mentioned, are therefore the $d \rightarrow d$ bands and are usually weak in intensity. They provide a great deal of information into the structure of the complex. We will be concerned solely with these bands in the remainder of this section.

Crystal field theory can offer a qualitative explanation of the d–d transitions in transition metal complexes. A d^1 system, such as Ti(III), should only show one such transition, corresponding to the excitation of the single electron from a t_{2g} to an e_g orbital. This band is seen at around 20,000 cm^{-1} in the spectrum of $[Ti(H_2O)_6]^{3+}$ and is a direct measure of Δ_0. The spectrum of a d^9 system, such as Cu(II), may also be simply interpreted. Here, however, it is said that the Cu(II) ion is a one positron ion (or one 'hole'), and that one band should be observed corresponding to the e_g–t_{2g} transition of the positron. A number of complicating features in these two cases will be discussed later.

The interpretation of the spectra of complexes of other d configurations is less straightforward. It is necessary to turn to the results of atomic spectroscopy and consider the energy levels derived from each d configuration for the free ion. For any particular d^n species ($2 < n < 8$) not all the arrangements of the electrons among the d orbitals are of equal energy. This is because the electrons repel each other differently in different orbitals. Thus, each d configuration will give rise to a number of energy levels or *terms*, each of which is usually degenerate. The full calculation of the relative energies of these terms is complex. They are described in terms of the Racah parameters B and C, which are in themselves composite quantities.

The terms are characterized by the total spin angular momentum and the total orbital angular momentum quantum numbers S and L respectively. (S and L for the term corresponds to s and l for the individual electron in an atomic orbital.) This involves the use of the Russell–Saunders coupling scheme which assumes that spin and orbital angular momenta do not interact. Terms with L values 0, 1, 2, 3, 4 ... are designated as S, P, D, F, G etc. and the classification is completed by reference to the spin multiplicity ($2S + 1$), the number of ways in which the unpaired electrons may be arranged. Thus if $S = 1$ the spin multiplicity is 3, a triplet. If $S = \frac{3}{2}$ and $L = 2$, the term is a ^4D term. The energy of the substates making up the term is given by the vectorial addition of L and S. This is represented by J. Thus, if $L = 2$, the four ways of arranging the three unpaired electrons in the ^4D term are $+\frac{1}{2}$, $+\frac{1}{2}$, $+\frac{1}{2}$; $+\frac{1}{2}$, $+\frac{1}{2}$, $-\frac{1}{2}$; $+\frac{1}{2}$, $-\frac{1}{2}$, $-\frac{1}{2}$, and $-\frac{1}{2}$, $-\frac{1}{2}$, $-\frac{1}{2}$ corresponding to the J values $\frac{7}{2}, \frac{5}{2}, \frac{3}{2}, \frac{1}{2}$. However, in the ^4D term L may take on any value between L and $-L$, so we also have to combine these S values with L values 1, 0, -1, -2 showing that the ^4D term contains 20 'degenerate' states, all in fact having slightly different energies. Similarly a ^3P term will have $3 \times 3 = 9$ substates.

The effect of the crystal field upon the Russell–Saunders terms of the free ion
The crystal field does not split the s or the p atomic orbitals but does split the d and f orbitals into two and three sets respectively. Similarly, the S and P terms of the ion are not split, but the D and F terms are split by the crystal field into two and three components. The Russell–Saunders states will always be affected in the same way, irrespective in what d configuration they occur. Table 2.3 gives the effect of the crystal field

TABLE 2.3 The splitting of Russell–Saunders states by the crystal field

Free ion	Crystal field states
S	A_1
P	T_1
D	$E + T_2$
F	$A_2 + T_1 + T_2$
G	$A_1 + E + T_1 + T_2$

upon the various Russell–Saunders states. The crystal field states are designated by group theoretical symbols. We recall that the five d orbitals were split into two sets irrespective of whether the crystal field were tetrahedral or octahedral, although the relative energies of the e_g and t_{2g} orbitals were inverted. Similarly, the Russell–Saunders D terms are split into two states by both tetrahedral and octahedral fields and so on, but again the energies of the two substates will be inverted in the two cases. It is usual to plot the Russell–Saunders terms on an energy scale and to plot the splitting against the crystal field energy Δ. This is done in Figure 2.9 for a d^8 ion (Ni^{II}) in an octahedral field. The splittings would be inverted in a tetrahedral field.

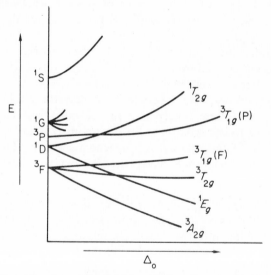

Figure 2.9 The effect of an octahedral crystal field upon the d^8 configuration.

We are now able to predict the electronic spectrum of a d^8 octahedral complex. Clearly a number of transitions may occur from the ground state to excited states. However, a quantum mechanical rule says that transitions may only occur between states of the same multiplicity. The ground state in the present case is a triplet state. We may therefore observe three transitions from the $^3A_{2g}$ ground state, to the $^3T_{2g}$, $^3T_{1g}(F)$ and $^3T_{1g}(P)$ excited states. The two different $^3T_{1g}$ states are characterized by (F) and (P), a reminder of the free-ion state from which they originate. The spectrum of a nickel(II) octahedral complex is given in Figure 2.10. It is possible to assign all these bands by comparison with the term-splitting

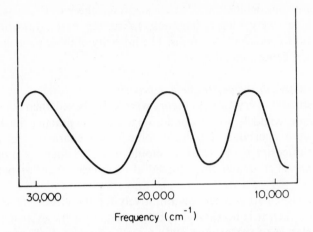

Figure 2.10 The electronic spectrum of $[Ni(en)_3]^{2+}$.

diagram. The lowest energy band, for example, will be the $^3A_{2g} \rightarrow {}^3T_{2g}$ transition, which in the present case is a direct measure of Δ_0.

It is possible to treat all d configurations in this way. The appropriate energy level diagrams are readily available. It should be noted that all crystal field states have the same multiplicity as the free ion term. Spin forbidden transitions may occur (i.e. between states of different multiplicity). These will be very weak in intensity and often can be satisfactorily assigned. Table 2.4 lists the Russell–Saunders states of various configurations. It may be seen for the d^5 configuration that only spin forbidden transitions may occur, as there is no state of the same multiplicity as the ground state. This explains why Mn(II) complexes are practically white in colour.

It may be noted that there is a symmetry in Table 2.4. This results from the fact that d^n systems are related to d^{10-n} systems by the electron/positron argument applied to the d^1 and d^9 systems. The splittings of the d^{10-n} states will, however, be inverted with respect to those of the d^n states.

TABLE 2.4 The Russell–Saunders states of the d configuration

d^1	d^2	d^3	d^4	d^5	d^6	d^7	d^8	d^9
^2D	^3F	^4F	^5D	^6S	^5D	^4F	^3F	^2D
—	^3P	^4P	^3H	^4G	^3H	^4P	^3P	—
—	^1G	^2H	^3G	^4F	^3G	^2H	^1G	—
—	^1D	^2G	—	^4D	—	^2G	^1D	—
—	^1S	^2F	—	^4P	—	^2F	^1S	—

This means that octahedral d^n species and tetrahedral d^{10-n} species will have similar energy level diagrams. In certain cases transitions may correspond to two electron jumps. The intensity of these bands is usually low, despite being spin allowed.

The weaknesses of the crystal field approach

If the postulates of crystal field theory were fully obeyed, then the intensity of d–d bands would be very low indeed. However, in practice the d orbitals are not pure d orbitals, but have some ligand character. As a result, a greater intensity is observed. It should be noted that the intensity of the d–d bands is greater for tetrahedral complexes than for octahedral complexes.

If it is wished to account quantitatively for the electronic spectra of complexes, then it is necessary that the energies of the excited states be known relative to the ground state for all values of Δ. This requires a knowledge of the Racah parameters B and C. They may be determined for the free ion from atomic spectra. However, to satisfactorily account for the experimental results, these terms B and C must be treated as parameters, their values being reduced below that of the free ion. This is a consequence of covalency as already described. All this is involved in the ligand field approach to electronic spectra. Values of B and Δ for the complexed ions may be obtained from Tanabe and Sugano diagrams which plot E/B against Δ/B for each state. Values of C may also be obtained.

Some general problems

The spectra of d^1 and d^9 complexes may be understood in simple terms, as already outlined. However, we also have to account for the breadth of these bands and for their lack of symmetry. The band widths in general are proportional to the slopes of the individual excited states in the energy level diagram. The ligand atoms are vibrating and so Δ is continuously changing over a narrow range. However, if the excited state slopes rapidly with change of Δ then a fairly wide range of energies will be associated with that transition.

One result of these broad bands is that it is difficult to note interactions of small energy. Spin-orbit coupling is such a factor. This can cause further splitting of the degenerate crystal field state. However, for the first transition series, the fine structure of a band resulting from this is not likely to be seen. This is more important for $4d$ and $5d$ elements.

A much more important effect in this context is the Jahn–Teller effect. This is associated with the presence of a crystal field of lower symmetry than O_h or T_d. The theorem states 'that any non-linear system which is in a degenerate state will undergo some kind of distortion to remove this degeneracy'. This means therefore that any states still degenerate in the crystal field will split, and implies that spin-orbit coupling has not done this. The Jahn–Teller splitting will result in further stabilization of the system.

The physical basis of the Jahn–Teller effect may be readily seen by considering the effect of unevenly filled e_g and t_{2g} shells on the symmetry of the complex. Let us consider a d^9 system and assume that we have one electron in the d_{z^2} orbital and two electrons in the $d_{x^2-y^2}$ orbital. Repulsion between ligand electrons and metal electrons will then be greater in the x–y plane than on the z axis, so producing a tetragonally distorted complex with short bonds on the z axis (Figure 2.11(a)). If the opposite arrangement of electrons holds, then the bonds on the z axis will be longer than the bonds in the plane. (This is usually the case.) This explains why the vast majority of copper(II) complexes are tetragonally distorted.

(a)   (b)

Figure 2.11   Jahn-Teller distortion of a d^9 octahedral complex.

It may readily be seen that for 6-coordinate complexes, uneven occupancy of the t_{2g} orbitals is not going to be as important as the uneven occupancy of the e_g orbitals in causing distortion. Table 2.5 lists those configurations for which a Jahn–Teller effect is expected to be observed on this basis. It should be noted that for high spin complexes, d^n and d^{5+n} configurations should have a similar effect. It is not possible to predict how large the distortion will be or its form, but it may be shown that the effect

TABLE 2.5 Jahn–Teller   distortions   of   octahedral complexes

	Observed	Predicted
d^1, d^6	slight distortion	Yes
d^2, d^7	slight distortion	Yes
d^3, d^8	no distortion	Yes
d^4, d^9	large distortion	Yes

is of importance in the spectra of octahedral complexes. Obviously, if the ground state is not subject to the Jahn–Teller effect then the excited state will be (unless a 2 electron jump occurs). If the ground state is split, then a band will appear in the near infrared corresponding to the transition between the two components of the ground state. However, it is unlikely that an effect will be seen in the visible region, as the upper of these two 'ground states' will not be sufficiently populated to allow an observable band to be seen for the transition to the excited state. If the Jahn–Teller effect is causing the splitting of the excited state, then transitions to both these may be seen. Clearly, however, the fact that degenerate terms are split by the Jahn–Teller effect, by energies of around a few hundred cm^{-1}, will provide another reason for the width of bands in electronic spectra.

Complexes of lower symmetry

In practice most complexes do not have six or four identical ligands around the metal ion and so only approximate to O_h and T_d symmetry, although their properties may often be understood satisfactorily on such a basis. We have discussed a number of other causes for lowered symmetry. In biological systems we often have to consider metal ion environments which are of very low symmetry. The electronic properties of these metal ions are difficult to explain.

As the symmetry of the complex is decreased beyond cubic, so the number of bands in the electronic spectrum will increase, as the crystal field states are split further, so allowing more transitions. The effect may only show itself in a broadening and splitting of the bands and often may be treated by regarding the low symmetry component as a perturbation on the main cubic field.

For complexes of non-cubic symmetry, it is necessary to introduce additional parameters. Clearly the effect of the crystal field cannot now be described by a single parameter. The energies of the five d orbitals will be split in a much more complex way. Thus a tetragonally distorted octahedral complex (of D_{4h} symmetry) requires in-plane and axial crystal fields to be considered. Two parameters are therefore needed. Each example of a distorted stereochemistry must be considered independently. Usually this is done in terms of molecular orbital theory.

MAGNETIC PROPERTIES OF TRANSITION METAL IONS

Both electrons and nuclei have associated magnetic properties. Those of the nuclei are manifested in the phenomenon of nuclear magnetic resonance, but are several orders of magnitude smaller than those of the electrons. In transition metal ions we are therefore specially concerned with the magnetic properties of unpaired electrons and with the information that may be derived from a study of these properties.

Molecules with closed shells of electrons only have no inherent magnetic properties as there are no degeneracies to be split by the magnetic field. The magnetic field will, however, induce a small opposing moment and so such molecules are repelled by the magnetic field. These molecules are termed *diamagnetic*. An unpaired electron has magnetic properties resulting from both its spin and orbital motion. Molecules with unpaired electrons are therefore attracted into a magnetic field and are said to be *paramagnetic*. The two effects, diamagnetism and paramagnetism, are opposed and in calculating the paramagnetic moment of a transition metal complex due allowance must be made for the diamagnetism of the ligand. It should also be emphasized that the following discussion refers to magnetically dilute species; that is, the individual paramagnetic centres do not affect each other.

If a substance is placed in a magnetic field H, the magnetic flux B (the field within the substance) is given by the equation:

$$B = H + 4\pi I$$

where I is the intensity of magnetization. This equation may be rewritten as:

$$B/H = 1 + 4\pi I/H = 1 + 4\pi K$$

where K is the magnetic susceptibility per unit volume. It is much more convenient, however, to work in grammes and it is usual to define the gramme susceptibility where

$$\chi = K/d$$

where d is the density of the substance. We may also write

$$\chi_M = \chi \cdot M$$

where M is the molecular weight and χ_M the molar susceptibility. A correction is then made for the diamagnetism of the ligand, giving χ_M^{corr}.

Curie's Law states that paramagnetic susceptibility is inversely proportional to the absolute temperature:

$$\chi_M^{corr} = C/T, \text{ where C is the Curie constant.}$$

It may also be shown statistically that

$$\chi_M^{corr} = N\mu^2/3kT,$$

where N is the Avogadro Number, k the Boltzmann constant and μ the magnetic moment. Hence

$$C = N\mu^2/3k, \quad \text{and} \quad \mu = \sqrt{(3k/N)} \cdot \sqrt{\chi_M^{corr} \cdot T}$$
$$= 2 \cdot 84 \sqrt{\chi_M^{corr} \cdot T}$$

This equation will allow the calculation of the magnetic moment of the paramagnetic species from the susceptibility. In accurate work a correction should also be made for Temperature Independent Paramagnetism (TIP) in χ_M^{corr} (see later).

Magnetic moments are expressed in terms of Bohr Magnetons, where 1 BM $= he/4\pi mc$. (e is the charge and m the mass of the electron, h is Planck's constant and c the velocity of light.)

Susceptibilities may be measured by the Gouy and Faraday methods and also by a NMR method (in solution). The advantages of the last two methods lie in the fact that only small amounts of material are required, an important point when considering metalloproteins. The non-heme protein ferredoxin has been studied in the solid state and solution by the Faraday and NMR methods, and this has shown that all seven iron atoms in the molecule are low-spin Fe(III). The Faraday method will allow the study of single crystals and hence the study of anisotropy in magnetic susceptibility. The most often used method is the Gouy method. The advantages here are that the equipment is simple and easy to operate, but comparatively large amounts of material are required. These methods are described in standard texts.

We have noted that there are two sources of the paramagnetic moment, i.e. spin and orbital contributions. If the full orbital contribution is made, then the magnetic moment will be given by the equation:

$$\mu = \sqrt{4S(S + 1) + L(L + 1)}$$

where L and S are the total orbital and spin angular momentum quantum numbers for the molecule. Clearly, if there is no orbital degeneracy (i.e. an S state ion, for which $L = 0$) there will be no orbital contribution. The remaining contribution will be the spin-only term. Table 2.6 lists spin-only

TABLE 2.6 Spin-only values

Unpaired electrons	S	μ
1	$\frac{1}{2}$	1·73
2	1	2·83
3	$1\frac{1}{2}$	3·81
4	2	4·93
5	$2\frac{1}{2}$	5·92

moments for various numbers of unpaired electrons. For other states, however, the orbital degeneracy could be lost or reduced by the environment of the other atoms. This is termed the quenching of the orbital contribution; quenching may be partial or complete. However, it may be shown that for an octahedral complex an orbital contribution is only to be expected if there is a ground state T_{1g} or T_{2g} term.

An orbital may be occupied by one electron with either $S = +\frac{1}{2}$ or $-\frac{1}{2}$. In the presence of a magnetic field these two arrangements will differ in energy and the more stable arrangement will be more favoured, although the actual occupancy will follow a Boltzmann distribution with respect to temperature. This stabilization of the ground state is the reason for the sample being drawn into the magnetic field. The splittings of spin (and orbital) degeneracy caused by the magnetic field are very small and as a result interactions too small to affect electronic spectra will have considerable effect upon the magnetic moment. In particular they affect the orbital contribution, because they help determine the ground state of the complex in the absence of a magnetic field. The presence of low symmetry crystal fields is one example, but in this case it is difficult to deal quantitatively with the problem. It should be noted, however, that if the ground state contains an odd number of unpaired electrons, then there is a degeneracy (Kramer's degeneracy) which a low symmetry field cannot remove. This will be split by a magnetic field.

One very important effect is that of *spin-orbit coupling*. Spin-orbit coupling may split states with both spin and orbital degeneracy into substates and can therefore have a marked effect upon the measured magnetic moment. In deriving the Russell–Saunders terms for the free ion, it is assumed that spin and orbital angular momenta will not interact and that their magnitude will determine the energy of a particular arrangement. In practice, however, they couple together to some extent, the result being that certain configurations of a nominally degenerate state are more stable than others.

The extent of spin-orbit coupling is given by the spin-orbit coupling constant λ, the measure of the coupling between resultant spin and orbital angular momenta. There is also a single electron spin-orbital coupling constant. Values for free ions may be obtained from atomic spectra and for complexes by magnetic measurements.

The effect of spin-orbit coupling for $d^1 - d^4$ ions is to reduce the measured magnetic moment (usually down to around the spin-only value) while for $d^6 - d^9$ ions the effect is to increase the measured moment. In Table 2.7 there are listed experimental moments for a number of species for different stereochemistries, together with calculated spin-only values. Orbital contributions will be expected for octahedral Co^{2+}, low spin Fe^{3+},

TABLE 2.7 Magnetic moments of transition metal complexes

Ion	Spin-only value	Measured moment	
		Octahedral	Tetrahedral
Cu^{2+}	1·73	1·7–2·2	—
Ni^{2+}	2·83	2·8–3·3	3·3–4·0
Co^{2+} h.s.	3·88	4·5–5·2	4·2–4·8
Co^{2+} l.s.	1·73	1·8	—
Fe^{2+} h.s.	4·90	5·1–5·7	5·3–5·5
Fe^{3+} h.s.	5·92	5·7–6·0	—
Fe^{3+} l.s.	1·73	2·0–2·5	—
Mn^{2+} h.s.	5·92	5·6–6·1	5·9–6·2
Mn^{2+} l.s.	1·73	1·8–2·1	—
Cr^{2+} h.s.	4·90	4·7–4·9	—
Cr^{2+} l.s.	2·83	3·2–3·3	—

Mn^{2+}, Mn^{3+} and Cr^{2+} and for tetrahedral Ni^{2+}. It is clearly possible to interpret magnetic data in a general way in the light of this information.

In summary we may see that for the formation of an octahedral complex the original degeneracies of the free ion will have been split successively by the crystal field, the tetragonal distortion, spin–orbit coupling and then by the magnetic field, each resulting in a successive stabilization of the ground state.

Temperature independent paramagnetism

There are a number of types of magnetic behaviour other than those of diamagnetism and paramagnetism. Ferromagnetism, ferrimagnetism and antiferromagnetism result from magnetically non-dilute species and will not be discussed further. The phenomenon of temperature-independent paramagnetism (T.I.P.) is a result of spin–orbit coupling, when a ground state (not split by spin-orbit coupling) is coupled by it to a degenerate excited state, well separated in energy. T.I.P. results because the magnetic field, as a result of spin-orbit coupling, pushes electron density into an excited state. The effect is a weak one and, as the thermal population of the excited state is essentially zero, does not vary with temperature.

Covalency

This may be allowed for by using the spin–orbit coupling constant as a parameter. Alternatively a new factor is introduced, the orbital contribution reduction factor, k, which as the name suggests, is incorporated with the orbital contribution, and usually has values around 0·7.

Electron paramagnetic resonance

EPR (or ESR) spectra provide a great deal of insight into iron and copper containing proteins and is a most valuable, although complex, technique. The spectra are usually only observed for ions which have an odd number of unpaired electrons in the ground state, i.e. those with a Kramers doublet. This degeneracy is removed by a magnetic field and the resulting transitions between the two components are observed in the EPR spectra. In view of the small energies involved, the spectra will clearly be sensitive to the various factors discussed above which result in splittings equivalent to or larger than those produced by the magnetic field.

From a more pictorial point of view, the unpaired spinning electron can be regarded as a magnet which will align itself with a magnetic field. However, absorption of energy could cause it to align itself against the magnetic field. Alternatively, if the system in question is subjected to a microwave beam and the strength of the magnetic field increased, it will be possible to note the particular strength at which the electron changes its alignment. This will be characteristic of the environment of the electron.

The unpaired electrons may interact with the nuclear spin I of the metal, giving $2I + 1$ bands. The separation associated with these is the hyperfine splitting, and will depend upon the stereochemistry of the complex. The presence of covalent interaction with a ligand will result in the appearance of a fine structure of $2I' + 1$ lines, where I' is the nuclear spin of the ligand. The greater the covalency the greater is the ligand hyperfine splitting.

The position of the EPR resonance gives the field at which it occurs. From this the g value (the splitting factor) may be calculated from the equation $hv = g\beta H$ where β is the Bohr magneton and H the field. The value of g is a measure of the contribution of the spin and orbital motion to the total angular momentum. For a free electron g = 2·0023. For inorganic complexes g values are higher than 2 and will be related to the arrangement of ligands around the metal ion. Inorganic complexes are often anisotropic, having g values dependent upon field direction. Values are therefore obtained for crystals in terms of g|| and g⊥. The anisotropy of the g value can give much information.

Organic free radicals will also give rise to EPR signals and these may prove a complicating factor in the interpretation of the EPR spectra of enzyme systems.

The distribution of electrons between two energy states is given by the Maxwell–Boltzmann expression. It is important that EPR measurements be made at low temperature to enhance the occupancy of the ground state at the expense of the excited state, so avoiding line broadening effects.

The value of EPR measurements for metalloproteins is that they will possibly tell us something about the oxidation state of the metal, e.g. for copper; they will confirm the presence of covalency and may help in the characterization of the ligand through the observation of the ligand hyperfine splitting.

ELECTRONIC PROPERTIES OF COMPLEXES OF COPPER AND IRON

These two transition metals figure prominently in later chapters. In order to provide a suitable basis for a discussion of the properties of their complexes, some relevant data are summarized at this point.

Copper

Copper(I) species, being of d^{10} configuration, will only show charge-transfer bands, Cu(II) $\rightarrow$ L. These have been observed up to 650 nm in the visible region. Copper(II) complexes (d^9) will also show charge-transfer bands, Cu(II) $\leftarrow$ L. Some complication may arise from the presence of ligand bands. The d–d spectra of copper(II) species are different to interpret in terms of the stereochemistry of the complex. The usual stereochemistry is that of a tetragonally distorted octahedral one. The effect of such a crystal field upon the degeneracy of the d orbitals has been seen in Fig. 2.4. As the ligands upon the z axis are withdrawn, so the energy of the d_z and d_{zx}, d_{yz} orbitals fall while that of the $d_{x^2-y^2}$ and d_{xy} will rise relative to these. Three positron transitions may then occur ($d_{x^2-y^2} \rightarrow d_{z^2}$; $d_{x^2-y^2} \rightarrow d_{xy}$; $d_{x^2-y^2} \rightarrow d_{xz,yz}$), but when the distortion is slight only one band will be seen as the energy separation between the split bands will be small. A similar state of affairs will hold for a grossly distorted square planar complex, while only for a medium distortion may three absorptions be seen. However, there may be other stereochemistries of low symmetry of importance in biological molecules and it may well be difficult to interpret their spectra.

Iron

Charge-transfer transitions may be observed for Fe(II) and Fe(III) complexes. For Fe(II), d^6 complexes in an octahedral stereochemistry the ^5D ground state is split into two states. One transition only therefore may be observed. At higher ligand fields it is possible that a singlet state may drop below the ^5D state and provide a new ground state. Fe(II) low spin complexes will, of course, be diamagnetic. For Fe(III), d^5 complexes with a ^6S ground state, spin forbidden bands only are possible as there are no other sextuplet states. The d–d bands will therefore be of low intensity. High spin complexes are usually observed with magnetic moments close to the spin-only value, as there is no orbital angular momentum contribution. Low spin complexes will have considerable orbital contribution and moments are usually raised above the spin-only value of 1·73 BM to about 2·3 BM. The d–d spectra of Fe(III) in hemoproteins have been a very useful guide to electronic structure and have also thrown light upon the nature of the axial ligand in a number of cases.

KINETICS AND MECHANISM OF REACTION OF TRANSITION METAL COMPLEXES

The reactions of transition metal complexes may be divided into the following three categories:

(a) *ligand substitution*, in which one coordinated ligand is replaced by another from solution;

(b) *redox processes* (electron transfer reactions), in which there is a change in the oxidation state of the metal ion; and

(c) *reactions of coordinated ligands*, which may be affected considerably by the presence of the metal ion. This topic is of wide fundamental interest and discussion will be reserved until the next chapter.

A useful classification in terms of general reactivity involves the division of complexes into inert and labile. Labile complexes are those whose reactions are over in less than one minute at 25° for concentrations around 0·1 mole litre^{-1}, while inert complexes are those whose reactions may be readily followed under these conditions, having half-lives greater than about one minute. Much work has been carried out on the reactions of inert complexes, particularly those of cobalt(III) and chromium(III). However, increasing attention is now being paid to the reactions of labile complexes. This partly reflects recent developments in the techniques for following very fast reactions in solution, particularly in the use of stopped-flow systems, relaxation techniques and magnetic resonance techniques. The use of relaxation techniques allows the study of reactions having first-order rate constants up to 10^{10} sec^{-1}. This avoids all the problems associated with the mixing of reactants by starting with a system already at equilibrium. The position of equilibrium depends upon a number of factors. If one of these, such as temperature, is changed very rapidly then there will be a time lag before the new equilibrium position is reached.

This time lag is related to the rate constants of forward and back reactions, which may then be calculated. In the temperature-jump (T-jump) method, the temperature rise is obtained by discharging a condenser through a few mls of reaction solution in a cell. The reaction is then followed spectrophotometrically or conductometrically, using instrumentation of the appropriate sensitivity.

Ligand substitution in octahedral complexes

Much of the earlier work involved octahedral cobalt(III) complexes with ethylenediamine or ammonia ligands. These reactions could be followed easily, while a wealth of background information on complexes of this type was available as a result of the work of Werner.

Ligand substitution is a nucleophilic substitution and attempts have been made therefore to classify these into S_N1 and S_N2 mechanisms as had already been done for substitution at carbon. It is better, however, to regard these as two extremes of mechanism dependent upon the timing of bond breaking and bond making.

Consider reaction (2) in which X^- is replaced by Y^-, A being a neutral, unidentate ligand.

$$[CoA_5X]^{2+} + Y^- \rightarrow [CoA_5Y]^{2+} + X^- \tag{2}$$

The S_N1 mechanism involves the rate-determining loss of X^-, giving a 5 coordinate transition state, followed by fast attack of Y^-.

$$[CoA_5X]^{2+} \xrightarrow{\text{slow}} [CoA_5]^{3+} + X^-$$
$$[CoA_5]^{2+} + Y^- \xrightarrow{\text{fast}} [CoA_5Y]^{2+}$$

This leads to the rate law (3)

$$\text{Rate} = k_1[CoA_5X^{2+}] \tag{3}$$

Alternatively, we may have a synchronous attack of Y^- upon the complex as X^- departs, with a 7 coordinate transition state. This is the bimolecular S_N2 mechanism, with a rate law (4).

$$\text{Rate} = k_2[CoA_5X^{2+}][Y^-] \tag{4}$$

The S_N1 and S_N2 mechanisms are sometimes termed dissociative and associative respectively.

The nature of rate laws (3) and (4) might suggest that a kinetic distinction between the two mechanisms is readily possible; however, there are complications. One of them involves the participation of the solvent. Many reactions have been studied in aqueous solution, but the water molecule is a good ligand and is obviously readily available in the outer coordination sphere. It is possible therefore that reaction (2) might occur *via* intermediate formation of an aquo complex, in which X^- was replaced in an S_N2 mechanism, followed by fast replacement of H_2O by Y^-, as in the following scheme

$$[CoA_5X]^{2+} + H_2O \xrightarrow{\text{slow}} [CoA_5H_2O]^{3+} + X^-$$
$$[CoA_5H_2O]^{3+} + Y^- \xrightarrow{\text{fast}} [CoA_5Y]^{2+} + H_2O$$

However, as the concentration of water is effectively unchanged, it will not appear in the rate law, which will therefore be:

$$\text{Rate} = k_1[CoA_5X^{2+}]$$

This corresponds to a unimolecular reaction, even though in practice the reaction is bimolecular.

Another complication is that of ion pairing. This is particularly important if the reactions are carried out in non-aqueous solvents. The cationic complex $[CoA_5X]^{2+}$ and the anion Y^- may well exist as the ion pair $[CoA_5X^{2+}]\ldots Y^-$ which may be the species undergoing reaction. In this case the rate law must include a dependence upon the concentration of the ion pair, which will in turn depend upon the concentration of complex *and* entering group, quite independently of the mechanism of the actual act of substitution.

The general approach to assigning mechanism has usually depended upon a comparison of the differing requirements of the S_N1 and S_N2 transition states with the actual experimental results. Thus, an S_N1 reaction should be speeded up by the presence of bulky ligands and the S_N2 reaction should be hindered, as the formation of the S_N1 transition state relieves overcrowding while the formation of the S_N2 transition state emphasizes it. Again, the presence of electron-releasing ligands should speed up an S_N1 reaction (as the leaving group has to leave with a pair of electrons) and the presence of electron-withdrawing groups should speed up the S_N2 reaction as then the nucleophilic attack of the entering group is assisted. On the basis of studies such as these it appears that there is substantial evidence for the S_N1 mechanism but that there is no uncomplicated evidence for the S_N2 mechanism. Many reactions do, however, appear to lie between these two extremes; thus the leaving group may almost have left before the entering group attacks.

Mention should be made, however, of one reaction, that of base hydrolysis, where it was thought that an S_N2 mechanism was operative.

$$[CoA_5X]^{2+} + OH^- \rightarrow [CoA_5OH]^{2+} + X^-$$
$$\text{Rate} = k_2[CoA_5X^{2+}][OH^-]$$

Much controversy has centred around this reaction but it appears that the reaction is actually S_N1 despite the bimolecular rate law: an S_N1 reaction undergone by the conjugate base of the complex, the concentration of which would depend upon both complex and the proton accepting base, in this case $[OH^-]$. This is the S_N1CB mechanism. (CB = conjugate base.)

$$[Co(NH_3)_5Cl]^{2+} + OH^- \rightleftharpoons [Co(NH_3)_4(NH_2)Cl]^+ + H_2O$$

$$[Co(NH_3)_4(NH_2)Cl]^+ \xrightarrow{slow} [Co(NH_3)_5(NH_2)]^{2+} + Cl^-$$

$$[Co(NH_3)_4(NH_2)]^{2+} + H_2O \xrightarrow{fast} [Co(NH_3)_5OH]2^+$$

A number of results indicate that this mechanism is almost certainly correct for base hydrolysis of cobalt(III) complexes.

The formation of complexes. The replacement of coordinated water

The replacement of coordinated water by an entering ligand group is a reaction of fundamental importance. For completeness we shall make brief reference here to the alkali metals and alkaline earth metals in addition to the transition elements. Most of their aquo complexes are six coordinate. The replacement of water and the exchange of water with solvent have been studied by the techniques we have mentioned earlier. The alkali metals form complexes faster than any other ions in the Periodic Table, but the complexes are weak and so the studies have been carried out at high concentration. They clearly show, however, that the rate of substitution varies linearly with the ionic radius of the cation (Table 2.8) suggesting that the rate-determining step is the loss of the water molecule (i.e. S_N1). The larger the metal ion (as the charge is constant), the poorer its 'hold' over the water molecule and the faster the rate. The S_N1 mechanism is confirmed by a very slight dependence of rate constant on the nature of the

TABLE 2.8 First-order rate constants for complex formation (25°)

	$10^{-7}\,k_1\mathrm{sec}^{-1}$		$k_1\mathrm{sec}^{-1}$		$k_1\mathrm{sec}^{-1}$
Li^+	4·7	Be^{2+}	10^2	V^{2+}	30
Na^+	8·8	Mg^{2+}	10^5	Cu^{2+}	10^8
K^+	15	Ca^{2+}	10^8	Mn^{2+}	3×10^7
Rb^+	23	Sr^{2+}	5×10^8	Fe^{2+}	3×10^6
Cs^+	35	Ba^{2+}	9×10^8	Co^{2+}	2×10^6
—	—	Cd^{2+}	5×10^9	Ni^{2+}	2×10^4
—	—	Hg^{2+}	3×10^9	Cu^{2+}	2×10^8
—	—	Pb^{2+}	6×10^8	Zn^{2+}	3×10^7

entering ligand. The rate constants for the reactions of the alkaline earth ions show a much wider spread. The reactions of the aquo complexes of Sr^{2+} and Ba^{2+} are very fast. Together with Ca^{2+} they appear to show a correlation between rate constant and ionic radius as observed in Group I. The rate differences between Ca^{2+} and Mg^{2+} are significant in helping to understand the different behaviour of these ions as enzyme activators.

The rates of exchange of coordinated water have been measured by n.m.r. for certain divalent transition metal ions. The results are as follows: $V^{2+} < Ni^{2+} < Co^{2+} < Fe^{2+} < Mn^{2+} < Zn^{2+} < Cr^{2+} < Cu^{2+}$. Crystal field stabilization energy effects are important here. Those ions that are highly stabilized in this way such as Ni(II) will be reluctant to undergo reaction. The transition state will be of different stereochemistry and so there will be a loss of C.F.S.E. in the transition state. This will be an added contribution to the activation energy. The greater the C.F.S.E. of an ion, the greater will be the activation energy in its reactions. However, as the stereochemistry of the transition state is not known with certainty, it is not possible to make many quantitative predictions, although this has been successfully done for certain reactions. The fast reactions of Cr^{2+} and Cu^{2+} species result from the Jahn–Teller effect, where the axial ligands are released more readily than the in-plane ligands. The sequence Co^{2+}, Fe^{2+}, Mn^{2+} is dependent upon the radii of the ions, so confirming the S_N1 mechanism.

Much study has also been carried out on the replacement of coordinated water in the reaction:

$$[ML_5(H_2O)]^{n+} \xrightarrow{Y^-} [ML_5Y]^{(n-1)+}$$

Here again the mechanism appears to be S_N1.

Substitution in square planar complexes

It is obviously more likely that substitution in these complexes would proceed by an associative mechanism, rather than a dissociative one. This is confirmed by the fact that the reaction rates of these complexes usually depend upon the nature and concentration of the reagent. The most stable square planar complexes are the d^8 systems, particularly Pt(II). These have received much attention, while complexes of Ni(II), Pd(II), Rh(I), Ir(I) and Au(III) have also been studied.

Reaction (5) usually proceeds with retention of configuration. Steric

$$[PtA_2LX] + Y^- \qquad [PtA_2LY] + X^- \qquad (5)$$

effects are important and 5 coordinate species have been isolated, confirming the suggested bimolecular mechanism. The kinetics are usually of the form:

$$k_{obs} = k_1[\text{Complex}] + k_2[\text{Complex}][Y]$$

This has been interpreted in terms of two simultaneous reactions, one with the solvent and the other with Y^-. In general $k_1 \lll k_2$. The five coordinate intermediate could be square pyramidal or a trigonal bipyramid. It appears that the latter is the important one despite the fact that crystal field considerations favour the square pyramid.

The trans-*effect*
In reaction (6) it is possible

$$[\text{PtLX}_3]^- + Y^- \rightarrow [\text{PtLX}_2Y]^- + X^- \tag{6}$$

to obtain a *cis* or *trans* product. This is largely determined by the nature of L. In general the lability of any group is determined by the ligand *trans* to it. This is the *trans* effect, and it is possible to write down an arrangement of ligands in order of increasing *trans* effect, such as

$$\text{H}_2\text{O} < \text{OH}^- < \text{NH}_3 < \text{Cl}^- < \text{Br}^- < \text{I}^- < \text{NO}_2^- < \text{CO} \sim \text{CN}^-$$

This effect is very useful in designing synthetic pathways to pure *cis*- and *trans*-isomers. This is best illustrated with an example.

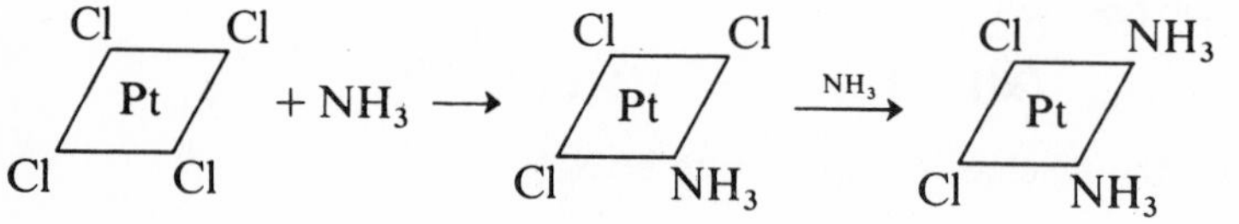

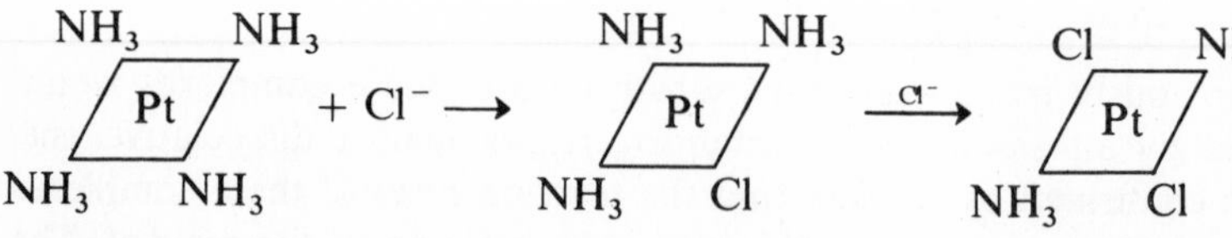

This effect is a kinetic effect, and must therefore be discussed in terms of the S_N2 mechanism. The presence of $\text{NO}_2^-, \text{CN}^-, \text{CO}$ at the high end of the above-listed series suggests that π-bonding is important in these cases. This is presumably by helping bond making by withdrawing the electron density from the metal and *trans*-ligand and hence aiding the nucleophilic substitution at that point. The remaining ligands must operate through a σ effect, presumably assisting bond breaking for the leaving *trans* ligand.

Redox reactions

Redox reactions of transition metal complexes are of two types. One type involves electron transfer with no change in the coordinated ligands, that is, only the oxidation state of the metal ion is changed. The other involves electron transfer coupled with group transfer. An example of the first type would be the reaction between ferrocyanide and ferricyanide, where $[Fe(CN)_6]^{4-}$ loses an electron and $[Fe(CN)_6]^{3-}$ gains one. That this reaction does occur very rapidly is shown by the use of isotopically labelled cyanide. An example of the second type is the oxidation of $[Cr(H_2O)_6]^{2+}$ by $[Co(NH_3)_5NCS]^{2+}$, giving eventually $[Cr(H_2O)_5NCS]^{2+}$. However, the initial product is the S-bonded thiocyanate complex, confirming that the transition state in the reaction involves a bridging thiocyanate group.

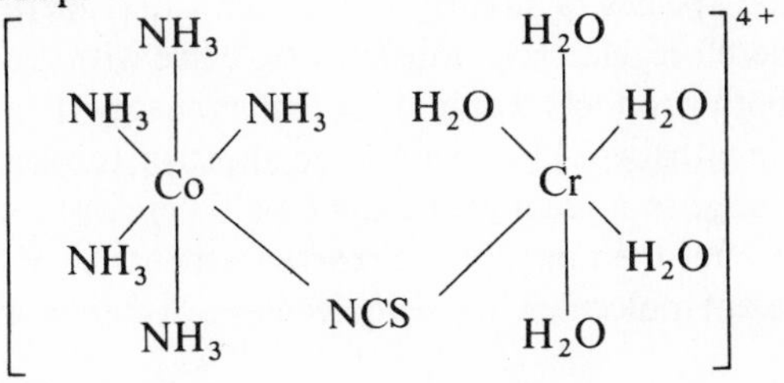

$$\left[\begin{array}{c} \underset{NH_3}{\overset{NH_3}{\underset{|}{NH_3}}} \quad Co \quad NH_3 \quad \quad \underset{H_2O}{\overset{H_2O}{\underset{|}{H_2O}}} \quad Cr \quad H_2O \\ NCS \end{array} \right]^{4+}$$

This particular reaction is slow, as the formation of the S-bonded thiocyanate is not favoured.

The actual mechanisms of redox reactions generally follow this same division. *Inner sphere mechanisms* are those in which the oxidant and reductant are connected by a bridging ligand common to the coordination shell of both metal ions, as in the reaction given immediately above. While group transfer usually occurs in this mechanism, it is not an essential feature of it, as we shall see shortly. The other mechanistic type is that of the *outer-sphere mechanism* in which electron transfer occurs through the intact coordination shell of both metal ions. The ferrocyanide-ferricyanide reaction proceeds by an outer sphere mechanism. It should be noted, however, that examples of outer sphere mechanisms are known where hydrogen ion transfer occurs as well. The following represents the transition state of a typical reaction of this type.

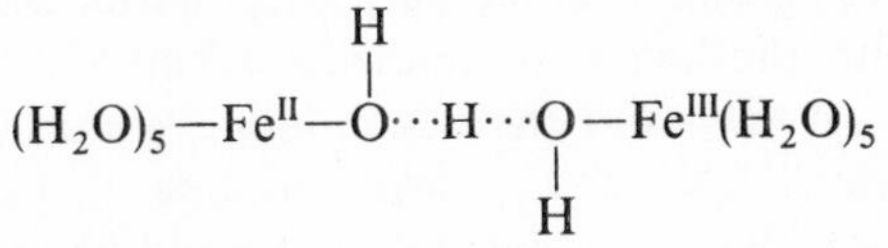

$$(H_2O)_5-Fe^{II}-\overset{\overset{\displaystyle H}{|}}{O}\cdots H\cdots \overset{\overset{\displaystyle H}{|}}{O}-Fe^{III}(H_2O)_5$$

Electron transfer by the outer sphere mechanism takes place more rapidly if the ligands are highly polarizable, such as cyanide or bipyridyl, so that the metal electrons may be delocalized over them. These reactions are very dependent upon the nature of the ligand.

The distinction between these two mechanisms is not a sharp one. An outer sphere mechanism may be confirmed if the rate of reaction is greater than the rate of ligand substitution into either coordination shell, as this obviously rules out the bridge mechanism. The rate law should also correspond to a transition state involving all the ligands of both metal ions.

One important principle that affects electron transfer reactions is the Franck–Condon principle which says that there must be no movement of nuclei during the time of electronic transition. This means that the geometry of the two species after reaction must be identical to that existing before the actual electron transfer. The Fe–CN bond length of the ferrocyanide species is slightly longer than that of the ferricyanide molecule. If, therefore, electron transfer took place with reactants in their ground state, both products would be of higher energy than the reactants, as their bond lengths would either be too short or too long. There is in fact no heat change in a reaction of this type. Therefore electron transfer will only occur after the appropriate rearrangements have taken place to give two equivalent molecules. These molecules will then be in vibrationally excited states.

This can provide a useful guide in assessing the relative rates of redox reactions. If substantial changes must take place before the configurations are alike, then the reaction will be slow. Other factors are also important, thus the rate of electron transfer between cobalt(II) and cobalt(III) species is much slower than the one we have just discussed. For cobalt(II) complexes are high spin and cobalt(III) complexes are nearly always low spin. The cobalt(III) complex prepared by electron loss from the cobalt(II) complex will then be a high spin complex and similarly the cobalt(II) species will be produced as a low spin complex. That is, the reaction is producing electronically excited products and the activation energy will therefore be appropriately higher.

Studies on electron transfer with group transfer reactions have been extensively carried out by Taube and his group, in particular with chromium(II) as the reducing agent. This is a good choice as chromium(III) is inert and hence group transfer may be confirmed. The reaction of chromium(II) with the following oxidizing agents has been examined: $CrCl^{2+}$, $FeCl^{2+}$, $AuCl_4^-$ and the pentammine species $[Co(NH_3)_5X]^{2+}$ where $X = NCS^-$, Cl^-, N_3^-, PO_4^-, acetate, oxalate etc. In the appropriate cases the use of ^{36}Cl has confirmed the transfer of Cl^- to the chromium ion, with electron transfer in the other direction. In addition, there is evidence for the S-bonded chromium(III) thiocyanate species, while intermediate bridged species have been isolated in other cases. Thus, in the oxidation of $Co(CN)_5^{3-}$ by $Fe(CN)_6^{3-}$, the species

$$[(NC)_5-Co-NC-Fe(CN)_5]^{6-}$$

has been isolated. Again, the rate of these reactions is very dependent upon the nature of the bridging ligand.

However, group transfer will not always occur. With $IrCl_6^{2-}$ oxidant hexa-aquochromium(III) is produced as the major product. This is a useful result as it shows that group transfer and electron transfer occur separately; that bridging serves to bring two metal ions together, but that once electron transfer has taken place, the bridge can break to give group transfer or no group transfer, depending upon the energetics of the overall reactions. This is illustrated in the following examples:

$$[Co(NH_3)_5Cl]^{2+} + [Cr(H_2O)_6]^{2+} \rightarrow$$
$$[(NH_3)_5Co-Cl-Cr(H_2O)_5]^{4+} \rightarrow$$
$$[Co(NH_3)_5H_2O]^{2+} + [CrCl(H_2O)_5]^{2+}$$
$$[IrCl_6]^{2-} + [Cr(H_2O)_6]^{2+} \rightarrow [Cl_5Ir-Cl-Cr(H_2O)_5] \rightarrow$$
$$[IrCl_6]^{3-} + [Cr(H_2O)_6]^{3+}$$

Much of the foregoing discussion is of relevance to biological oxidations and will be discussed further in the appropriate chapters. Thus it is of interest to note that we can have atom transfer in a redox reaction that does not involve the fission of the metal–ligand bond. Hydrogen transfer in the ferrous–ferric reaction discussed earlier is such an example. A second example of especial significance is given below, and is a possible mechanism for biological hydroxylation reactions.

$$
\begin{array}{ccc}
\text{H} & & \text{H} \\
\diagdown & & \diagup \\
& \text{O--O} & \quad\rightarrow\quad \text{Fe}^{2+}-\text{R} \\
\diagup & & \\
\text{Fe--R} & &
\end{array}
\qquad
\begin{array}{cc}
\text{OH} & \text{OH} \\
| & | \\
\text{Fe}^{2+}-\text{R} &
\end{array}
$$

A further point of interest is that there are a number of examples where electron transfer occurs *via* extended bridges, and in which the transfer of electrons through the bridge becomes rate-determining. It is obviously of interest to compare the rate of electron conduction through different ligands. Not unexpectedly the presence of unsaturated and polarizable ligands speeds up the reaction.

Taube and his associates have studied the reaction

$$(NH_3)_5Co-O-\overset{\overset{\displaystyle O}{\parallel}}{C}-R^{n+} + Cr^{2+} \longrightarrow$$

$$Co^{2+} + 5NH_3 + (H_2O)_5Cr-O-\overset{\overset{\displaystyle O}{\parallel}}{C}-R^{n+}$$

where R is a wide range of compounds with differing donor atoms for coordination to Cr(II). One striking feature is the increased rate observed for fumarate, oxalate, formate and maleate, attributable to the ready coordination of Cr(II) to the free carboxylate group and the enhancement of electron drift to cobalt by the unsaturated nature of the ligand. In the latter three examples it is suggested that the electron is transferred from Cr(II) to the ligand which remains coordinated as a radical anion, while finally the electron is transferred to the cobalt(III), with the regeneration of the ligand. There are a number of lines of evidence which support the radical anion theory and for the attack by Cr(II) on a remote position away from the carboxyl group coordinated to the cobalt.

Complementary and non-complementary reactions

The redox reactions discussed so far involve two reactants undergoing one electron change. These are complementary reactions. Non-complementary reactions are those where the oxidizer and reducer undergo reactions involving different numbers of electrons. In general, complementary reactions are faster than non-complementary reactions due either to the low probability of termolecular reactions (for example, in $2Fe^{2+} + Tl^{3+} \rightarrow 2Fe^{3+} + Tl^{+}$) or to the necessity of forming unstable oxidation states.

Whenever two-equivalent redox reactions occur, there are two possibilities, either that there is a two-electron transfer mechanism, with a doubly bridged intermediate, or that two one-electron steps occur with the formation of intermediate unstable oxidation states. In the reactions of Tl^{3+} the latter mechanism would involve the intermediate formation of Tl^{2+}. At present, there is little evidence for intermediate oxidation states. One non-complementary reaction of biological importance is the reaction of oxygen with iron(II).

$$2Fe^{II} + O_2 \rightarrow 2Fe^{III} + O_2^{2-}$$

This is a one step termolecular reaction.

Higginson has suggested the following rules:

(a) transition metal complexes will react with each other by a series of univalent changes,

(b) non-transition element compounds will react with each other in a series of bivalent changes unless at least one of the reactants is a free radical, in which case univalent changes occur,

(c) species derived from a transition element and a non-transition element will react with each other by either univalent or bivalent changes, univalent changes being more common.

These rules say nothing about the mechanism of the reaction.

A redox mechanism for base hydrolysis of cobalt (III) complexes

Two mechanisms have already been discussed for base hydrolysis of cobalt(III) complexes. These are the S_N2 and S_N1CB mechanisms. At this stage it is possible to return to this problem and put forward a third mechanism, involving electron transfer, that may well have some relevance to biochemical systems. Hydrolysis of cobalt(III) complexes by OH^- has certain special features. Thus this reaction differs from those of the analogous chromium(III) and rhodium(III) complexes, in that base hydrolysis is very much faster than acid hydrolysis, and that stereochemical change occurs frequently. In addition, ligand substitutions carried out in basic solution usually give equilibrium concentrations of possible products through ligand scrambling, implying the formation of labile cobalt(II) complexes. A number of reactions of cobalt(III) complexes are known to involve redox catalysis. All this suggests that base hydrolysis may also proceed *via* a cobalt(II) intermediate and generation of a radical.

In the following scheme Y represents any base.

$$[L_5Co^{III}X]^{n+} + Y^- \rightleftharpoons \{[L_5Co^{III}X]Y\}^{(n-1)+} \text{ Ion pair}$$

$$\{[L_5Co^{III}X]Y\}^{(n-1)+} \rightleftharpoons \{[L_5Co^{II}X]Y^{\cdot}\}^{(n-1)+} \text{ Ion pair}$$

$$\{[L_5Co^{II}X]Y^{\cdot}\}^{(n-1)+} \rightleftharpoons [L_5Co^{II}Y^{\cdot}]^{n+} + X^-$$

$$Co^{II} + 5L + Y^{\cdot} \qquad\qquad [L_5Co^{III}Y]^{n+}$$

The overall reaction pathway will then depend upon the lability of the cobalt(II) intermediate and the oxidizing power of the radical $Y^{\cdot}$. For $Y^- = OH^-$, the rate-determining step is the electron transfer from OH^- to Co(III), presumably to an excited state of Co(III) to avoid the generation of a low spin cobalt(II) state.

This scheme will allow an explanation of the special features of base hydrolysis of cobalt(III) complexes discussed above. Cobalt(III) differs from chromium(III) and rhodium(III) in the ease of reduction of the metal to the bivalent state, so explaining why base hydrolysis is so much faster than acid hydrolysis for cobalt(III). In general, it should be noted that the larger the ligand field strength, the smaller the differences in rates of acid and base hydrolysis for cobalt(III). Stronger ligand fields stabilize cobalt(III) over cobalt(II), (C.F.S.E. effect) and hence increase the redox potential for cobalt(II)–cobalt(III). Such an electron transfer reaction has been suggested for $Co^{3+} + {}^-OOH \rightarrow Co^{2+} + \cdot OOH$.

FURTHER READING

C. S. Phillips and R. J. P. Williams, *Inorganic Chemistry*, Oxford University Press, 1965

F. A. Cotton and G. Wilkinson, *Advanced Inorganic Chemistry* (2nd Ed.), Interscience, New York, 1966

J. Lewis and R. G. Wilkins (Eds.), *Modern Coordination Chemistry*, Interscience, New York, 1960

S. F. A. Kettle, *Coordination Compounds*, Nelson, London, 1969

B. N. Figgis, *Introduction to Ligand Fields*, Interscience, New York, 1966

A. E. Earnshaw, *Introduction to Magnetochemistry*, Academic Press, New York, 1969

D. Sutton, *Electronic Spectra of Transition Metal Complexes*, McGraw-Hill, London, 1968

F. J. C. Rossotti and H. S. Rossotti, *The Determination of Stability Constants*, McGraw-Hill, London, 1961

F. Basolo and R. G. Pearson, *Mechanisms of Inorganic Reactions*, (2nd Ed.), Wiley, New York, 1967

J. O. Edwards, *Inorganic Reaction Mechanisms*, Benjamin, New York, 1964

A. McAuley and I. Hill, 'Kinetics and Mechanism of metal-ion Complex Formation.' *Quart. Rev. Chem. Soc.*, 1969, 18

R. D. Gillard, 'A Possible Mechanism for Base Hydrolysis and Other Reactions of Cobalt(III) Complexes,' *J. Chem. Soc. (A)*, 1967, 917

CHAPTER THREE

THE STUDY OF METALLOPROTEINS AND OTHER METAL-CONTAINING BIOLOGICAL MOLECULES

Reference has already been made to a number of possible roles for metal ions in biological systems. These are summarized again for convenience in Table 3.1. In recapitulation, it may be seen that there is a relationship

TABLE 3.1 Some roles of metal ions in biological processes

Metal ion	Role	Metal binding strength
Na^+, K^+	Charge carriers	Weak
Mg^{2+}, Ca^{2+}	Control mechanisms	Medium
	Enzyme activators	
Zn^{2+}	Strong Lewis acid in hydrolytic enzymes	Strong
Fe, Co, Cu, Mo	Redox catalysts	Very strong

between the role of the metal ion and the binding strength of the protein for the metal. Thus, for the alkali metals, whose major role is that of charge carrier, interaction with the protein is slight, whereas the metal ions of the redox metalloenzymes are very firmly bound indeed, usually by macrocyclic molecules of the porphyrin type or by strongly polarizable donors. While in the following discussion much emphasis is laid upon the role of the metal ion, its importance must always be kept in perspective. The catalytic efficiency of the enzyme is an overall effect of the whole enzyme. Thus, the role of zinc in certain hydrolytic enzymes is one of a Lewis Acid, but the rate enhancement of these enzymatic reactions, compared with non-enzymatic reactions, is such that it is very unlikely that this is the only factor. This has been demonstrated[1] for bovine carboxypeptidase and the complex with glycyl-L-tyrosine[1b] by X-ray diffraction studies. This has shown that very large conformational changes occur in the presence of this substrate. The enzyme catalyses the hydrolysis of C-terminal peptide links in peptides and proteins and contains one mole of zinc per mole of protein. The substrate is bound through the peptide carbonyl group to the zinc. This serves to orientate the substrate so that an interaction occurs between its carboxyl group and a positively charged arginine residue in the protein, resulting in this residue moving by 0·2 nm towards the substrate. However, this movement is magnified through the effect of the protein structure so that the phenolic hydroxyl group of a tyrosine residue moves about 1·2 nm with a twisting of the carbon–carbon chain, so that the OH group is in the vicinity of the substrate's peptide link. The zinc will also polarize the carbonyl group while the tyrosine could donate its hydrogen to the peptide amide nitrogen thus facilitating cleavage of the link. This is depicted schematically in Figure 3.1. Further aspects of this reaction will be discussed later.

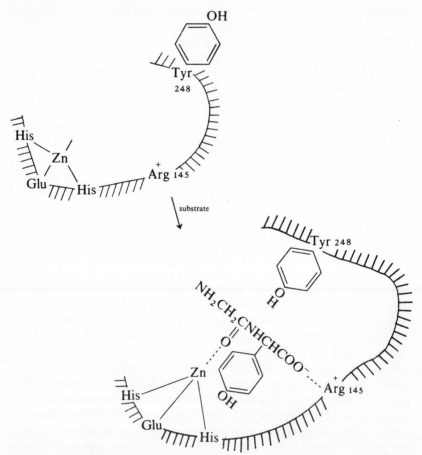

Figure 3.1   Conformational changes induced in carboxypeptidase A by substrate binding.

In the study of metalloproteins all the factors involved in the study of enzymes in general are important. In this chapter we are concerned particularly with the function of the metal. The first step towards elucidating the function of the metalloprotein must involve the characterization of the metal–protein interaction, in terms of the nature of the binding groups and the strength of the interaction with the metal. This will then lead on towards an understanding of the structure of the active site and the associated electronic properties that are responsible for the catalytic function of the enzyme.

The identification of metal binding groups in metalloproteins is a complex problem in view of the large number of potential binding groups present. In certain cases, as noted in Chapter 1, the situation is somewhat simpler, as some of these groups have a much greater affinity for the metal ion than others; the imidazole group of histidine and the —SH group of cysteine are particularly important. However, in other metalloenzyme systems, where the binding is much weaker, the active site will be defined with much greater difficulty. It should also be emphasized that the description of the binding site in terms of the immediate binding groups will not allow the explanation of phenomena resulting from long range effects.

In this chapter it is intended to survey briefly the methods for measuring metal ion–ligand interaction, to show what broad generalizations can be applied to the behaviour of different metals and ligands, and then to use this information with that provided by other approaches (such as spectroscopy, redox potentials, pH titrations) to indicate how the binding of the metal may be studied. Much emphasis will be laid upon the value of replacing main group metals by either one or a series of transition metals in order to throw further light on this problem. The use and design of model compounds will be discussed together with the information that may be obtained from a study of their differences and similarities when compared with the *in vivo* systems. Mention will also be made of more specialized techniques which will eventually doubtless supersede some of the approaches described here. These techniques are those of NMR, EPR, Mossbauer spectroscopy, ORD/CD, and others such as the application of relaxation techniques in kinetic studies.

An increasing number of enzymes, including metalloenzymes, have been examined by X-ray diffraction methods.[1d] These, especially when coupled with the results of chemical amino acid sequence determination, will give much valuable information. Studies at 0·5–0·6 nm resolution, while allowing the examination of overall protein structure, will not necessarily allow the positive identification of metal-binding groups, but will still allow certain conclusions to be made. Thus, a substantial weight of evidence had been accumulated on carboxypeptidase indicating that zinc-binding was through RS$^-$ and RNH$_2$ groups. However, a 0·6 nm resolution study[2] indicated that the peptide chain ended about 25 nm away from the zinc. This cannot be reconciled with the suggestion that the RNH$_2$ group binds zinc, although it should be noted that other workers[3] had produced chemical evidence against RNH$_2$ involvement in zinc binding. The more recent 0·2 nm resolution study[1] has shown that the zinc atom has three protein ligands, two imidazoles (His 69, His 196) and a glutamic acid residue (Glu 72). There is no evidence for RS$^-$ binding.

Another recent X-ray study involves rubredoxin, a non-heme iron protein, isolated from *C. Pasteurianum* and other species. Here the environment of the iron atom is tetrahedral, the iron being bound to four cysteine sulphur atoms.[4] In Table 3.2 are listed the metalloenzymes whose

TABLE 3.2 X-ray diffraction studies

	Resolution (nm)	Binding groups/metal	Footnote
Oxymyoglobin	0·20	Porphyrin, imidazole, O$_2$/Fe	a
Deoxymyoglobin	0·28	Porphyrin, imidazole, /Fe	b
Oxyhemoglobin	0·28	Porphyrin, imidazole, O$_2$/Fe	c
Deoxyhemoglobin	0·55	Porphyrin, imidazole, /Fe	d
Erythrocruorin	0·25	Porphyrin, imidazole, /Fe	e
Ferricytochrome *c*	0·28	Porphyrin, imidazole, methionine, /Fe	f
Rubredoxin	0·25	Four cysteine S atoms/Fe	g
Carboxypeptidase	0·20	Two imidazoles, glutamic acid, H$_2$O/Zn	h
Carbonic Anhydrase	0·55	Not —SH or —NH$_2$, probably three imidazoles, H$_2$O/Zn	i

[a] J. C. Kendrew, R. E. Dickerson, B. E. Strandberg, R. G. Hart, D. R. Davies, D. C. Phillips and V. C. Shore, *Nature*, 1960, **185**, 422.

[b] C. L. Nobbs, H. C. Watson and J. C. Kendrew, *Nature*, 1966, **209**, 339.

[c] M. F. Perutz, H. Muirhead, J. M. Cox, L. G. Goaman, F. S. Matthews, E. L. McGandy and L. E. Webb, *Nature*, 1968, **219**, 29.
M. F. Perutz, H. Muirhead, J. M. Cox and L. G. Goaman, *Nature*, 1968, **219**, 131.

[d] W. Bolton, J. M. Cox and M. F. Perutz, *J. Mol. Biol.*, 1968, **33**, 283.

[e] R. Huber, O. Epp, W. Streigemann and H. Formanek, *Eur. J. Biochem.*, 1971, **19**, 42.

[f] R. E. Dickerson, T. Takano, D. Eisenberg, O. B. Kallai, L. Samson, A. Cooper and E. Margoliash, *J. Biol. Chem.*, 1971, **246**, 1511.

[g] J. R. Herriot, L. C. Sieker, L. H. Jensen and W. Lovenberg, *J. Mol. Biol.*, 1970, **50**, 391.

[h] G. N. Reeke, J. A. Hartsuck, M. L. Ludwig, F. A. Quiocho, T. A. Steitz and W. N. Lipscomb, *Proc. Nat. Acad. Sci. U.S.*, 1967, **58**, 2220.
W. N. Lipscomb, G. N. Reeke, J. A. Hartsuck, F. A. Quiocho and P. H. Bethge, *Phil. Trans. Roy. Soc. Lond.*, 1970, **B257**, 177.

[i] K. Fridborg, K. K. Lannan, A. Liljas, J. Lundin, B. E. Strandberg, R. Strandberg, B. Tilander and G. Wiren, *J. Mol. Biol.*, 1967, **25**, 505.
See S. Lindskog, *Structure and Bonding*, 1971, **8**, 153.

three-dimensional structures have been determined at the time of writing, together with details of the resolution and, where possible, the binding groups. It is unlikely that sufficient resolution will be achieved in many cases to allow a full assessment of metal–protein interaction, and in any case studies on the electronic states of the different groups will still be required.

Perutz had reviewed[1d] the results of X-ray analysis of enzymes, and has attempted to relate these to their function. The structures of the different enzymes are very complex, and from an overall point of view have little in common with each other, apart from the fact that polar residues are excluded from the interior (except when specially required for the function of the enzyme). In terms of detailed structure, as noted in Chapter 1, there is a cleft for substrate binding together with the appropriate groups for providing the necessary binding interactions. The hydrophobic interior of the protein facilitates these interactions. The oxygen carriers hemoglobin and myoglobin have received particular attention, and in Chapter 7 we will see how the work of Perutz and his group has made a very significant contribution to the understanding of the mechanism of action of hemoglobin.

Another approach to studying binding groups is illustrated by the following study[5] of model compounds for the non-heme iron proteins, in which an attempt was made to identify the chromophore in those proteins by comparing EPR and adsorption spectra with those of a number of related model compounds. The mixing of iron(III) chloride with an equal amount of sodium sulphide and excess 2-mercaptoethanol at pH 9 gave a solution of an iron mercaptoethanol sulphide complex with similar spectroscopic properties to those of the actual protein. A Job's plot (see later) indicated that a 1:2 complex with mercaptoethanol had been formed, so it was suggested that the environment around the iron in the model compound is represented by Figure 3.2, with an octahedral

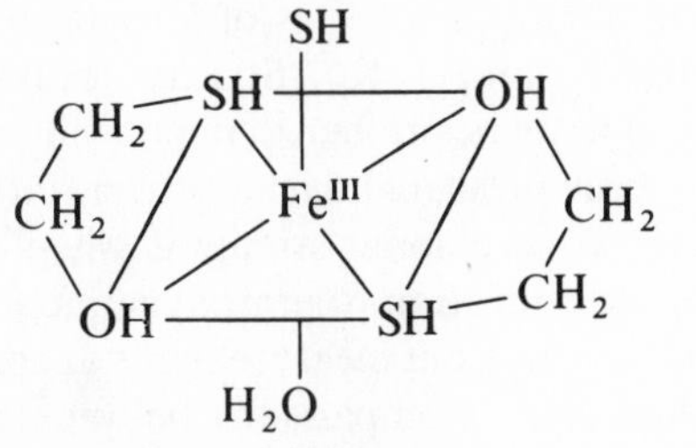

Figure 3.2   The structure of a model complex for metal-binding in non-heme iron proteins.

stereochemistry. The environment of the iron in the protein is then assumed to be similar, although here there are a number of iron atoms per mole of metalloenzyme and a bridged structure has been postulated. The model compound itself is inherently unstable due to reduction of Fe^{3+} to Fe^{2+} by mercaptoethanol, and the reduced form is readily reoxidized by molecular oxygen. Its properties do therefore justify an analogy with the non-heme iron proteins.

FORMATION CONSTANTS (STABILITY CONSTANTS)

These provide a quantitative measure of the extent to which a metal will complex with any particular group or ligand. There are now a number[6] of very useful compilations of formation constants including those for a range of metals with many important biological molecules, such as amino acids, peptides, purines, phosphates, nucleic acids, etc.

If we have a metal ion M and a monodentate ligand L, then, provided precipitation does not occur, a series of stepwise equilibria will come into existence for which formation constants may be written, the higher the value of K the greater being the concentration of the complex species. 'Stepwise' stability constants are illustrated as follows:

$$M + L \rightleftharpoons ML \qquad K_1 = [ML]/[M][L]$$

$$ML + L \rightleftharpoons ML_2 \qquad K_2 = [ML_2]/[ML][L]$$

$$ML_{(N-1)} + L \rightleftharpoons ML_N \qquad K_N = [ML_N]/[ML_{N-1}][L]$$

An alternative way of expressing these results involves the use of overall formation constants.

$$M + L \rightleftharpoons ML \qquad \beta_1 = [ML]/[M][L]$$

$$M + 2L \rightleftharpoons ML_2 \qquad \beta_2 = [ML_2]/[M][L]^2$$

$$M + NL \rightleftharpoons ML_N \qquad \beta_N = [ML_N]/[M][L]^N$$

It may be shown that the two types of formation constants are interrelated by $\beta_N = K_1 . K_2 . K_3 \ldots K_N$. Strictly speaking, they should be expressed in terms of activities rather than concentrations. While formation constants have been measured over a concentration range in order to allow extrapolation to zero ionic strength, when the activity will be represented by the concentration (thermodynamic formation constants), it is more customary to carry out measurements at constant ionic strength using an added electrolyte. Comparisons between systems at the same ionic strength are then meaningful. These are stoicheiometric stability constants.

The measurement of formation constants

This subject has been discussed in detail in a number of publications.[7] Most measurements are made at 25°. The analytical method must be one that will not disturb the equilibrium. For the formation of a 1:1 complex, provided the initial concentrations of metal and ligand are known, then it is only necessary to know the concentration of one species at equilibrium to allow the calculation of the formation constant. Usually, however, a variety of complex species will exist in stepwise equilibria and a number of formation constants have to be determined, so the situation is obviously much more complex. The calculations themselves are tedious and computing techniques are usually employed.

One widely used method, that of potentiometric titration, may be applied if the ligand may be protonated and the pK_a of its conjugate acid is known. The accuracy of this method falls away rapidly as the pK_a drops towards 2. The method itself involves setting up a competition for the ligand between metal ion and hydrogen ion. Usually a known solution of metal salt and mineral acid is titrated with a standard solution of ligand while the pH is read at appropriate intervals. Alternatively, a mixture of ligand and acid may be titrated with metal salt, or the metal ion and ligand may be titrated with alkali. The difference between total added acid and free acid (pH) is a measure of the amount of protonated ligand. From the pK_a the amount of free ligand may be calculated. The amount of ligand complexed by a known amount of metal ion is then derived from the equation

$$L_{total} = LH^+ + L + LM^{n+}$$

It is then possible to work out $\bar{n}$, the average number of ligand molecules complexed with each metal atom. A plot of $\bar{n}$ against pA, where A is the concentration of the free ligand, is known as the formation curve. A curve may level off, for example, at $\bar{n} = 2$. This indicates that a 1:2 species is the highest species formed in solution. By interpolating at $\bar{n} = 1$, $\bar{n} = 2$, etc. it is possible to obtain rough values of formation constants. Accurate values of overall stability constants are usually obtained by a curve fitting procedure. Measurements have to be carried out over a range of metal ion concentration to ensure that dimerization is not occurring.

This method is probably the quickest and most widely applicable of all methods (provided that the pK_a of the ligand allows its application) but there are many others. These include a variety of optical methods, the use of emf measurements, polarography and solvent extraction.

Job's method may often prove to be useful in determining the stoicheiometry of complex formation in solution and also in obtaining the formation constants in favourable cases. This method only applies where the complex absorbs in a region of the spectrum where the ligand and metal ion do not. A number of solutions are prepared in which the concentrations of the two reagents are continuously varied, keeping the total molarity constant. The optical density of each solution is measured at λ_{max} and plotted against solution composition. The absorbance (and hence the concentration of complex) will be a maximum when the two reagents are present in the same stoicheiometry as the complex. The stoicheiometry of the solution at maximum Optical Density will therefore correspond to the stoicheiometry of the complex. Figure 3.3 illustrates the formation of 1:1 and 1:2 complexes. The method becomes less practical at higher stoicheiometries.

Some formation constant data is presented in Table 3.3.

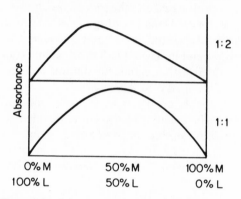

Figure 3.3 Job's method of continuous variation.

A number of general points must be borne in mind when considering formation constant data.

(1) Successive formation constants will decrease in the absence of any special effect. This is a statistical phenomenon reflecting the decreasing number of available coordination sites as more ligands complex.

(2) Steric repulsions between ligands may result in a lowered stability.

(3) It is customary to speak of a 'chelate effect', in which chelate complexes are supposedly more stable than analogous complexes with unidentate ligands, e.g. $[Ni(en)_3]^{2+}$ and $[Ni(NH_3)_6]^{2+}$, en = $NH_2CH_2CH_2NH_2$. The non-existence (or at least the non-importance) of the chelate effect has been demonstrated.[8] Comparisons in support of it are usually based upon formation constants of different dimensions. It may be shown, however, that five-membered ring chelates are more stable than complexes with six-membered rings and that increasing ring size leads to increased instability.

TABLE 3.3[a] Logarithms of formation constants in aqueous solution at 25°C

Ligand	Ionic strength	Constant	Mg^{2+}	Ca^{2+}	Mn^{2+}	Fe^{2+}	Fe^{3+}	Co^{2+}	Ni^{2+}	Cu^{+}	Cu^{2+}	Zn^{2+}
Glycine	→0	K_1	3·44	1·38	3·44	4·3	—	5·23	5·77	—	8·62	5·52
	→0	K_2	3·01	—	—	3·5	—	4·02	4·80	—	6·97	4·44
Cysteine	0·01	K_1	<4	—	4·1	—	—	9·3	—	19·2	—	9·86
	0·01	K_2	—	—	—	—	—	7·6	19·3	—	—	8·84
	→0	β_2	—	—	—	11·77	32·1	—	—	—	—	—
Serine	Varied	K_1	—	~0·5	—	7·0	—	8·0	—	—	—	—
	Varied	β_2	—	—	—	—	—	—	—	—	14·54	—
Arginine	0·15	K_1	—	—	—	—	—	3·87	—	—	—	—
	0·15	K_2	—	—	—	—	—	3·20	—	—	—	—
	0·15	K_3	—	—	—	—	—	2·08	—	—	—	—
	0·01	β_2	—	—	—	—	—	—	—	—	13·90	7·80
Aspartic acid	0·10	K_1	2·43	1·60	3·74	—	—	5·90	7·12	—	8·57	5·84
	0·10	K_2	—	—	—	—	—	4·28	5·27	—	6·78	4·31
Histidine	0·01	β_2	—	—	7·74	9·3	—	13·86	15·9	—	18·33	12·88
Lysine	0·01	β_2	—	—	2·0	4·5	—	6·8	8·8	—	13·7	7·6
Glutamic acid	Varied	K_1	1·9	2·05	3·3	4·6	—	5·06	5·90	—	7·85	5·45
		K_2	—	—	—	—	—	3·40	4·44	—	6·55	4·01
Glutamine	0·16	K_1	—	0·18	—	—	—	—	—	—	—	—
Asparagine	0·01	β_2	~4·0	—	~4·5	6·5	—	8·13	10·6	—	14·9	8·7
Glycylglycine	→0	K_1	1·06	1·24	2·15	—	—	3·49	—	—	6·04	3·80
		K_2	—	—	—	—	—	2·39	—	—	5·62	2·77
Glycylserine	0·01	K_1	—	—	—	—	—	—	—	—	3·7	3·7
Alanylglycine	0·01	K_1	—	0·66	—	—	—	3·15	—	—	3·0	3·0
Glycylalanine	0·01	K_1	—	—	—	—	—	2·58	—	—	—	4·1
	0·01	K_2	—	—	—	—	—	—	—	—	—	—

TABLE 3.3 *continued*

Ligand	Ionic strength	Constant	Mg^{2+}	Ca^{2+}	Mn^{2+}	Fe^{2+}	Fe^{3+}	Co^{2+}	Ni^{2+}	Cu^{+}	Cu^{2+}	Zn^{2+}
Glycylglycylglycine	→0	K_1	—	—	1·41	—	—	2·98	—	—	5·41	3·33
	—	K_2	—	—	—	—	—	2·61	—	—	5·15	2·99
Imidazole	Varied	K_1	—	0·08	—	—	—	—	2·94	—	4·20	2·58
		K_2	—	—	—	—	—	—	2·41	—	3·47	2·19
		K_3	—	—	—	—	—	—	1·99	—	2·84	2·41
		β_2	—	—	—	—	—	—	—	10·87	—	—
Proline	0·03	β_2	<4	—	5·5	8·3	—	9·3	—	—	16·8	10·2
1,10 Phenanthroline	Varied	K_1	—	—	—	—	—	—	—	—	6·30	—
	—	K_2	—	—	—	—	—	—	—	—	6·15	—
	—	K_3	—	—	—	—	—	—	—	—	5·50	—
	—	β_3	—	—	7·35	21·0	14·10	—	—	—	—	—
2-methyl,1,10 phenanthroline	—	β_3	—	—	—	10·8	—	—	—	—	—	—
Ammonia	—	K_1	0·23	-0·2	—	—	—	2·11	2·79	5·93	4·15	2·37
	—	K_2	-0·15	-0·6	—	—	—	1·63	2·24	4·93	3·50	2·44
	—	K_3	-0·42	-0·8	—	—	—	1·05	1·73	—	2·89	2·50
Ethylenediamine	1·0M	K_1	0·37	—	2·73	4·28	—	5·93	7·66	—	10·72	5·92
	—	K_2	—	—	2·06	3·25	—	4·73	6·40	—	9·31	5·15
	—	K_3	—	—	0·88	1·99	—	3·30	4·55	—	—	1·86
1,3 diaminopropane	0·15	K_1	—	—	—	—	—	—	6·98	—	9·77	—
	—	K_2	—	—	—	—	—	—	4·93	—	7·17	—
Ethanolamine	→0	β_2	—	—	—	—	—	—	—	—	6·68	—
2-mercapto ethylamine	1·0	K_1	—	—	—	—	—	9·38	10·96	—	—	8·07

TABLE 3.3 continued

Ligand	Ionic strength	Constant	Mg^{2+}	Ca^{2+}	Mn^{2+}	Fe^{2+}	Fe^{3+}	Co^{2+}	Ni^{2+}	Cu^{+}	Cu^{2+}	Zn^{2+}
N,N dimethyl ethylenediamine	0·1	K_1	—	—	—	—	—	—	—	—	9·23	—
	—	K_2	—	—	—	—	—	—	—	—	6·73	—
N,N' dimethyl ethylenediamine	0·1	K_1	—	—	—	—	—	—	—	—	9·69	—
	—	K_2	—	—	—	—	—	—	—	—	6·65	—
2-methyl thiourea	Varied	K_1	—	—	—	—	—	—	3·23	—	5·42	—
	—	K_2	—	—	—	—	—	—	2·79	—	5·11	—
Acetate	Varied	K_1	0·51	0·41	—	—	—	—	0·67	—	2·16	1·03
Oxalate	—	β_2	2·55	3·0	3·82	4·7	—	4·7	—	—	8·5	7·02
Salicylic acid	—	K_1	—	0·36	—	—	15·2	—	—	—	10·6	—
Malic acid	→0	K_1	1·55	2·66	—	—	—	—	—	—	—	3·32
Uramil-N-N diacetic acid	→0	K_1	3·1	5·2	4·0	—	—	3·2	3·3	—	—	3·2
Nitrilo triacetic acid	0·1	K_1	7·0	8·17	7·44	8·84	15·87	10·6	—	—	12·68	—
	—	K_2	—	3·43	—	—	8·45	3·9	—	—	—	—
EDTA	0·1	K_1	9·12	11·0	14·04	13·9	25·1	16·2	18·56	—	18·79	—
Hydroxide	—	K_1	2·6	1·1	2·8	3·2	—	3·6	3·8	—	6·5	4·7
Azide ion	0	K_1	—	—	—	—	5·06	—	—	—	2·56	—

ᵃ Taken from 'Stability Constants', Special Publication of the Chemical Society No. 6, 1957. Further data are contained in Chapter 8. A few constants were not measured at 25°C.

FACTORS AFFECTING THE STABILITY OF METAL COMPLEXES

It is clearly of great importance to be able to assess the factors that influence complex stability. If we are given certain metals and a range of donor groups, it should then be possible to make sensible predictions regarding the mode of metal binding. A consideration of the available results on complex stabilities allows us to make certain broad generalizations and also to observe more refined trends superimposed upon these.

Class (a) and Class (b) metals

Chatt and Ahrland have divided metal ions into these two types. Class (a) metals include the alkali and alkaline earth metals, zinc and the early members of the first transition series (up to Cr). These form more stable complexes with ligands having donor atoms from the first short period (N, O, F), than with analogous ligands having donor atoms from the second short period (P, S, Cl). Class (b) behaviour does not extend to many metals of biological interest but the later transition elements (Mn on) are border line in behaviour between class (a) and class (b). The reverse stability situation is true. For class (b) metals, the stability of the complex will be greater when the donor atom is from the second short period. More generally the stability will decrease with donor atom in the sequence $S > I > Br > Cl > N$. For class (a) metals there will be a useful correlation between the base strength of the ligand and complex formation, that is regarding H^+ and M^{n+} as being equivalent. For class (b) elements, simple consideration of this type plays a less important role and other factors will be more important; thus they will form π bonds with ligands having empty π orbitals such as CO and CN^-. Crystal field effects will also be important as have already been discussed.

'Hard' and 'Soft' acids and bases

This classification is becoming increasingly more important. 'Hard' metal ions or acids are like the proton, small and not easily polarized ($\equiv$ class (a)), while 'Soft' acids are large and easily polarized. Ligands with highly electronegative donor atoms are hard bases while polarizable ligands are soft. A general rule is that stable complexes are those formed between hard acids and hard bases, and soft acids and soft bases. Table 3.4 classifies some acids and bases. While we will reserve our summary of conclusions until later, it may be seen that the information contained in Table 3.4 will allow a number of phenomena to be appreciated. Sulphur donors are soft bases

TABLE 3.4 'Hard' and 'Soft' acids and bases

Acids				Bases			
Hard		*Intermediate*	*Soft*	*Hard*		*Intermediate*	*Soft*
H^+ Mn^{2+}		Zn^{2+} Sn^{2+}	Cu^+ Pd^{2+}	H_2O RNH_2		pyridine	$-SCN^-$ R_3P
Li^+ Cr^{3+}		Cu^{2+} Pb^{2+}	Ag^+ $Pt^{2+,4+}$	ROH $-NCS^-$		Br^-	$-CN^-$ R_3As
Na^+ Fe^{3+}		Ni^{2+}	Au^+ Cd^{2+}	R_2O Cl^-		N_3^-	RSH H^-
K^+ Co^{3+}		Fe^{2+} Group V	Tl^+	OH^- PO_4^{3-}		NO_2^-	R_2S $S_2O_3^{2-}$
Group IIA/IIIA				OR^- SO_4^{2-}			RS^-
				NH_3			

while nitrogen donors are hard. It is not surprising therefore that the thiocyanate ion is nitrogen bonded to the elements of the first transition series but sulphur bonded to those of the second and third.

The Irving–Williams series of stability

It is often valuable to prepare a series of model compounds with varying transition metal ions and to compare their formation constants with those of the apoenzyme of the system in study with the same series of metals. Normally in model compounds, the formation constants of complexes of the bivalent ions of the first transition series follow the Irving–Williams sequence of stability (Mn(II) < Fe(II) < Co(II) < Ni(II) < Cu(II) > Zn(II). The complexing power of a metal ion will vary with the charge/radius ratio of that ion, i.e. polarizing power. For the above ions this ratio varies in the sequence Mn(II) < Fe(II) < Co(II) ⩽ Zn(II) ⩽ Ni(II) > Cu(II). The departure of the results from this sequence reflects the varying C.F.S.E. contributions of each ion. However, the situation with regard to the metalloenzymes may not always be straightforward, stability constant data may or may not follow the same pattern of behaviour as model compounds and this must be accounted for in the mechanistic scheme. This may be of use in testing a suggested mechanism.

Some conclusions from formation constant data

The general trends discussed above and the detailed formation-constant data allow useful conclusions to be drawn. It is possible to predict which ligands will combine with a wide range of metal ions and those which will be much more selective in their action. Much of this information provides the underlying reasons for the choice of certain analytical reagents for certain operations. The hard-soft concept seems to be particularly useful in rationalizing the data. It is also possible to understand some data in terms of steric hindrance in the ligand and hence a selectivity for metal ions based upon the preferred stereochemistry of that metal. Thus o-phenanthroline is a good ligand for both Fe^{2+} and Cu^+, which favour octahedral and tetrahedral stereochemistries respectively. The presence of methyl groups adjacent to the nitrogen donor atoms in a phenanthroline makes it a selective reagent for copper in the presence of iron. The methyl groups prevent the ligand satisfying the stereochemical requirements of Fe^{2+}.

Considering the data from the point of view of the metal ion, it may be seen that interesting parallels and differences between certain biologically important metal ions emerge. This data is important when we come to consider metal for metal replacement, as metal ions must bind to the same extent and to the same binding groups in order for successful replacement to be feasible. There is a good parallel between Mg^{2+} and Mn^{2+}, and manganese will replace magnesium in many biological systems. However, Mg^{2+} and Ca^{2+} are rather more different. The ligand preferences for certain metal ions are listed in Table 3.5.

TABLE 3.5 Preferred ligand binding groups for metal ions

Metal	Ligand groups
K^+	Singly-charged oxygen donors or neutral oxygen ligands
Mg^{2+}	Carboxylate, phosphate, nitrogen donors
Ca^{2+}	$\equiv Mg^{2+}$ but less affinity for nitrogen donors, phosphate and other multidentate anions
Mn^{2+}	Similar to Mg^{2+}
Fe^{2+}	—SH, NH_2 > carboxylates
Fe^{3+}	Carboxylate, tyrosine, $—NH_2$, porphyrin (four 'hard' nitrogen donors)
Co^{3+}	Similar to Fe^{3+}
Cu^+	—SH (cysteine)
Cu^{2+}	Amines ≫ carboxylates
Zn^{2+}	Imidazole, cysteine
Mo^{2+}	—SH
Cd^{2+}	—SH

METAL ION INTERACTION WITH AMINO ACIDS, PEPTIDES AND PROTEINS

A discussion of metal ion interaction with some model ligands is given later in this chapter. In view of the importance of metal–protein interactions we will discuss separately model metal complexes with amino acids, peptides and proteins and consider the implications of these results in terms of metal ion interactions in metalloproteins. This has been the subject of a comprehensive review.[9]

Amino acids and peptides

Each of the twenty or so naturally occurring $L - \alpha$ amino acids may form a stable five membered chelate ring with a metal ion. If there are no complicating donor side chains, then the donor groups will be the amino and carboxylate groups. Under other pH conditions this may not be the case; thus at lower pHs the amino acid may coordinate as a neutral ligand. When the carboxylate group is not part of such a chelate ring, then often four-membered rings are formed in which both oxygen atoms of the carboxylate are coordinated, alternatively the carboxyl group may bridge the metal atoms.

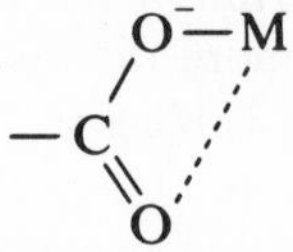

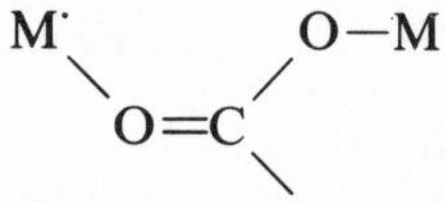

$$-C\overset{\displaystyle O^--M}{\underset{\displaystyle O}{\Big\backslash}}$$

$$M\diagdown_{\displaystyle O=C}\diagup^{\displaystyle O-M}\diagdown$$

Simple peptides (again other than those with side chains) will combine less strongly with metal ions than do the amino acids, as is indicated by formation constant data. In order to avoid the formation of large rings it is not likely that the terminal groups alone would be utilized. Much work, including X-ray studies,[10] has been carried out on metal complexes of amino acids and peptides. Presumably the conclusions reached in X-ray studies also hold in solution. These show the importance of planar peptide links in metal peptide complexes. It appears that non-deprotonated N peptide atoms are not used as metal-binding sites. These would involve a tetrahedral environment. When a metal is bonded to three donor groups of a single peptide molecule (the central one of which is the deprotonated nitrogen atom), then the three donor atoms will be in the same coordination plane. A number of peptide complexes are known in which a tautomeric form of the peptide is stabilized by coordination

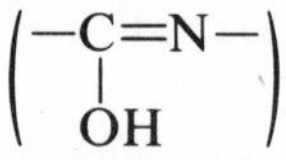

$$\left(\begin{array}{c}-C=N-\\ |\\ OH\end{array}\right)$$

so allowing the coordination of the nitrogen atom in a planar environment.

The presence of side chains will markedly affect the behaviour of amino acids and peptides. Histidine and cysteine are both examples where strong affinities for metals may be explained by the side chains (imidazole and SH), although these groups may remain uncoordinated in certain cases. The common nitrogen donors found in proteins lie in the spectrochemical series: $N(imidazole) < N(peptide) < -NH_2$.

Much work has been done on copper complexes. As the ligand field of the four in-plane donor atoms increased, so the axial interactions become weaker. The Pauling electroneutrality principle states that a complex is most stable when the charge on each atom lies between -1 and $+1$. In the present case, as the charge transferred to the metal in the plane increases, so there is less need for transfer of charge from axial ligands. Five coordination could thus be favoured by the presence of four strongly bound donor atoms (e.g. three peptide donors and a water molecule). The fifth position would probably be filled by a side chain carboxyl group or a water molecule. Such a structure could well be of importance in metalloenzymes. It would, for example, allow the approach of a sixth ligand towards the other axial position and hence the replacement of the fifth ligand—as its link to the metal is weakened. Five-coordinate peptide complexes with other metals are known, but not for other biologically important metals. A number of peptide complexes of nickel(II) have been examined but these are not important as there is little evidence for nickel–protein interaction in natural systems.

Cobalt(II) peptide complexes appear to involve amino nitrogen, side chain and oxygen donors and no deprotonated peptide nitrogen atoms. These latter donors are important in cobalt(III) complexes. This is because the higher ligand field of the deprotonated N(peptide) link will allow the C.F.S.E. advantage of cobalt(III) over cobalt(II) to be utilized, so giving stable cobalt(III) complexes. The stability of a number of cobalt(II) and cobalt(III) complexes with amino acids may be understood on this basis.

Metal–protein complexes

A number of additional points must be borne in mind when comparing complexes of amino acids and peptides with those of proteins. It is reasonable that the measured first formation constant for a metal–protein system should relate to that of the simple model approximating to that of the binding group in the protein. Second and third formation constants etc., are not so meaningful as coordination of these groups may be determined by what groups are brought near by the coordination of the first group. Despite this, however, in general the behaviour of model metal–protein complexes is like that of peptide complexes, that is they are normal coordination complexes. This is not necessarily true of naturally occurring metalloproteins.

The presence of side chains (carboxyl, sulphur, imidazole groups) may well dominate coordination by proteins, with formal peptide links playing little part. Charges on the protein will be important as also will be the basic stereochemistry of the protein molecule, as this may result in certain groups being presented in a favourable position for coordination and other groups in unfavourable positions. This too will depend upon the favoured coordination number and stereochemistry of the metal ion and upon the size of the metal ion.

A statistical factor will also be operative. This will depend upon the number of different donor groups available in the protein in addition to the reactivity of those groups for the metal ions. If there are many more less-reactive groups than reactive ones, then this will result in a greater possibility of their coordination despite their lower reactivity.

In natural systems there will also be competition for the donor site with H^+. Protonation of a donor group prevents its coordination. Metal binding will be favoured by a large acid dissociation constant for the protonated ligand, ligandH^+. This may well determine which of two ligands will be coordinated to a particular metal ion at any pH. We will illustrate this with reference to ammonia and imidazole. These have a similar affinity for Zn^{2+} as is shown by the following formation constant data: $\log K_1$ for $Zn:NH_3 = 2.37$, $\log K_1$ for $Zn:$imidazole $= 2.58$. However, the pK_a values for NH_4^+ and imidazole H^+ are 9.26 and 7.03 respectively. At pH 7, therefore, less than 1% of ammonia is present as NH_3, while over 50% of imidazole is unprotonated. Thus, on a statistical basis, metal–imidazole interaction is favoured by a factor of 10^3. As the pH increases over 7, coordination of ammonia relative to imidazole will become more effective.

THE STUDY OF METALLOPROTEINS AND OTHER METAL-CONTAINING SYSTEMS

Metal binding

The environment of the metal in metal enzyme systems may include ligand groups from one or both of the protein and substrate. The simplest function of the metal ion, that of bringing the reacting groups into the correct orientation, must involve coordination of both protein and substrate. Such behaviour may usually be understood on the basis of formation constants and the stereochemistry of model compounds. This is often the case for systems involving Mg^{2+}, Mn^{2+} and Ca^{2+}.

We will briefly describe the application of the methods we have just considered to the problem of elucidating metal-binding sites, prior to considering other methods for examining metalloproteins in more detail. It should be emphasized, however, that these methods cannot lead to definite conclusions regarding the full details of metal-binding sites. The results of X-ray structural studies in every case studied so far have not been in accord with the suggestions put forward on the basis of chemical evidence, even though features of schemes have been confirmed. Data on metalloproteins have been reviewed by Vallee and Wacker.[10a]

Metal for metal replacement. Metal ions as probes

The techniques to be described in the following sections that depend upon the electronic properties of the metal ion may be made available to enzymes having metal ions without these electronic properties by metal for metal replacement. Thus, a main group metal may be replaced by a transition metal ion or an NMR or Mossbauer active metal ion. There is a tremendous interest at present in the use of probes of this type. For successful use of a probe metal ion it must fill the same site as the original metal ion, i.e. the principle of isomorphous replacement must hold. A useful guide here is whether or not the biological activity of the molecule is maintained. If this is so, then the new metal ion is assumed to be bound at the same site. The ionic radius of the metal ion is an important consideration. Thus K^+ may be replaced by NH_4^+ or Tl^+. The probe metal must also have similar requirements in terms of preferred stereochemistry of the metal-binding site and the nature of the binding ligands. Table 3.6 lists some probes that

TABLE 3.6 Metal ion probes[a]

Native cation	Ionic radius (nm)	Prove, with ionic radius and electronic property utilized
K^+	0·133	Tl^+ (0·140, NMR, fluorescence), NH_4^+ (0·145) Cs^+ (0·169, NMR)
Mg^{2+}	0·065	Mn^{2+} (0·080, EPR, paramagnetism) Ni^{2+} (0·069, d–d spectra[b]), other late $3d$ elements, but enzyme inactive in these cases
Ca^{2+}	0·099	Mn^{2+} (0·080), Eu^{2+} (0·112, paramagnetic Mossbauer), Lanthanide^{3+} ions, (0·115–0·093, range of probe properties)
Fe^{3+}	0·053	Gd^{3+} (paramagnetic[c]), Tb^{3+}, Eu^{3+}, Er^{2+}, Ho^{3+} (fluorescence[d])
Zn^{2+}	0·069	Co^{2+} (0·072, d–d spectra[e]), other $3d$ elements

[a] A fuller account of probes for Group I and Group II native cations is given in Chapter 8.
[b] E. J. Peck and W. J. Ray, *J. Biol. Chem.*, 1969, **244**, 3748.
[c] R. C. Woodworth, K. A. Morallee and R. J. P. Williams, *Biochemistry*, 1970, **9**, 839. J. Reuben, *Biochemistry*, 1971, **10**, 2834.
[d] C. K. Luk, *Biochemistry*, 1971, **10**, 2834.
[e] See Chapter 4; Review by S. Lindskog, *Structure and Bonding*, 1971, **8**, 153.

have been used in the study of biological systems. These will be discussed again in Chapter 8. The lanthanide elements are particularly useful as there is the possibility of having a series of probes differing only slightly in ionic size.

In addition to this aspect of metal ion replacement, it is often of value to replace a metal by a series of metal ions, as outlined earlier, and compare catalytic efficiency with the changing properties of the metal ions in the series.

Chemical testing

In a number of cases, the chemical testing of metalloenzyme and apoenzyme for the presence of specific groups has been of value. The underlying principle here is that a binding group for a metal in a metalloenzyme is bound so firmly that it does not react with the analytical reagents. Thus, the —SH content of proteins may be found by titration with a variety of reagents. Titration with Ag^+ shows[11] that hemocyanin, the oxygen-carrying copper protein, does not have any free —SH groups, while the results for the copper-free apoenzyme show the presence of free —SH. Hence it is suggested that copper is bound by —SH in the enzyme. On the other hand there are examples where this chemical testing method has given incorrect results, possibly because a group is inaccessible in the metalloenzyme and accessible in the apoenzyme as a result of changes in the protein structure.

pH methods

Another approach[9] involving a comparison of metalloprotein and apoenzyme is that of hydrogen ion titration. A large number of groups are involved in protonation equilibria but some attempt has been made to determine pK_as of metal binding groups by a comparison of titration curves in the presence and absence of the metal. However, it is known that pK_a values of protein groups may be different from those of the group in isolation and so while the pK_as of the binding group may be calculated, this does not necessarily allow the identification of the binding group.

Thermodynamic measurements

Formation constants may be measured for reversible metal–protein interaction by potentiometric methods and compared with those of model compounds. It is certainly possible to predict the favoured groups out of those available on the protein but the statistical factor may cause complication. Thus the data in Table 3.3 show that Cu(II) has a much stronger tendency to coordinate with nitrogen donors than oxygen donors. We would expect, therefore, that Cu(II) would be bound to imidazole rather than to a carboxylate group. But carboxylate groups are usually more plentiful, so in practice we may find Cu(II) combined with both groups, even though formation constant data would not suggest this.

Hydrogen ion competition is often a useful guide alongside this particular approach.

Comparisons between the reactivity of different metals for the same site are useful but are only valid when 'isomorphous replacement' occurs. Irving and Williams have shown[12] that the slope of the plot of formation constant against atomic number of the metal is greater for nitrogen and sulphur ligands than for oxygen, reflecting the different C.F.S.E. contributions. The review article of Dennard and Williams[13] cites a number of examples where this approach has been used, including one example, zinc in carboxypeptidase, where it led to the incorrect suggestion that RS^- was a binding group. Again, Lindskog and Nyman[14] have measured formation constants for a number of metal carbonic anhydrase complexes and by a suitable comparison with data for the corresponding ethylene-diamine complexes were able to suggest that the probable binding of zinc in carbonic anhydrase was through three nitrogen bases and did not involve RS^-.

The redox potentials of transition metal ion couples will provide information on the nature of the ligand groups as has already been discussed. The redox potentials of metalloenzymes are often rather different from those observed for model compounds. The copper 'blue' proteins provide an example of this behaviour. This is associated with the irregular stereochemistry of the metal binding site, as an unusual geometry will usually destabilize the higher oxidation state of an element more than the lower.

Absorption spectra

The band pattern and band intensity in the d–d spectra of transition metal complexes will give information regarding the stereochemistry of the site. For metalloenzymes this has led to the conclusion that the geometry is usually rather different from that observed for complexes of the metal. Thus for cobalt(II) carbonic anhydrase (i.e. Zn^{2+} replaced by Co^{2+}) the visible absorption spectrum indicates the presence of a distorted tetrahedral site, a proposition that is supported by X-ray studies. In addition the spectrum is pH dependent in a way that may be correlated with enzymatic activity. By contrast,[15] cobalt(II) complexes of the metal activated enzymes yeast and rabbit muscle enolases, pyruvate kinase and β methyl aspartase do not show this distorted spectrum. The role of the metal ion here will be more difficult to elucidate.

The band positions in the d–d spectra, together with a knowledge of the spectrochemical series, will allow the postulation of reasonable binding groups, while it may also be possible in the appropriate cases to note something of the nature of the spin state of the metal ion and hence, perhaps note indications of bond length changes (alongside Mossbauer and EPR spectroscopy and magnetic susceptibility data where possible). An example of the use of the position and intensity of bands lies in the work of Brill, Martin and Williams[13,16] on the visible spectrum of the copper protein erythrocuprein. It is suggested that the cupric ion is surrounded by at least four nitrogen donors while the high extinction coefficient (for d–d bands) is attributed to a degree of asymmetry such as is observed for complexes of the ligand bis(salicylaldehyde)ethylenediimine.

Charge-transfer bands will also help define the ligand, knowing the metal ion. An often quoted example[17] of the use of charge-transfer bands in model systems involves the Fe(III) catalysis of the decarboxylation of dimethyloxaloacetic acid. A purple colour appears during the reaction which is characteristic of a charge-transfer band involving a ferric enolate group. This shows that keto-enol tautomerism is involved. The scheme was given in Chapter 1.

Vallee[18] has used charge-transfer bands in the study of metallothionein. This contains zinc and cadmium and has an intense band at 250 nm, similar to that observed for cadmium mercaptide complexes. Again, cadmium can replace zinc in liver alcohol dehydrogenase giving a band at 250 nm. On the other hand, if cadmium in metallothionein is replaced completely by zinc, then a charge-transfer band at 215 nm is observed. The zinc-mercaptoethanol complex has a band at 215 nm. This is good evidence for metal binding by —SH in these proteins.

Optical rotatory dispersion and circular dichroism. The Cotton effect
Optical rotatory dispersion (ORD) is concerned with the variation of optical activity (i.e. the angle of rotation α) with the wavelength of the incident light. The ORD curve is a plot of α against wavelength. If the medium through which the light passes exhibits circular dichroism then it will absorb right and left circularly polarized light unequally. A CD spectrum involves a plot against wavelength of the difference in extinction coefficient for left and right hand circularly polarized light $(E_l - E_r)$. The two phenomena are collectively known as the Cotton effect.

The Cotton Effect is depicted in Figure 3.4. In the region of an optically active absorption band the ORD curve becomes more negative or positive. It reaches a maximum near the absorption band, reverses its direction to a minimum and then gradually increases again. A Cotton Effect is positive when the peak is at the longer wavelength and negative when the trough is at the longer wavelength.

The measurement of circular dichroism is concerned only with light absorption and so is of particular relevance to optically active chromo-

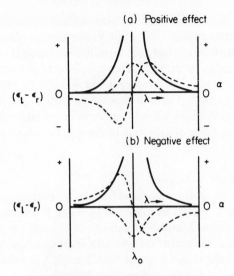

Figure 3.4   The Cotton effect.

phores, such as chelate complexes. Much has been done on correlating the configuration of optically active complexes with their Cotton effect and also in applying this technique to the assignment of electronic spectra. Cotton effects have been observed for a number of metalloproteins, indicating that the metal-binding sites are asymmetric.

ORD allows the study of general protein structure and the configuration of particular areas such as the enzyme active site. Vallee and Ulmer[19] have reviewed the application of ORD to iron proteins. The Cotton effects of proteins basically arise from the asymmetry of the polypeptide skeleton and will be sensitive to the secondary and tertiary structure of the protein. These have been termed *Intrinsic Cotton effects*. Superimposed upon this will be *side chain Cotton effects*, resulting from amino acid residues in asymmetric environment, and *Extrinsic Cotton effects*,[20] resulting from the asymmetric interaction of the protein or peptide with other groups such as coenzymes, inhibitors, metal ions or substrates. The first extrinsic Cotton effect was observed in the visible spectrum of ferricytochrome-*c*. The results were attributed[21] to the presence of the iron atoms, as heme itself is essentially symmetrical in structure. The Cotton effect will therefore be a reflection of the asymmetry of the metal-binding site, either as a direct result of the conformations of the ligands or as a result of their interaction with metals. Alternatively, the binding of an additional (not necessarily optically active) group to the metal may produce the asymmetry.

These effects have been used for comparison purposes in the case of ferredoxins and rubredoxin,[22,23] while Vallee and Riordan[24] have used the extrinsic Cotton effect to study conformational changes occurring during the action of carboxypeptidase. The coupling of p-azobenzenearsonate with tyrosyl residues of carboxypeptidase A leads to the presence of an extrinsic Cotton effect. On addition of substrate glycyl-L-tyrosine the band at 530 nm disappears while lower wavelength bands are also affected. This shows that binding of substrate to the enzyme alters the conformation of azotyrosine.

Magnetic resonance techniques

Nuclear magnetic resonance and electron paramagnetic resonance spectroscopy have proved of immense value in the study of biologically important molecules. Many illustrations of the use of these techniques may be found in some recent reviews and symposia reports.[16,25,26]

EPR. This technique was discussed in Chapter 2. It will give information regarding the redox state of the metal (although this may be ill-defined), covalency of the metal–ligand bond, the nature of the ligands and the stereochemistry around the metal ion. The spectrum itself will be defined in terms of g values, hyperfine structure constants and the area under the absorption curve. For metalloproteins EPR measurements have to be made on frozen solutions, rather than a crystal, and there is, therefore, an averaging out of all orientations of the molecules compared with the fixed orientation of a single crystal. The interpretation of such EPR data is difficult and often rather empirical conclusions have been reached for metalloproteins, mainly by comparison with the spectra of model compounds.

Studies have been carried out on iron, copper and molybdenum proteins. Thus EPR studies have shown that 40–50 % of the copper in ceruloplasmin is present as copper(II), confirming a result already obtained by chemical analysis and by magnetic susceptibility measurements. Unusual results have been obtained for the oxygen-carrying hemocyanins, both oxy and deoxy forms of *C. Magister* showing no signal. The implication of these results is bound up with the general problem[27] of assigning oxidation states to copper in these proteins.

An example of the use of EPR in detecting the binding groups for a metal is the study of a single crystal of copper-substituted insulin. It has been suggested,[28] on the basis of the ligand hyperfine structure, that the spectra are typical of copper interacting with two magnetically equivalent nitrogen donor atoms in a complex of trigonal symmetry. Associated titration studies involving a comparison between experimental and computed titration curves lead to the suggestion that the cupric ion is in fact bound to identical nitrogens, either imidazole or the amino groups of N-terminal phenylalanines. A second example[5] is that already given at the beginning of this chapter involving the binding sites in ferredoxin.

A number of workers have applied EPR spectroscopy to iron and molybdenum species. In particular much work[16,25,26] has been done on flavoproteins and non-heme iron proteins, together with model compounds for them. In the case of reduced non-heme iron proteins, it has been noted that signals centred around g = 1·94, observable at low temperature, seem to be a common feature of their EPR spectra. Much attention has been paid to this, and attempts have been made to obtain model compounds that show this feature, such as[28a] reduced $[Fe(CN)_5NO]^{2-}$ and $[Fe(MeCS-CH_2-CSMe)_2Cl_2]^{2+}$. A recent study[28b] has examined the EPR spectra of the products of sulphide and dithionite reaction with $[Fe(CN)_5NO]^{2-}$ and $[Fe(NO)(H_2O)_5]^{2+}$ and illustrates the general use of models in such problems.

EPR studies have also been carried out on enzymes which contain a probe paramagnetic ion. Cohn and her colleagues have pioneered the use of Mn(II) in this connection and have been able to make substantial contributions to the understanding of the behaviour of a number of Mg(II) catalysed enzymes, usually associated with phosphate groups. An alternative approach[28c] is to use a 'spin label' which is a stable free radical such as a nitroxide which will bind to a particular site on the enzyme. It is not always possible specifically to label an amino acid. EPR has been used to detect conformational changes induced by metal ions and substrates in creatine kinase. The study of enzymes containing Mn(II) and a spin label has been of particular value, as the interaction between these two centres may be detected in the EPR spectrum. Thus in creatine kinase the geometry at the active site has been clarified in that it has been possible to estimate that the distance between the Mn(II) ion and the spin-label in the Mn–ADP complex is 0·8 ± 0·02 nm. These will be discussed further in Chapter 4.

NMR.[28d] Proton magnetic resonance techniques have been extensively applied to the study of the detailed structure of proteins and nucleic acids, particularly since high-resolution spectrometers have been available. The use of computer averaging has helped improve the signal/noise ratio so that low concentrations can now be studied. Carbon-13 and phosphorus-31 are also important nuclei while calcium-43 and sodium-23 have also been studied (see Chapter 8). In addition, potassium can be replaced[29] by Tl^+ (or Cs^+) which are possible NMR probes.

Williams[28e] has discussed the application of proton magnetic resonance in the study of conformational changes in metalloproteins. For example, it is known that when the sixth ligand in myoglobin is fluoride, iron(III) is effectively 100 % high spin; when the sixth ligand is cyanide, the complex is low spin. Now the NMR spectrum of myoglobin is considerably affected by this changeover in spin state (the metal–ligand bond length will change by 0.01 nm). It is suggested that shifts in spin state balance in other heme proteins will be associated with conformation changes in the protein. Similarly, the uptake of CO or O_2 by heme proteins causes a change from high to low spin Fe(II); this too will be reflected in conformational changes, as observed by NMR. Results such as these are obviously of importance in understanding the role of transition metal ions in redox reactions, particularly as many of the enzymes involve several metal atoms.

One very important application of NMR lies in the study of the effect of a paramagnetic ion upon the spectrum of a metal-containing system. Metal for metal substitution has allowed this approach to be applied quite widely. One example[30] involves the preparation of an active manganese carboxypeptidase complex and the demonstration that one molecule of water is bound by the metal ion and that this is displaced by the inhibitor β-phenylpropionate.

Cohn[31] has looked in particular at the use of Mn(II) as a paramagnetic probe. She has shown that the effect of the paramagnetic ion upon the relaxation rate of the NMR signal of water protons is very sensitive to the type of metal complex formed, and that it can, in appropriate cases, provide much useful information in terms of the role of the metal ion and the nature of the metal–enzyme and metal–enzyme–substrate complexes, including their binding constants. She has applied this technique in particular to enzymes which catalyse phosphoryl transfer reactions of the type $ATP^{4-} + XH \rightleftharpoons ADP^{3-} + XPO_3^{2-} + H^+$.

A 'spin label' may be used as an alternative to a paramagnetic probe. In general, the spin label is less effective in causing nuclear relaxation as there will only be one free unpaired electron in the free radical.

Mossbauer spectroscopy

The general background of this effect has been reviewed and a number of illustrative examples presented.[16,25] The elements Fe, K, I are all Mossbauer nucleides of potential biological interest but Fe^{57} is the most easily used and at present the only one used extensively. Mossbauer spectra are presented in terms of three parameters; the isomer shift, the quadruple splitting and the magnetic hyperfine splitting. Isomer shifts reflect the electronegativities of the nearest neighbour atoms and hence the ionic character of bonds. Quadruple splittings reflect the distribution of electric charge in atoms, while the hyperfine splitting often enables a distinction to be made between the different possible spin states. Low temperature is essential for observing the hyperfine splitting while the variation of the other parameters with temperature provides useful information. Most iron-containing systems, however, involve too dilute an iron concentration to observe the hyperfine splitting. The non-heme iron proteins provide one exception here. The low natural abundance of Fe^{57} may necessitate Fe^{57} enrichment of the iron containing protein by appropriate methods. Mossbauer studies have been carried out on hemoglobin and myoglobin[32] and on non-heme proteins. Thus oxidized spinach ferredoxin (containing 2 iron atoms per 13,000 molecular weight) has been shown by studies at 4·6°K to involve identical iron atom environments with iron as Fe(III). After reduction with excess dithionite the Mossbauer spectrum indicated that one half of the iron had been reduced to Fe(II). The quadruple splitting confirmed that this was high spin Fe(II). Chemical studies had already shown that ferredoxin was a single electron transporter.

A similar examination[34] of the oxidized and reduced forms of the non-heme iron in xanthine oxidase showed that there was little distinction between the spectra of the two forms. It was concluded that the spectra arose from a diamagnetic state in both oxidized and reduced form, although conclusions were tentative as the spectra were of poor quality. This result agreed with magnetic susceptibility work but not with EPR results. This serves to illustrate an important point to bear in mind when using those two techniques as complementary tools. EPR is very sensitive to the presence of small amounts of paramagnetic species. Mossbauer spectroscopy will be sensitive to all the iron atoms present and so will not emphasize paramagnetic forms. This effect may be most important in examples such as the one above, where the quality of the spectrum is low.

The use of europium(II) as a probe allows a further application of Mossbauer spectroscopy.[34a]

Table 3.7 lists some conclusions that have been drawn concerning the nature of the metal-binding groups in metalloproteins on the basis of the results of applying the techniques described in the preceding sections.

The stereochemistry and oxidation states of metals in metalloenzymes

It has already been emphasized that the stereochemical requirements of the protein molecule may result in an abnormal site symmetry for the metal, and that, for redox metalloenzymes, this is often associated with unusual redox potentials and difficulties in defining the oxidation state of the metal. When in a subsequent section we come to talk about model compounds, we will then be placed in a difficulty inasmuch as it is not possible to obtain models that duplicate these factors. Vallee and Williams[35,36] have considered this problem and have put forward the

TABLE 3.7[a] Some results of the application of chemical and instrumental methods (other than X-rays) to the determination of metal binding groups in metalloproteins

Metalloprotein	Binding groups and experimental method	Comments
Carboxy-peptidase (Zn)	RS^- and NH_2: chemical tests on apo-enzyme and enzyme for the SH group; a comparison of stability constants for a range of metal carboxypeptidases with those for complexes of mercaptoethyl-amine	X-ray studies have shown that —SH and NH_2 are *not* binding groups
Carbonic Anhydrase (Zn)	Three nitrogen bases: a comparison of stability constant data for some metal-carbonic anhydrases and metal ethyl-enediamine complexes; EPR data on the Cu enzyme and model complexes; chemical evidence against —SH binding	X-ray studies indicate binding *via* three imidazole groups
Ferredoxin (Fe)	Inorganic sulphide and —SH: require-ment for inorganic sulphide for refor-mation of the enzyme from the apo-enzyme; chemical studies on enzyme and apoenzyme for binding *via* —SH	
Alkaline phosphatase (Zn)	Three imidazoles of histidine: photo-oxidation of native and apoenzymes shows that three histidyl residues are protected in the native enzyme and are not oxidized, the treated apoenzyme has a reduced zinc-binding capacity	
Copper proteins	There is much evidence for —SH binding of Cu in a number of copper proteins. This has involved titration of native and apoenzyme for —SH groups. The EPR spectra of copper conalbumen and transferrin (formed by adding the metal to the apoenzyme) shows ligand hyperfine splitting in-dicating binding *via* nitrogen donors.	
Transferrin and Conalbumin (Fe)	Tyrosine and histidine residues: pH studies indicate that the metal binding groups have pK_a values about 11·2 and 7; also EPR studies on copper substi-tuted proteins (see above)	
Cytochrome *c* (Fe)	The sixth ligand has been suggested to be histidine, lysine and methionine by different workers	X-ray studies show it to be methionine

[a] For a review of these properties see *The Proteins* (Second Edition), Ed. H. Neurath. Volume 5, 'Metalloproteins,' by B. L. Vallee and W. E. C. Wacker (Academic Press, New York, 1970).

general suggestion that the active metal site of the enzyme is in a geometry approaching that of the transition state of the appropriate reaction and as such is uniquely fitted for catalytic action. They have termed this an entatic state. In later chapters some reference will be made to this. Table 3.8 lists a number of metalloproteins for which irregular physical properties have been observed. Some have already been discussed in the text.

TABLE 3.8 Irregular binding sites for metals in metalloenzymes

Protein	Metal	Stereochemistry/Coordination number	Techniques
(a) Native			
Hemoglobin	Fe^{2+}	5-coordinate	X-rays
Myoglobin	Fe^{2+}	5-coordinate	X-rays
Hemoglobin	Fe^{3+}	Iron displaced out of	
Myoglobin	Fe^{3+}	porphyrin plane. Strong axial field	X-rays/EPR
Cytochromes	Fe^{3+}	Rhombic	EPR
Ferredoxins	Fe^{3+}	EPR g-value $= 1.94$ not understood	EPR
Rubredoxin	Fe^{3+}	Asymmetric	EPR/CD
Conalbumen Transferrin	Fe^{3+}	Very low symmetry	EPR/CD
Vitamin B_{12} coenzymes	Co^{3+}	5-coordinate	d–d/NMR
Copper blue proteins	Cu^{2+}	Low symmetry	d–d/EPR/CD
Oxyhemocyanin	Cu^{2+}	Distorted orthorhombic	d–d
(b) Substituted			
Carbonic anhydrase	Co^{2+}	Distorted, 5-coordinate ?	d–d
Carboxypeptidase	Co^{2+}	Irregular	d–d
Alkaline phosphatase (two atoms Co^{2+})	Co^{2+}	Distorted	d–d
Phosphogluco-mutase	Co^{2+}	Irregular, 5-coordinate ?	d–d

Studies with inhibitors

A study of the role of inhibitors in enzymatic reactions has contributed greatly to an understanding of enzyme mechanisms. In this section the inhibition of metalloenzymes by added chelating agents will be discussed. One method of showing the presence of a metal in an enzyme is to demonstrate inhibition by added chelating and other coordinating agents. Those widely used include EDTA, nitrilotriacetic acid, Tiron (4,5-dihydroxy *m*-benzene sulphonic acid), ethylenediamine, dimethylglyoxime, mercaptoethanol, dipyridyl, dithizone and the inorganic ions azide and cyanide. It is assumed that any effect is due solely to coordination to the metal ion and that there is no other interaction between chelating agent and protein, substrate or coenzyme, i.e. that the inhibitors will preferentially replace the protein groups binding the metal. This problem really involves a comparison of formation constants, but there are severe complications in this. Thus the metal binding groups may not be necessarily known. Even if they are known, then the reactivity of the metal ion towards the inhibitor may not be that expected as the basis of the study of model compounds. Structural features in the protein may prevent the operation of the chelating agent, while the strength of any metal-inhibitor binding will reflect the stereochemistry and strength of the binding of the metal to the protein. Quantitative comparison is therefore difficult, but despite this much of value can come from a study of metalloenzyme-inhibitor interaction.

The inhibiting action may either involve complete removal of the metal leaving the apoenzyme, which may or may not be reactivated by addition of the original metal ion, or the replacement of some of the protein groups only, so giving a mixed enzyme–inhibitor metal complex. Prevention of inhibition by having free metal ion present, with which the chelate may react, is usually taken as confirmation that the chelating properties of the inhibitor are responsible for the inhibitory effect. If a mixed complex is formed then the inhibition should be reversed by adding excess metal ion to compete for the inhibitor. This reversal of inhibition can be used to demonstrate the presence of the mixed complex.

The measurement of the stoicheiometry of metalloenzyme–inhibitor complex is also of value here. Job's method may be applied, using either a maximum of light absorption or of ORD to monitor the system. If a 3:1 complex is formed with a bidentate inhibitor, then clearly the metal ion has been removed completely from the protein. The observation of the formation of a 1:1 complex would suggest the presence of the mixed complex.

Inhibition by chelating agents will throw light on mechanism in a number of ways. Kinetic studies on inhibition can be of value, while in general the effect of inhibition on the physical properties of the system allows the correlation between inhibition and the change in those properties. In Chapter 4 we will consider the way in which the addition of inhibitors to cobalt carbonic anhydrase at different pH values causes changes in the *d–d* spectrum of the cobalt which may be correlated with inhibitor function and hence the mechanism of action of carbonic anhydrase.

THE STUDY OF METALLOPROTEINS

Ulmer and Vallee[37] have shown that the presence of a chelating agent in a mixed complex can result in optical activity. This is an example of an extrinsic Cotton effect. The study of the carboxypeptidase-β-phenylpropionate inhibitor complex (by X-ray diffraction and NMR) have shown that the carboxyl group is coordinated to the zinc, or probe atom in the NMR case. This, too, will be discussed in Chapter 4.

KINETIC STUDIES ON
MODEL SYSTEMS AND METALLOENZYMES

In this section we are concerned particularly with studies on model systems for metalloenzyme reactions and also with studies involving the metal in metalloenzymes. Again, the theme of the value of *comparing* metal ions is stressed. For a quantitative comparison it is necessary to have kinetic data as well as formation constant data.

In many model systems, the catalytic effect of various metal ions follows the Irving–Williams series, that is, it is the same as the sequence of formation constants. This is of interest mechanistically. Most metal aquo ions form complexes in aqueous solution by fast diffusion of a ligand molecule to the aquo complex forming an outer sphere complex, while the rate-determining step involves the replacement of the aquo group by the ligand (by the appropriate mechanism). The rate of replacement[38] of water in octahedral complexes is $Ca^{2+} > Cu^{2+} > Zn^{2+} > Mn^{2+} > Fe^{2+} > Co^{2+} > Mg^{2+} > Ni^{2+}$, which is of course different from the Irving–Williams series. It follows therefore that in model systems the rate-determining step is not the formation of the complex. Rather, as the catalytic order follows the Irving–Williams order, it may mean that the rate-determining step is one involving a rearrangement of the metal–substrate complex in which the metal ion assists in orientating the substrate for reaction. The rate, from metal ion to metal ion, will then depend upon the extent to which complex formation has taken place, in which case there should be a direct relationship between the catalytic rate constant, k, and the formation constant K for the different metals. If the formation constant doubles, so will the rate constant.

$$M + S \overset{fast}{\rightleftharpoons} MS \qquad K$$

$$MS \xrightarrow{slow} \text{Products } k$$

Alternatively, the activity of the metal may be one of a Lewis acid. The catalytic effect should then depend as before on the extent of complex formation, together with the differing Lewis acidity of the various metal ions. As Lewis acidity also follows the Irving–Williams series, the overall effect follows this series, but now there will not be the direct correlation between rate constant and formation constant observed for the first case. The rate constant will increase more rapidly than the formation constant. It may be seen therefore that variation of the metal ion and a comparison of rate constant and stability constant may allow some insight into the role of the metal. An example[39] of such Lewis acid catalysis lies in the metal ion catalysed decarboxylation of acetone dicarboxylic acid, where a comparison of rate constant and formation constant over a range of metal ions clearly shows the Lewis acid requirement for the metal ion to catalyse the removal of CO_2.

An increasing number of studies on the kinetics of complex formation in model systems have been reported. These are very fast reactions and are usually studied by the temperature-jump relaxation technique. An early example[40] is the study of the kinetics of complex formation between magnesium(II) aquo ions and 8-hydroxyquinoline ('oxine'). In order to attempt to represent enzyme–metal–substrate complexes more closely in the matter of the presence of ligands other than water in the coordination shell, these workers studied oxine substitution into magnesium ATP, magnesium polytriphosphate and magnesium uramil-NN-diacetate complexes in addition to hexaaquomagnesium(II). This meant that a statistical correction had to be introduced to allow for the binding of different numbers of water molecules to the metal ion in the different cases. It appears, however, that the first ligand had comparatively little influence on the rate at which the metal aquo complex reacts with a second ligand.

On the basis of the calculated rate constants for the different stages of the reaction, an attempt was made to rationalize the differing catalytic powers of Ca^{2+} and Mg^{2+}. While not differing very much in terms of formation constant, their catalytic powers can be quite different. Thus Ca^{2+} will inhibit the reactions of some Mg^{2+} activated enzymes and vice versa. It appears that the rate constants for the formation of a complex are about 1000 times larger for Ca^{2+} than for Mg^{2+}. As the formation constants are fairly similar, this must mean that calcium complexes dissociate about 1000 times more rapidly than magnesium complexes, as $K = k_f/k_b$. In other words, the extent of complex formation stoicheometrically may be the same, but the lifetime of the complexes is much shorter for Ca^{2+}.

The relative catalytic properties of the two ions may then be understood by a scheme in which an internal rearrangement of the complex E–M–P (P = product) may or may not be necessary before the final product is liberated by complex dissociation. If the rearrangement *must* occur (e.g. a conformational change of the protein) then the necessary changes must take place in a time shorter than the lifetime of the complex. For calcium, whose complexes have a shorter lifetime, this cannot occur, but a comparison of reaction rates suggests that it may be possible for magnesium. For such a reaction Ca^{2+} will be an inhibitor.

On the other hand, if there is no need for the rearrangement, the product will be produced more quickly for Ca^{2+}, as the rate of dissociation is faster. Ca^{2+} will therefore be a more effective catalyst of such reactions.

It should be noted that it is possible to explain the differing catalytic efficiencies of Mg^{2+} and Ca^{2+} purely in terms of their preferences for different binding sites (i.e. binding groups and site symmetry). This seems a more reasonable explanation.

A second study[41] with oxine has involved measuring the kinetics of its complex formation with molybdate. This has been used as a model system for enzyme–substrate binding in xanthine oxidase. This is a 290,000 molecular weight protein which catalyses the oxidation of xanthine to uric acid by molecular oxygen. EPR studies have shown that electron transfer from substrate to oxygen proceeds *via* molybdenum, FAD (flavin adenine dinucleotide) and iron. A Mo(V) signal is observed within 10 msec reaction time between substrate and enzyme, so indicating that molybdenum forms part of the active site of the enzyme. The formation of a complex between molybdate and oxine will serve as a model system to indicate the feasibility of this suggestion.

The results in the pH region 7·9–8·9 show that a 1:1 complex is formed, to an extent dependent upon the pH, reflecting protonation both of the molybdate and oxinate anions. The enzymatic reaction has been studied, and it appears that the overall forward rate constant for the oxine–molybdate reaction is about 4–5 orders of magnitude lower than that[42] estimated for the enzyme–substrate binding reaction. However, there is EPR evidence[25] that xanthine oxidase involves a proton in association with the molybdenum, so it may be more meaningful to compare with oxine substitution into $[MoO_3OH]^-$, data for which were calculated.[41] The rate constants are in fact fairly close to the estimated one for the enzyme–substrate binding reaction and are therefore compatible with the suggestion that molybdenum is bound in the suggested way in xanthine oxidase (Figure 3.5).

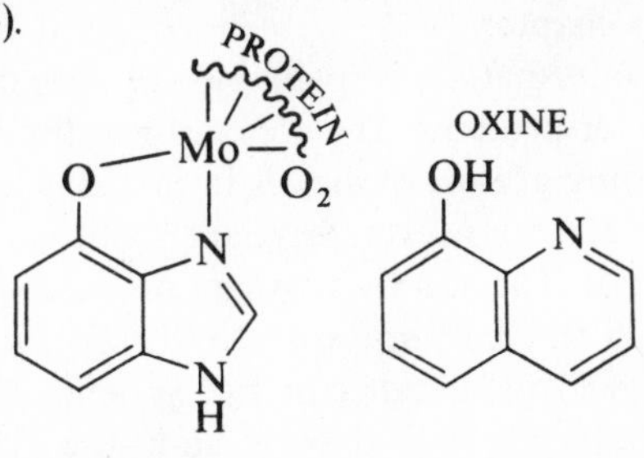

Figure 3.5 Suggested binding of xanthate to molybdenum in xanthine oxidase.

To turn from model compounds to metalloenzymes, kinetic studies of particular interest here include metal for metal exchange rates. These are very much slower than those for model compounds, the rate-determining step probably being the relaxation of the protein.

The binding of zinc by the apoenzyme of carbonic anhydrase has been studied.[42a] The recombination is accelerated by the dissociation of protons from groups with apparent pK_a values of 5·4 and 7·2. Again the rate constants are orders of magnitude smaller than for the formation of zinc complexes where the rate-determining step is dissociation of water from the first coordination sphere. The kinetic parameters also differ in the two cases.

THE USE OF MODEL COMPOUNDS

The complexity of biological macromolecules, and the associated problems in their study, means that it is often worthwhile to examine the behaviour of simpler molecules that contain what appear to be the essential features of the original molecule, uncomplicated by other features of that molecule. The model compounds may also have other useful properties such as solubility, not possessed by the original species. Provided that the comparison is a real one, then the conclusions reached for model compounds may be extrapolated to the parent molecule and some light thrown upon its reactions and functions. On the other hand, differences between the behaviour of the models and the molecule may also have some significance, as will be demonstrated for certain hydrolytic enzymes.

Despite the fact that good chemical models are available for most enzymatic reactions, it should still be emphasized that a comparison of rates, in general, shows that the reactions undergone by the models are very much slower than the enzymatic reactions. It is possible to attempt to correct for certain catalytic features in the enzyme but even so, the enzymatic reactions may be up to 10^{11} times faster than the model reactions. This serves to re-emphasize the cautionary note expressed in the introduction to this chapter.

A great deal of information is available on amino acids and simple peptides as models for proteins. The conclusions discussed earlier in this chapter on the binding groups of metals in proteins are based upon this type of information. Here, however, there are problems which we have not yet discussed in detail. The reactivity of amino acid side chains (such as pK_a) may be changed on incorporation into the protein, due to the hydrogen bonding or other interactions in the protein. Often the catalytic function of an enzyme may well depend upon the protein environment generating an abnormal physical property for some group or grouping.

Such a feature obviously cannot be incorporated into the design of model compounds. Again, other protein groups may be rendered inaccessible by the protein structure, and so the reactivity for that group implied by model compounds may not mean anything in that particular example. This explains why chemical tests for the presence of certain residues in the protein may not function. —SH groups are well known for demonstrating lowered reactivity on incorporation into the protein.

We have already discussed the unusual stereochemistry of the metal ions in many metalloenzymes. This cannot be readily represented by model compounds, so their value is limited. The order of catalysis of hydrolysis in model compounds by various metal ions is not always followed in enzyme systems for the same metals. Again the relationship to the models must be examined closely. Finally, when considering models for redox enzymes, it must be emphasized that the electronic state of the model compound must be known before it can be compared with the metalloenzyme.

The use of model compounds has allowed the build up of much information regarding the reactivity of coordinated ligands. The presence of a metal ion often has a very marked effect upon the reactivity of the group. The Lewis acid behaviour of metal ions will lower the basic properties of coordinated groups and so enhance attack by nucleophiles or cut down the possibility of electrophilic attack. The use of model compounds also illustrates the role of metal ions as templates and their possible role in biosynthesis of macrocyclic molecules. There are many examples in model compounds of the condensation of amines and carbonyl groups to give Schiff bases being aided by metal ion templates. Other condensations include[43] the formation of square planar nickel (II) chelates in which α-diketones condensed with β-mercapto amines.

Without the metal ion, the yield is very poor, as the amine and mercapto groups act as competitive nucleophiles. The chelate may then be closed by reacting with an appropriate difunctional molecule.

Another well known example is the formation[44] of stable Ni(II) and Cu(II) complexes via the condensation of coordinated ethylenediamine with acetone. Tris(ethylenediamine)nickel(II) reacts with dry acetone under certain conditions to give a quadridentate macromolecule. The reaction has been extended to include other metals, amines, ketones and also aldehydes, so that a variety of different macrocyclic compounds may be prepared. The reaction is accelerated by base and retarded by water. The first step in the example cited involves the condensation of acetone with the suitably orientated coordinated ligand.

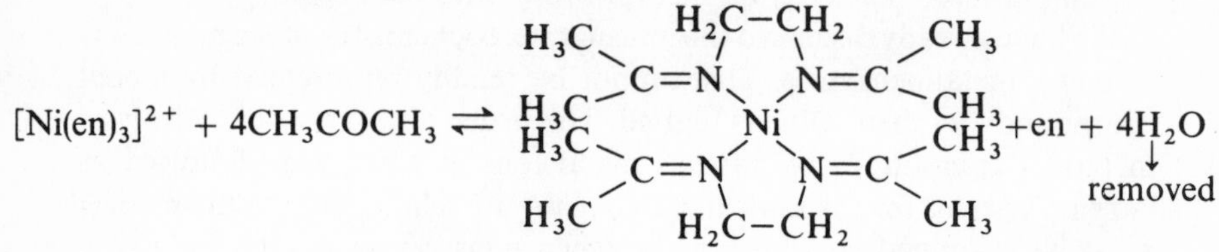

$$[Ni(en)_3]^{2+} + 4CH_3COCH_3 \rightleftharpoons$$

It is suggested that the released ethylenediamine then serves as a base to catalyse the following reaction.

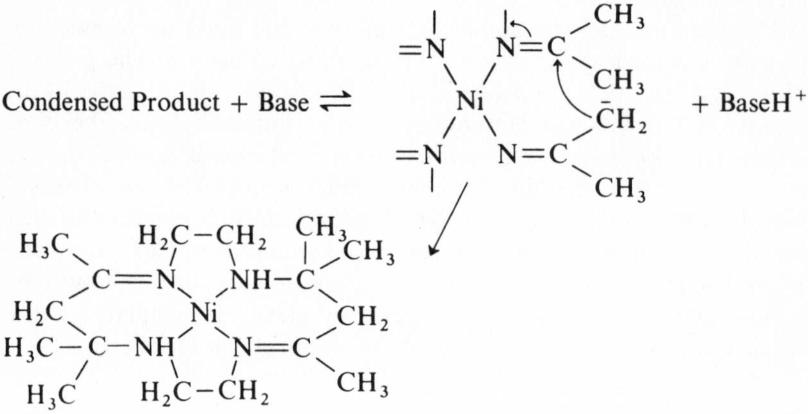

Condensed Product + Base $\rightleftharpoons$

This complex may then be readily reduced to give the chelate of the saturated amine. The nickel ion may be removed by reaction with cyanide, leaving the macrocyclic molecule.

Applications of this type of behaviour may well be important in the biosynthesis of certain natural products and have in fact been used in the synthesis[45] of corrin complexes, as noted in Chapter 1.

Models for vitamin B_{12} coenzymes

The corrin and porphyrin ring systems are of great biological importance. They are shown in Figure 3.6.

The porphyrins[46] are derived from porphin, varying according to the nature of the substituents. They are all intensely coloured and highly con-

Corrin   Porphin

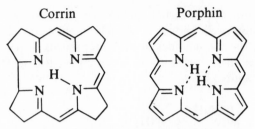

Figure 3.6   Corrin and porphin ring structures.

jugated. The group itself is tetradentate with four pyrrole-like nitrogens surrounding a central site for metals. On coordination the two N–H protons are lost, as is the NH proton in corrin. To illustrate the importance of the tetrapyrrole group it may be noted that iron complexes of the porphyrins are the hemes, and that depending upon the nature of the protein in the hemoprotein and the substituent on the porphyrin, we have (i) the hemoglobins and myoglobins in which the iron(II) hemoprotein will combine reversibly with oxygen without oxidation to the iron(III) state; (ii) peroxidases, which catalyse the hydrogen peroxide oxidation of substrates, and (iii) catalases, which catalyse the decomposition of hydrogen peroxide. On the other hand, if magnesium lies at the centre of the hydroporphin nucleus, we have the chlorophylls, while if cobalt is the central atom of the corrin ring system we have vitamin B_{12} and the series of cobamide coenzymes, depending in both cases upon the nature of the substituents.

From the chemical point of view the 'organometallic chemistry' of cobalt in vitamin B_{12} and its derivatives is unusual and interesting. This may be parallelled quite remarkably in many cases by some quite simple model compounds.

The corrinoid ring systems provide a square planar set of donor atoms, while the cobalt usually has axial ligands above and below the plane, although sometimes one only. If one of the axial ligands is the benzimidazole of the associated nucleotide, the complex is a cobalamin; if this is replaced by a water molecule, then we have the cobinamides. Cobalt may be formally regarded as cobalt(III). Vitamin B_{12} is cyanocobalamin (i.e. the sixth group is cyanide, a result of the isolation procedure), B_{12a} aquocobalamin, B_{12c} nitrito cobalamin. The B_{12a} form may be reduced first to B_{12r} and then to B_{12s}. B_{12r} is a low spin cobalt(II) complex, while B_{12s} is apparently a cobalt(I) species.

The 5-deoxyadenosyl cobalamin has the adenosyl group carbon bonded to the cobalt. This was the first naturally occurring Co–C bonded structure to be recognized. Previously only a small number of highly unstable compounds with cobalt-carbon bonds had been known. Alkyl-cobalt systems are now well known in models based upon the vitamin B_{12} structure. For the cobinamides in particular the alkyl species exist partly as 5-coordinate structures, with the water molecule displaced. Structures are given in Figure 3.7.

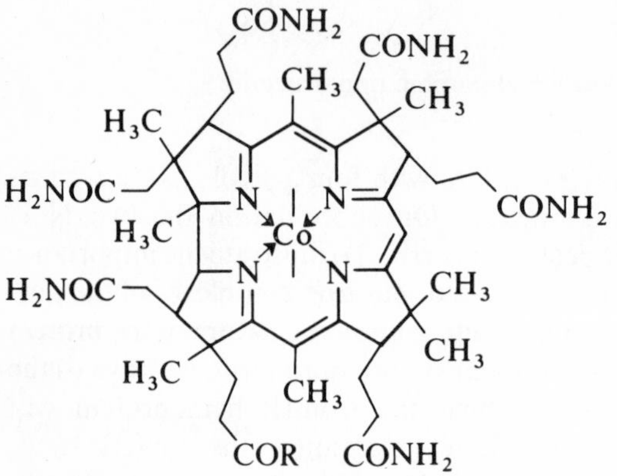

Cobinamide $R = NHCH_2CH(OH)CH_3$

Cobalamin $R = NHCH_2CHCH_3$

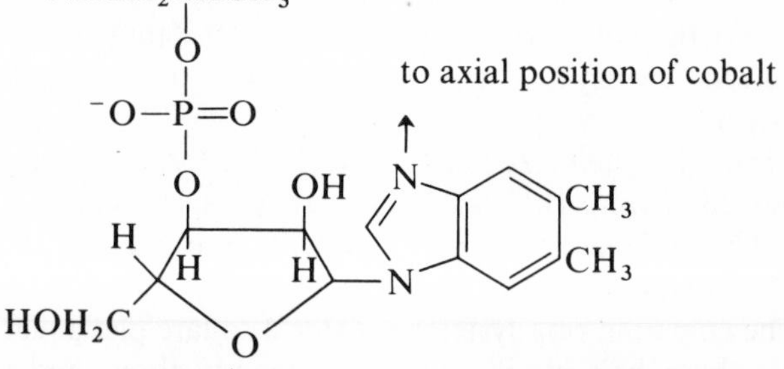

Figure 3.7   The structure of cobalamin and cobinamide.

Porphyrins have been used as models for corrin ring systems. They will add groups readily in the axial positions. This is not typical of all porphyrins. Thus Cu(II) and Ni(II) porphyrins have a low affinity for extra ligands, while Mg(II), Zn(II) and Cd(II) porphyrins add on one ligand giving a square pyramidal five coordinate structure. Fe(II), Mn(II) and Co(II) porphyrins all add on two moles of ligand. Cobalt porphyrins[46] can be converted into organic derivatives, but they cannot be reduced to the cobalt(I) state in aqueous solution. The reduction is usually carried out with a Grignard reagent in non-aqueous solution.

However, bis(dimethylglyoxime)cobalt complexes (the cobaloximes) (Figure 3.8) show very many of the reactions of the cobalt atoms in corrins. They too add on axial groups and form stable organic derivatives readily, and also they can be reduced to cobalt(I) species. These, though, are very sensitive to oxygen and only exist in alkaline pH. In acid solution, like vitamin B_{12s}, they decompose liberating hydrogen and giving cobaloxime.

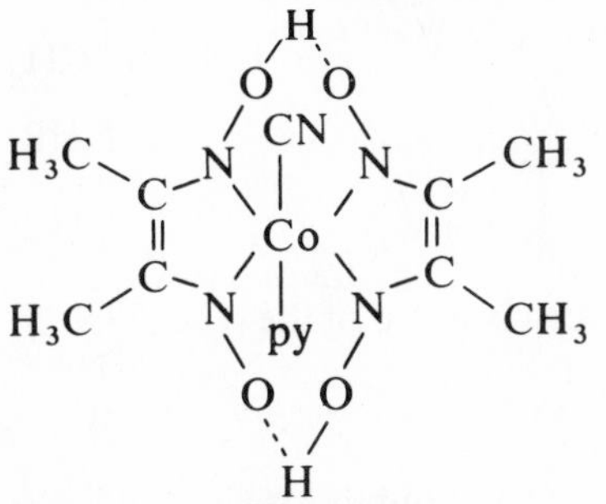

py = pyridine

Figure 3.8 Cyanopyridine cobaloxime.

The comparison between cobaloximes and B_{12s} has contributed to an understanding of the latter. It appears that the close similarity between cobalamins and cobaloximes is due to the presence of an in-plane ligand of similar strength and is independent of the axial ligands. This is supported by physical methods[49-51] while the similarity is also justified by theoretical methods.[52]

In order to illustrate the comparison with the model compounds, it should be noted now that the reactions catalysed by the cobamide coenzymes include

(1) methyl group transfer in which the cobalt(III) presumably is the methyl-accepting group,

(2) reduction reactions, and

(3) rearrangement reactions.

The first type of behaviour is that shown in the biosynthesis of methionine:

$$\begin{array}{c} CH_3 \\ | \\ Co \end{array} + \text{homocysteine} \longrightarrow (Co^I)^- + \text{methionine}$$

(methylcobalamin)

Methylcobaloximes will similarly methylate homocysteine, although the reaction is not reversible. Demethylation is possible, however, provided it is first converted to the S-adenosyl derivative.

Both vitamin B_{12} coenzymes and cobaloximes will also catalyse reduction reactions involving the synthesis of N-methyl groups from formaldehyde and amines in the presence of a reducing agent such as molecular hydrogen.

$$HCHO + NH_2C_6H_5 \rightleftharpoons C_6H_5NHCH_2OH$$

$$Co \Big| H_2$$

$$C_6H_5NHCH_3 + H_2O$$

Intermediates have been isolated of the type

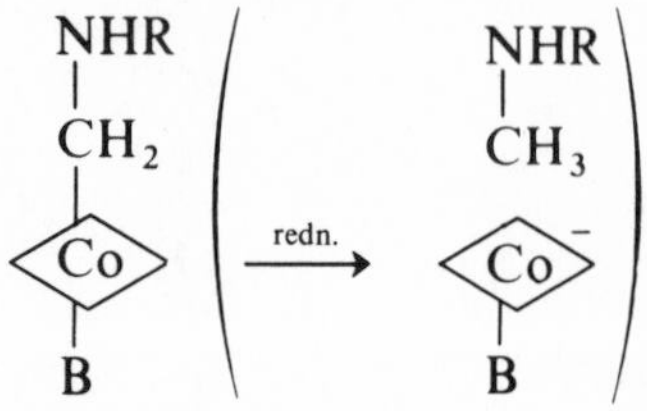

$$
\begin{array}{c}
NHR \\
| \\
CH_2 \\
| \\
\langle Co \rangle \\
| \\
B
\end{array}
\quad \xrightarrow{\text{redn.}} \quad
\begin{array}{c}
NHR \\
| \\
CH_3 \\
| \\
\langle Co \rangle^- \\
| \\
B
\end{array}
$$

Other aspects of model compounds for vitamin B_{12} will be discussed later; the work of Schrauzer and his group has, however, shown the value of a comparison of alkycobalamins and alkylcobaloximes.

GENERAL CONCLUSIONS

The methods outlined in the preceding sections indicate the general approaches that may be followed in the study of the binding and function of metals in biological macromolecules. A number of examples have been given for each method. In the following chapter we will examine a number of hydrolytic enzymes in detail to show how these methods may be integrated together. We have chosen to study, in more detail, the role of zinc in carboxypeptidase and carbonic anhydrase in view of the experimental information, particularly X-ray results, available. We will also comment in more general terms on other hydrolytic enzymes.

REFERENCES

1. (a) G. N. Reeke, J. A. Hartsuck, M. L. Ludwig, F. A. Quiocho, T. A. Steitz and W. N. Lipscomb, *Proc. Nat. Acad. Sciences*, 1967, **58**, 2220
   (b) W. N. Lipscomb, G. N. Reeke, J. A. Hartsuck, F. A. Quiocho and P. H. Bethge, *Phil. Trans. Roy. Soc. Lond.*, 1970, **B257**, 177
   (c) W. N. Lipscomb, *Proc. 8th Int. Congress of Biochemistry*, 1970, 128
   (d) M. F. Perutz, *Eur. J. Biochem.*, 1969, **8**, 455
2. W. N. Lipscomb, J. C. Coppola, J. A. Hartsuck, M. L. Ludwig, H. Muirhead, J. Searl and T. A. Steitz, *J. Mol. Biol.*, 1966, **19**, 423
3. T. Ando and H. Fujioka, *J. Biochem. Tokyo*, 1962, **52**, 363

4. J. R. Herriot, L. C. Sieker, L. H. Jensen and W. Lovenberg, *J. Mol. Biol.*, 1970, **50**, 391
5. C. S. Yang and F. M. Huennekens, *Biochemistry*, 1970, **9**, 2127
6. Stability Constants, Special Publication 17, *The Chemical Society, London*, 1964
7. F. J. C. Rossotti and H. Rossotti, *The Determination of Stability Constants*, McGraw Hill, 1961
8. G. Beech, *Quart. Rev. Chem. Soc.*, 1969, **23**, 410
9. F. R. N. Gurd and D. E. Wilcox, *Advances in Protein Chemistry*, 1956, **11**, 311
10. H. C. Freeman, *Advances in Protein Chemistry*, 1968, 342
10. (a) B. L. Vallee and W. E. C. Wacker, *The Proteins*, (2nd. Edit.) 1970, Vol. V. Academic Press
11. A. Ghiretti-Magaldi, C. Nuzzulo and F. Ghiretti, *Biochemistry*, 1966, **5**, 1943
12. H. M. N. H. Irving and R. J. P. Williams, *J. Chem. Soc.*, 1953, 3192
13. A. E. Dennard and R. J. P. Williams, 'Transition Metal Ions as Reagents in Metalloenzymes', *Transition metal chemistry*. (Ed. Carlin) Vol. 2. Arnold
14. S. Lindskog and P. O. Nyman, *Biochim. Biophys. Acta*, 1964, **85**, 461
15. Cited by E. W. Westhead, *Proc. 8th Int. Congress of Biochemistry*, 1970, 131
16. A. S. Brill, B. R. Martin and R. J. P. Williams, in *Electronic Aspects of Biochemistry*, Ed. B. Pullman, Academic Press, 1964, 554
17. F. H. Westheimer, *Proc. Chem. Soc.*, 1963, 253
18. J. H. R. Kagi and B. L. Vallee, *J. Biol. Chem.* 1961, **236**, 2435; R. Druyan and B. L. Vallee, *Fed. Proc*, 1962, **21**, 247
19. B. L. Vallee and D. D. Ulmer, *Non Heme Iron Proteins. Role in Energy Conversion*, Ed. San Pietro, The Antioch Press, Yellow Springs, Ohio, 1965, 43
20. E. R. Blout, *Biopolymers Symp. No.* 1, 1964, 397
21. G. L. Eichorn and J. F. Cairns, *Nature*, 1958, **181**, 994
22. A. N. Atherton, K. Garbett, R. D. Gillard, R. Mason, S. G. Mayhew, J. L. Peel and J. E. Stangroom, *Nature* 1966, **212**, 590
23. R. D. Gillard, E. D. McKenzie, R. Mason, S. G. Mayhew, J. L. Peel and J. E. Stangroom, *Nature*, 1965, **208**, 769
24. B. L. Vallee and J. F. Riordan, *Brookhaven Symp. Biol.*, 1968, **21**, 91
25. *Magnetic Resonance in Biological Systems*, Eds. A. Ehrenberg, B. G. Malmström and T. Vanngard. MacMillan, 1967
    A. J. Bearden and W. R. Dunham, *Structure and Bonding*, 1971, **8**, 1
26. A. San Pietro, (Ed), *Non Heme iron proteins. Role in Energy Conversion*, The Antioch Press, Yellow Springs, Ohio, 1965
27. D. C. Gould and A. Ehrenberg, *Physiology and Biochemistry of Hemocyanins*. Ed. F. Ghiretti, Academic Press, 1968, 95
28. A. S. Brill and J. H. Venable, *The Biochemistry of Copper*, (Eds. J. Peisach, P. Aisen and W. E. Blumberg), Academic Press, 1966, 67
    (a) K. Knauer, P. Hemmerich and J. D. W. Van Voorst, *Angew. Chem. Int. Ed.*, 1967, **6**, 262
    (b) B. A. Goodman and J. B. Raynor, *J. Chem. Soc. (A)*, 1970, 2038
    (c) H. M. McConnel and B. G. McFarland, *Quart. Rev. Biophysics*, 1970, **3**, 91
    (d) K. Wuthrich, *Structure and Bonding*, 1971, **8**, 53
    (e) R. J. P. Williams, *Proc. 8th Int. Congress of Biochemistry*, 1970, 126
29. J. P. Manners, K. G. Morallee and R. J. P. Williams, *Chem. Comm.* 1970, 965
30. R. G. Shulman, G. Navon, B. J. Nyluda, D. C. Douglass and T. Yamane. *Proc. Nat. Acad. Sci. U.S.*, 1966, **56**, 39

31. M. Cohn, *Biochemistry*, 1963, **2**, 623; *Quart. Rev. Biophysics*, 1970, **3**, 61; M. Cohn and J. Reuben, *Acc. Chem. Res.*, 1971, **4**, 214
32. W. Marshall and G. Lang, *Proc. Phys. Soc. (Lond)*, 1966, **87**, 3
33. A. J. Bearden and T. H. Moss, *Magnetic Resonance in Biological Systems*, Eds. A. Ehrenberg, B. G. Malmström and T. Vanngard, MacMillan, 1967, p. 391
34. C. E. Johnson, R. C. Bray and P. F. Knowles, ibid, p. 417
34. (a) R. J. P. Williams, *Quart. Rev. Chem. Soc.*, 1970, **24**, 355
35. B. L. Vallee and R. J. P. Williams, *Proc. Nat. Acad. Sc. U.S.A.*, 1968, **59**, 498
36. B. L. Vallee and R. J. P. Williams, *Chemistry in Britain*, 1968, **4**, 397
37. D. D. Ulmer and B. L. Vallee, *Adv. Enzymol*, 1965, **27**, 37
38. M. Eigen, *VII Int. Conf. on Coord. Chem.*, Butterworths, 1963, 97
39. J. E. Prue, *J. Chem. Soc.*, 1952, 2331
40. D. N. Hague and M. Eigen, *Trans. Far. Soc.*, 1966, 1236
41. P. F. Knowles and H. Deibler, *Trans. Far. Soc.*, 1968, 977
42. H. Gutfreund and J. M. Sturtevant, *Biochem. J.*, 1959, **73**, 1
42. (a) R. W. Henkens and J. M. Sturtevant, *J. Amer. Chem. Soc.*, 1968, **90**, 2669
43. M. C. Thompson and D. H. Busch, *J. Amer. Chem. Soc.*, 1962, **84**, 1762
44. N. F. Curtis, *J. Chem. Soc.*, 1960, 4409; 1962, 1204, 3016; *Coord. Chem. Rev.*, 1968, **3**, 3
45. A. Eschenmoser, R. Scheffold, E. Bertele, M. Pesaro and H. Gschwend, *Proc. Roy. Soc.*, 1965, A **288**, 306
46. J. E. Falk, *Porphyrins and Metalloporphyrins*, Elsevier Pub. Co. 1964
47. G. N. Schrauzer, *Acc. Chem. Res.*, 1968, **1**, 97
48. D. Dolphin and A. W. Johnson, *Chem. Comm.*, 1965, 494
49. G. Lenhert, *Chem. Comm.*, 1967, 890
50. G. N. Schrauzer, R. J. Windgassen and J. Kohnle, *Chem. Ber.*, 1965, **98**, 3324
51. G. N. Schrauzer and L. P. Lee, *J. Amer. Chem. Soc.*, 1968, **90**, 6541
52. G. N. Schrauzer, L. P. Lee and J. W. Sibert, *J. Amer. Chem. Soc.*, 1970, **92**, 2997

CHAPTER FOUR

HYDROLYTIC METALLO- AND METAL ACTIVATED ENZYMES

A number of recent reviews[1-4] have considered certain aspects of metal dependent hydrolytic enzymes. Table 4-1 lists some such enzymes. The metal ions that are important here are the Group II cations Mg^{2+}, Ca^{2+} and Zn^{2+}. A number of techniques, including the use of tracers, may of

TABLE 4.1 Some hydrolytic enzymes[a]

Enzyme	Reaction catalysed	Metal ion
Carboxypeptidase	Hydrolysis of C terminal peptide residues	Zn^{2+}
Collagenase	Hydrolysis of collagen in native configuration	Zn^{2+}, Ca^{2+}
Dipeptidase	Hydrolysis of dipeptides such as glycylglycine	Zn^{2+}
Alkaline phosphatase	The hydrolysis of phosphate esters; also a phosphotransferase	Zn^{2+}
Carbonic anhydrase	$CO_2 + HO^- \rightleftharpoons HCO_3^-$	Zn^{2+}
Phospholipase C	Hydrolysis of phospholipids	Zn^{2+}
Neutral protease	Hydrolysis of peptides	Zn^{2+} (Ca^{2+})
α-amylase	Hydrolysis of glucosides	Ca^{2+} (Zn^{2+})
Protease (pseudomonas)	Hydrolysis of peptides	Ca^{2+}
ATPase	$ATP + H_2O \rightarrow ADP + \text{ortho-phosphate}$	Mg^{2+}
Inorganic pyro-phosphatase	pyrophosphate $\rightarrow$ orthophosphate	Mg^{2+}
Various phosphatases (e.g. fructose diphosphatase)	Phosphate hydrolysis reactions (Fructose 1-6 diphosphate $\xrightarrow[H_2O]{Mg}$ fructose 6 phosphate)	Mg^{2+}
Aminopeptidase	Hydrolysis of N-terminal peptide residues	Mg^{2+} (Mn^{2+})

[a] The hydrolytic enzymes are thoroughly reviewed in *The Enzymes* (3rd Edition) Volumes III, IV and V. (Ed. Boyer) Academic Press, New York, 1971.

course be applied to these enzymes but in general metal ion replacement by transition metal ions will be necessary to obtain the full benefit of modern techniques.

In this chapter we will consider first, in detail, the role of zinc in the enzymes carbonic anhydrase and carboxypeptidase. These are enzymes which catalyse the dehydration of carbonic acid and the hydrolysis of the carboxyl terminal peptide bond in peptides respectively. It is obviously better to begin by looking at those metalloenzymes for which X-ray diffraction data of high resolution is available, as this eliminates some of the doubts associated with deciding the nature of the metal binding groups. We will, however, look at the chemical and other evidence that has been collected on this matter in order to see how successful those methods have been in determining the nature of the zinc ligands in these enzymes. We will examine the conclusions obtained from the use of models and consider also how the techniques discussed in Chapter 3 have contributed to our understanding of the mode of action of these enzymes. Finally, we will survey other hydrolytic enzymes in more general terms.

PRELIMINARY COMMENTS

(1) Carboxypeptidase A

This is a zinc-containing enzyme of molecular weight 34,600, having some 300 amino acid residues, including three methionyl residues. The molecule has been cleaved at these positions by treatment with cyanogen bromide. The amino acid sequence of the resulting four fragments has been studied by Neurath and coworkers[5] and three have now been determined in full. The carboxypeptidase zymogen, that is, the precursor, exists as a trimer or dimer with other units. Carboxypeptidase A is generated by appropriate enzymatic action on the zymogen in which a peptide bond is cleaved, releasing an N-terminal fragment of some 60 amino acid residues from the precursor. Four different carboxypeptidases are generated depending upon the size of the fragment released; these are termed α, β, γ and δ. Details are given in Table 4.2. All four forms have essentially similar properties although varying in heat stability and solubility. X-ray studies have been carried out on CPAα.

TABLE 4.2 Forms of carboxypeptidase

Enzyme	No. of residues	N-terminus
CPAα	307	Alanine
CPAβ	305	Serine
CPAγ	300	Asparagine
CPAδ	300	Asparagine

The activity of the enzyme is directed specifically towards carboxyl

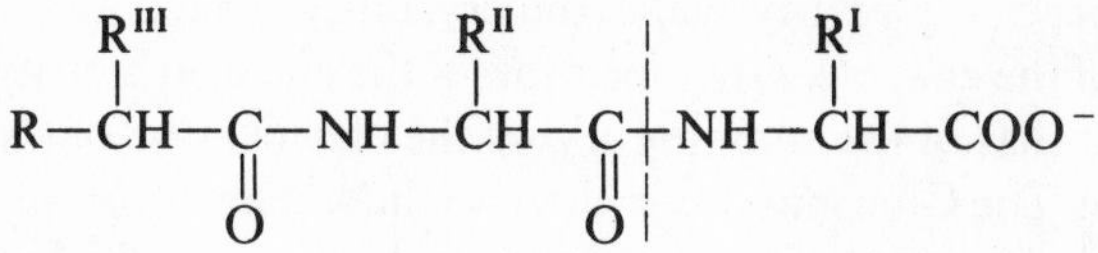

$$
\begin{array}{ccccccc}
R^{III} & & R^{II} & & & R^{I} & \\
| & & | & & & | & \\
R-CH-C-NH-CH-C{+}NH-CH-COO^{-} \\
\parallel & & \parallel & & & \\
O & & O & & & \\
\end{array}
$$

terminal peptide bonds where the C-terminal residue is aromatic or branched aliphatic of L configuration. It appears that at least the first five terminal residues affect enzyme activity. The enzyme also possesses esterase activity towards model compounds in which the peptide nitrogen is replaced by an oxygen. The mode of action of carboxypeptidase has been much studied by Vallee and coworkers.[6]

(2) Carbonic anhydrase

This enzyme, which occurs in human and animal blood, is essential for respiration as the non-enzymatic dehydration of carbonic acid is not fast enough for physiological purposes. In addition to catalysing this reaction carbonic anhydrase also catalyses the hydrolysis of certain esters.

$$ HCO_3^- \rightleftharpoons CO_2 + OH^- $$

The enzyme contains one atom of zinc per molecule and has a molecular weight of 30,000. It has been less extensively studied than carboxypeptidase. The molecule contains some 260 amino acid residues with a high proline content and no disulphide bridge. Only limited sequence studies have been carried out. The enzyme occurs in three forms (A, B and C), form C being 30 times more active than form B.

X-RAY STUDIES

Both molecules are ellipsoidal in shape and show a cavity or cleft which is associated with the active site, the zinc atom being inside the cavity.

Carboxypeptidase A

X-ray studies have been carried out[7,8,9] at successively increased resolution (0·6, 0·28, and 0·2 nm) on the native enzyme and heavy atom derivatives. The molecular dimensions are about $5 \times 4·2 \times 3·8$ nm, with the zinc bound by three protein groups, two imidazoles of histidine and one glutamic acid residue. The remaining coordination position is filled by a water molecule which is replaced by the substrate on formation of the enzyme-substrate complex.

Carboxypeptidase forms an extremely stable enzyme-substrate complex with the dipeptide glycyl-tyrosine, the crystals of which are isomorphous with those of the enzyme. A comparison of the electron density map of the enzyme with that of the complex allows the clarification of the binding of the substrate. The C-terminal side chain of the substrate fits into the pocket or cleft of the enzyme. There appears to be no specific binding group involved here in accord with the lack of high specificity for the nature of the side chain. The terminal carboxylate group interacts with the positively charged guanidinium group of Arginine 145, while the carbonyl oxygen of the peptide link which is to be split is probably bound to the zinc, with resulting displacement of the coordinated water group.

It has already been noted that the enzyme undergoes dramatic conformational changes when Gly–Tyr is bound to the active site. This was schematically represented in Figure 3.1. The overall result is that the phenolic OH group of the residue Tyrosyl 248, which was known to be implicated in the catalysis, moves about 1·2 nm and the OH group comes from the surface of the molecule to within a close distance (0·27 nm) of the peptide bond that is to be cleaved. A number of other conformational changes also occur, but the overall result is this bringing together of the components of the active site. This is the first demonstrated example illustrating Koshland's 'induced fit' mechanism.[10]

Carbonic anhydrase C

The structure of this enzyme and of its complex with the inhibitor acetoxymercurisulphanilide have been determined[11] to 0·55 nm resolution using four heavy atom derivatives. The dimensions of the molecule are approximately $4 \times 4·5 \times 5·5$ nm. At the active site the molecule has the cleft previously mentioned with the zinc atom bound at the bottom. One part of the cleft is a narrow slit where the inhibitor is bound to the zinc. The SH group of the protein molecule is too far away from the zinc to be involved in metal binding. Although the resolution is not very high, the chain folding and one terminus has been determined (close to the metal ion), while the α-helix content and the other terminus have been determined to a lower degree of probability. A recent[12] report has confirmed that the environment around the zinc atom is a distorted tetrahedral one, in accordance with electronic spectra results on the cobalt substituted enzyme. The zinc appears to be bound by three protein ligands, probably all imidazole groups of histidine residues, while the fourth site is accessible to other groups such as inhibitors. In the absence of inhibitors, the site would be filled by a water molecule or an hydroxide ion.

MODELS FOR HYDROLYTIC METALLOENZYMES

Much information is available[1,13] on the hydrolysis of simple peptides and esters, so providing a suitable basis for the consideration of the possible mode of action of carboxypeptidase. It has been shown that the hydrolysis of esters and peptides is subject to both acid and base catalysis. Particular attention has been paid to the role of metal ions in the hydrolysis of small peptides and correlations drawn between metal ion-peptide stability constants and rate constant over a range of metal ions. The order of reactivity of metal ions in such model reactions is not necessarily paralleled by the order of metal ion activity in a series of metalloenzyme catalysed reactions, but even so some useful points may be noted. The role of the metal ion as a Lewis acid has been clearly demonstrated.

Bender and Turner[13a] have studied the copper(II) catalysed hydrolysis of esters of α-amino acids. On the basis of oxygen-18 exchange data and kinetic studies they have suggested a mechanism involving the formation of a complex in which the ester is chelated *via* the amino and carbonyl groups. Polarization of the carbonyl group then results in enhanced susceptibility of the carbon to nucleophilic attack by a base. Nucleophilic attack is also assisted by the departure of the bonding around the carbonyl group from planarity.

$$\begin{array}{c} X \\ \diagdown \\ C = O + M^{2+} \rightarrow \\ \diagup \\ Y \end{array} \quad \begin{array}{c} X \\ \diagdown \\ C \overset{\delta+}{-} \overset{\delta-}{O} \rightarrow M^{2+} \\ \diagup \\ Y \end{array} \xrightarrow[\text{base}]{H^+-OH^-} XH + YCOOH + M^{2+}$$

X = RNH or EtO

Results such as these obtained with labile complexes are difficult to interpret and other mechanisms cannot be excluded. Of greater significance[13b] is work involving inert cobalt(III) complexes, in particular bisethylenediamine complexes with coordinated glycine esters, *cis*-[Co(en)$_2$(NH$_2$CH$_2$COOR)Cl]Cl$_2$, where R = CH$_3$, C$_2$H$_5$ and i-C$_3$H$_7$. In these complexes the ester is coordinated *via* the amino group, with the ester group free. They are stable in aqueous solution for several hours with respect to hydrolysis. In this case the carbonyl group is not co-ordinated. This is consistent with the mechanism outlined above.

In acid solution Hg(II) reacts with the complex, with hydrolysis of the ester and formation of a glycinatobis(ethylenediamine)cobalt(III) complex. Spectrophotometric studies show that the coordinated halide is removed by mercury(II) in the first step (a well known catalyst of aquation) and that hydrolysis of the ester occurs in a subsequent step. Infrared spectroscopy suggests that the intermediate is the chelated ester complex.

The mercury(II) ion induced aquation of the chloro complex results in the formation of a five coordinate species, the sixth position then being filled by the ester carbonyl group. In the absence of other nucleophiles water

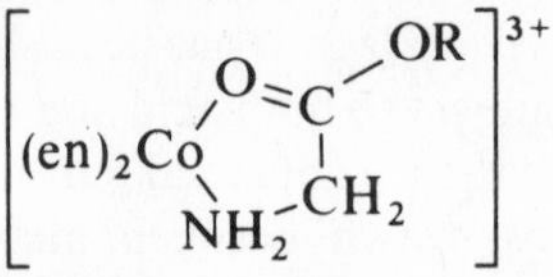

$$\left[(en)_2Co \begin{array}{c} O{=}C{-}OR \\ | \\ NH_2{-}CH_2 \end{array} \right]^{3+}$$

attacks the positive centre of the polarized carbonyl group. The hydrolysis reaction is subject to general base catalysis.

More recently[13c] further oxygen-18-tracer studies have been carried out on the base hydrolysis of glycine esters coordinated to cobalt (III), in cis-$[Co(en)_2X(glyOR)]^{2+}$, $(X = Cl, Br)$ and β_2-$[Co(trien)Cl(glyOC_2H_5)]^{2+}$. Evidence has been produced for two competing processes, intermolecular hydrolysis of the chelated ester and intramolecular attack of coordinated OH^- at the carbonyl group of the monodentate ester. Both reaction pathways arise from competition for the five coordinate deprotonated intermediate formed by loss of the halide ion via an S_N1CB mechanism.

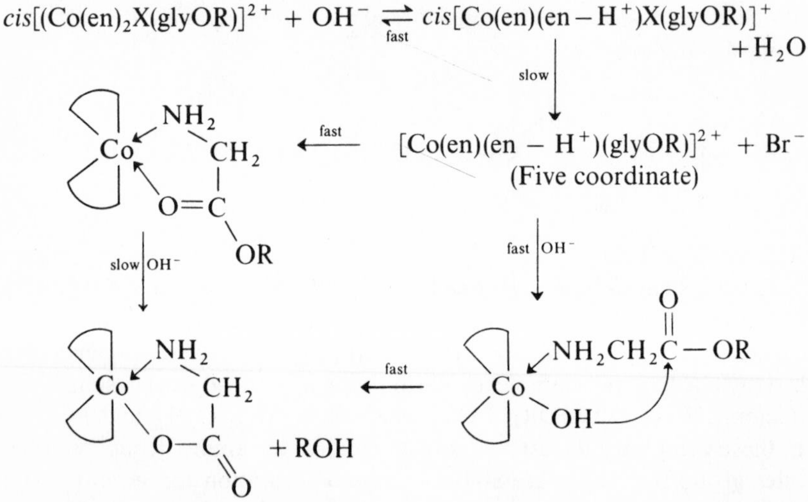

$$cis[(Co(en)_2X(glyOR)]^{2+} + OH^- \underset{fast}{\rightleftharpoons} cis[Co(en)(en-H^+)X(glyOR)]^+$$

$$+H_2O$$

However, these reactions could not be observed directly as they are fast compared to hydrolysis of coordinated halide. Therefore a study[13d] of the Co(III) promoted hydrolysis of glycine amides has been of particular value, as base hydrolysis of chelated glycine amides is much slower than that of ester. This allows the hydrolysis of the chelated amide to be observed following loss of Br^- in cis-$[Co(en)_2BrglyNR_1R_2]^{2+}$ and hence the possibility of confirming intramolecular hydrolysis by coordinated OH^-

Evidence for the existence of an intermediate of reduced coordination number, which may be competed for by species in solution, was provided by showing that azide ion could be incorporated into the complex during hydrolysis of cis-$[Co(en)_2Br(glyNH_2)]^{2+}$ giving $[Co(en)_2N_3(glyNH_2)]^{2+}$. No evidence could be found for the formation of cis-$[Co(en)_2N_3(glyO)]^+$ where glyO is the nitrogen bound monodentate glycinate anion, so implying that hydrolysis of the monodentate amide does not occur before or during loss of Br^-. Oxygen-18 studies indicate that the two possible products resulting from competition for the sixth coordination position (i.e. by the carbonyl oxygen of the amide or the OH^- group) are formed about equally. Oxygen-18 studies also show that reaction via the chelating amide pathway involves attack of solvent OH^- without opening the ring, and the second pathway to occur via intramolecular attack of coordinated OH^- at the carbonyl carbon of the monodentate amide.

It is clear that cobalt(III) is very effective in inducing hydrolysis of amino acid ester and amides. The cobalt(III) catalysed intramolecular hydrolysis reaction in cis-$[Co(en)_2OH(glyNH_2)]^{2+}$ is about 10^7 times faster than the uncatalysed reaction. This mechanism appears to be more efficient than that resulting from direct metal ion polarization of the carbonyl group. For it to occur it is necessary that the coordinated water molecule is acidic enough to exist in the hydroxo form.

This work is particularly relevant to carbonic anhydrase in which a zinc bound hydroxide group converts CO_2 to bicarbonate ion, and also carboxypeptidase where two mechanisms have been suggested; the 'zinc-carbonyl' mechanism and the 'zinc-hydroxide' mechanism. These will be discussed later, but for the present it is assumed that the first of these two mechanisms is operative.

The reactions of CO_2 as models for carbonic anhydrase have been discussed in detail by Dennard and Williams[1] in their review. Here, the reactions of CO_2 may be understood in terms of the charge arrangement in the molecule, i.e. $\overset{\delta-}{O}=\overset{\delta+}{C}=\overset{\delta-}{O}$ with the CO_2 molecule adding its carbon to a negative centre. Thus the reaction with a Grignard reagent (1) may be

$$CO_2 + R\text{--}Mg\text{--}X \longrightarrow RCO_2^-Mg^+ - X \tag{1}$$

understood in terms of the polar Mg–R link polarizing the carbon dioxide molecule. Such a reaction involves combined acid–base attack.

A number of decarboxylation reactions[14] have been examined. These are the back reactions corresponding to the overall reaction (2) and therefore involve the same transition state.

$$CO_2 + RCOCH_3 \longrightarrow RCOCH_2COO^- + H^+ \tag{2}$$

Here too the catalytic role of metal ions as Lewis acids has been clearly demonstrated.

Dennard and Williams[14a] have studied the rate of reaction between carbon dioxide and water over a wide pH range, together with the catalytic effect of a number of anions. In contrast to the results of earlier workers they found that there did not appear to be any relationship between the pK_a of the conjugate acid of the anion and its catalytic efficiency. The effect of added anions was measured over a range of pH by using buffers. If catalysis did occur it was then possible to confirm that this was only through the anion by comparing the experimental variation of rate constant with pH and a theoretical curve constructed on the basis of the pK_a of the acid, assuming that the anion was the catalytic species. Good catalysts were species such as arsenite, sulphite, hypochlorite, hypobromite, while carbonate, bicarbonate, nitrite, nitrate, sulphide, sulphate, phosphate are typical of non-active anions. It is suggested that the requirements for catalytic activity are: (I) that the bases should be oxyanions of non-metals in their lower oxidation states, having at least one long pair, thus HPO_3^{2-} is non-catalytic having no lone pair of electrons; (II) if the oxyanion involves a high oxidation state of a non-metal then there must be no equivalent oxygen atom in the anion to that from which the proton has been removed. This condition prevents the possibility of charge distribution away from the oxygen function involved in catalysis.

Certain mechanisms for base catalysis of CO_2 hydration such as scheme (A) may be eliminated as they would require a correlation between activity and pK.

$$(A) \quad \rightarrow HCO_3^- + HOX$$

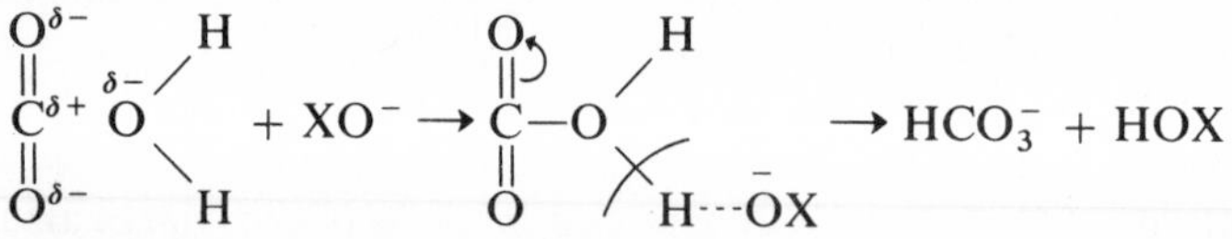

Dennard and Williams suggest that scheme (B) is a feasible alternative, for here depending upon the stability of the first intermediate there may or may not be a dependence of k upon pK.

$$(B) \quad \rightarrow \text{Products}$$

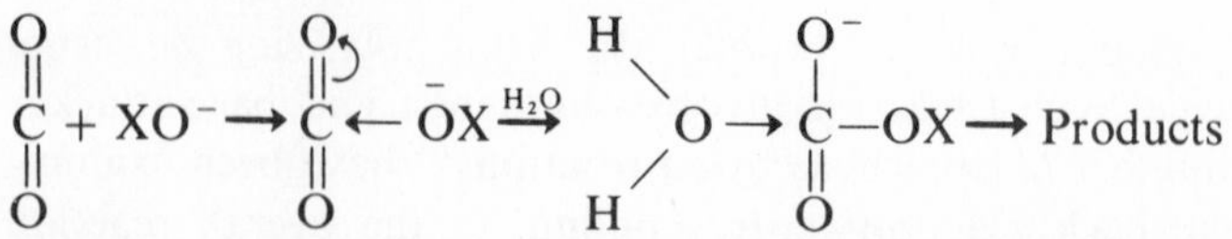

Another mechanism could involve catalytic function through X rather than the oxygen atom, in accord with the requirement (I) for catalytic activity. This is illustrated in scheme (C), one involving concerted acid–base attack upon CO_2.

$$(C) \quad \begin{matrix} O \\ \| \\ C \\ \| \\ O \end{matrix} \xrightarrow[XO^-]{H_2O} \quad \begin{matrix} O \quad\quad H \\ \| \quad\quad / \\ C\text{------}O \\ \| \quad\quad\quad H \\ O_{\cdots} \\ \quad X-O^- \end{matrix} \rightarrow \begin{matrix} O \quad\quad H \\ \| \quad\quad / \\ C-O \\ | \quad\quad\quad H \\ O \\ \quad X-O \end{matrix} \rightarrow \text{Products}$$

Bromine exerts a large catalytic effect upon carbon dioxide hydration and dehydration. This has been studied recently[14b] by 0–18 techniques and the catalysis shown to result from hypobromous acid, with a rate law:

$$\text{Rate} = k[\text{HOBr}][\text{HCO}_3^-]$$

It has been suggested that the catalytic efficiency is such that it implies that the compound functions as a brominating agent rather than as a general acid catalyst. The HOBr is regenerated.

$$\text{HO}\overset{\frown}{-}\text{Br} \overset{H}{\underset{|}{O}}-\overset{O}{\underset{\|}{C}}-O^- \rightarrow \text{HOBr} + CO_2 + OH^-$$

The reverse reaction, that of hypobromite catalysis of CO_2 hydration, is among those considered by Dennard and Williams and discussed above. Scheme (C) involves interaction of HOBr and bicarbonate ion in the reverse reaction, but obviously differs from the mechanism involving bromination.

In concluding the section on model compounds it should be noted that infrared studies on the CO_2 molecule bound in the hydrophobic slit at the active site of carbonic anhydrase have indicated that CO_2 is not coordinated to the metal but that the bicarbonate ion is bound to zinc *via* its negatively charged oxygen atom. This will be discussed in more detail when the enzyme itself is considered, but it is clear that the zinc ion itself in carbonic anhydrase can still be put forward as the Lewis acid partner in *concerted* acid–base attack, provided that hydroxide is the other partner, as this will generate bound bicarbonate ion. It will not involve specifying CO_2 coordinated to zinc.

CARBOXYPEPTIDASE

A number of approaches have been adopted in the study of the function of this enzyme. These have included enzyme kinetics, including the pH-rate profile; the chemical modification of differing protein groups and a study of the resulting effect upon the activity of the metalloenzyme;

the replacement of zinc by other metal ions; and the use of specific instrumental techniques upon the apoenzyme, native enzyme, appropriate metalloenzyme or enzyme-substrate, and enzyme-inhibitor complexes.

It is important to note that both chemical enzyme modification and metal for metal substitution result in increased esterase activity towards the substrate hippuryl-DL-phenyllactate (HPLA) and decreased activity towards peptide substrates such as carbobenzoxyglycyl-L-phenylalanine (CGP). It has been assumed that the presence of some group is necessary for peptide hydrolysis and that it retards ester hydrolysis. Concommitant decrease in peptidase activity and increase in esterase activity is therefore due to the elimination of this group by the various modifications.

If the argument based on models is reasonable, we must expect any mechanism to provide an H^+ and to promote attack of water or some other base at the carbon atom of the carbonyl group of the appropriate peptide link, the carbonyl group being bonded to the metal, and polarized accordingly. This assumes that the 'zinc-hydroxide' mechanism is not operative.

In the following discussion it may be seen that although chemical approaches had not succeeded in identifying the binding groups of the zinc, they had allowed a substantial insight into a number of features of the mechanistic scheme suggested by the X-ray diffraction studies.

It is assumed that the results obtained for the binding of the substrate glycyl-L-tyrosine to the enzyme, as outlined earlier, also hold for other peptides, although it should be noted that if the H atom of the NH group to be cleaved is replaced by other substituents then the peptidase activity of the enzyme is either inhibited or greatly reduced.

The X-ray studies show that there are only certain protein groups near enough to the substrate in the complex to be reasonably implicated in enzyme activity. These are listed as follows, together with the probable function of the group.

Arg 145—interaction with carboxyl terminus of the peptide.

Glu 270—this group will be very close to the carbonyl carbon of the hydrolysable bond and could act in promoting general base catalysis by a water molecule or by direct nucleophilic attack at this carbon atom.

Zn^{2+} —Lewis acid.

Tyr 248 —To donate a hydrogen bond to the NH of the hydrolysable bond and possibly eventually to provide the H to be incorporated in the $-NH_2$ group of the product.

While certain of these functions appear to be well defined, there are a number of problems left unsolved and there are a number of alternatives not yet discussed. After a discussion of the more chemical approaches to studying the function of carboxypeptidase, we will briefly reconsider some of these problems.

There is one other point to be emphasized. The formation of the metal-substrate complex results in the loss of the zinc water molecule, and the expulsion of water from the enzyme cavity before it is closed off by the conformational changes of the Tyr 248 group. The resulting hydrophobic region will enhance the effectiveness of the zinc in polarizing the carbonyl group and this may well provide the real driving force of the reaction.

Chemical studies on the nature of the zinc binding groups

A comparison[15] of stability constants for a series of metallocarboxy-peptidases with those of bidentate metal complexes suggested that the metal ion was bound by nitrogen and sulphur ligands. The change in relative stability for the metalloenzymes is very similar to the change in stability from metal ion to metal ion when the ligand is mercaptoethyl-amine but is different from the behaviour with other ligands. In addition, the apoenzyme gave[16'] the normal tests for the presence of an —SH group and an RNH_2 group but the tests were negative on the zinc enzyme, suggesting that the metal ion was bound to these groups so modifying their reactivity towards the reagents used to test for their presence. The reagents for the thiol group were Ag^+, p-mercuribenzoate and ferri-cyanide. The results appeared convincing in that the zinc content of the enzyme and the Ag^+ titratable groups always gave a constant molar total over a range of zinc concentrations.

Despite this evidence, it is quite clear from X-ray studies that a thiol group is not implicated in zinc-binding in carboxypeptidase. The chemical evidence may be explained if it is assumed that conformational changes occur in the protein on the binding of the zinc by the apoenzyme and that the —SH group, once accessible to the reagents, becomes inaccessible in the metalloenzyme as a result of this conformational change.

Metal for zinc substitution

The zinc in carboxypeptidase has been replaced by cobalt, nickel, man-ganese, cadmium, iron(II), mercury, rhodium, lead and copper. The results of the last section have illustrated one use of a series of metal substituted enzymes. The peptidase activity of the cobalt enzyme is greater than that of the zinc enzyme, but only the nickel, iron and manganese enzymes of the remaining metalloenzymes show peptidase activity. This order of activity is quite different from the order of activity observed in model systems. The cadmium, mercury, rhodium and lead enzymes do, however, show esterase activity towards HPLA. That of the cadmium enzyme was inhibited[6] by peptide substrates, implying that peptides are bound to the cadmium enzyme, although not hydrolysed by it.

A general feature of zinc metalloenzymes appears to be the fact that the zinc can be effectively replaced by cobalt, presumably because these two metal ions prefer similar geometries, and that the electronic spectrum of the resulting cobalt enzyme is indicative of a distorted environment around the metal ion. Similarly, cobalt carboxypeptidase has a spectrum that is not typical in either band position or intensity of either simple tetrahedral species or octahedral species. Circular dichroism studies have confirmed the asymmetry of the cobalt environment. The general phenomena of unusual stereochemistries for metalloenzymes has already been discussed. It appears that the active site of carboxypeptidase could also be in such an 'entatic' state.

NMR studies have been carried out[17] on manganese carboxypeptidase and have demonstrated that water or hydroxide is bound to the metal and that this is displaced on formation of the enzyme-inhibitor complex with β-phenylpropionate.

Chemical modification of protein groups

Vallee and his coworkers[6] have carried out a series of elegant studies on the effect of the modification of amino acid side chains on the peptidase and esterase activity of carboxypeptidase. The implication of tyrosyl residues in the catalytic function has been shown by modification[6] through acetylation, nitration, iodination and coupling with 5 diazo-IH tetrazole or p-azobenzene-arsonate.[18] In all cases enzymatic activity was altered, usually peptidase activity being reduced while esterase activity is increased.

When acetylation is carried out in the presence of the inhibitor, β-phenylpropionate, two tyrosine residues are protected, suggesting that these two residues are of particular importance. The X-ray studies have shown that a second tyrosine (Tyr 198), in addition to Tyr 248, is in the general area of the active site. A great deal of work has been carried out on the modification of these two groups and has been discussed in detail.[6b] We will only give one further example. The coupling of the tyrosyl group with p-azobenzenearsonate[18] introduces an asymmetric centre into the protein. On addition of either inhibitor or glycyl-L-tyrosine, the circular dichroism spectrum is changed, so indicating that the conformation of the azotyrosyl residue is altered on substrate/inhibitor binding.

In his original mechanism, (1963), Vallee suggested that the C-terminal carboxyl group specificity requirement of the enzyme could be understood in terms of an interaction of this group with a positively charged group, such as a lysyl or arginyl residue. No evidence could be obtained by chemical modification for the implication of lysyl groups, but subsequent to the earlier X-ray studies implicating Arg 145, Vallee and Riordan[69] were able to modify arginyl residues in carboxypeptidase by use of a 150 fold excess of diacetyl and to demonstrate that peptidase activity was lost. Esterase activity was increased, so indicating that the arginyl residue was not essential for the binding of the ester. It has also been shown chemically that the modification of a histidyl residue resulted in decreased peptidase activity. However, apart from the zinc binding groups, there is no histidine residue near enough to the zinc to be implicated in catalytic activity, so it is rather difficult to interpret this result at present.

Studies on inhibitor and substrate binding

The binding of substrates and inhibitors to the apoenzyme and the various enzymatically inactive metal substituted enzymes has been well studied and has contributed a great deal to the understanding of the mode of action of carboxypeptidase.

The apoenzyme will bind peptide substrates but will not bind esters. It is reasonable to believe therefore that the interaction between apoenzyme and peptide must be through the peptide nitrogen group *via* hydrogen bonding to a base, presumably the tyrosyl group already implicated by the work of Vallee. The binding of ester substrates by, for example, the copper enzyme, suggests that in this case the interaction is through metal ion and carboxylate group.

The inhibitor β-phenylpropionate is also only bound by the metalloenzyme; this too probably involves carboxylate metal ion interaction. It has already been mentioned that, in the presence of this inhibitor, chemical modification of tyrosyl groups does not affect enzyme activity, although it does have a marked effect in the absence of the inhibitor. It appeared, therefore, that the inhibitor interacts also with the tyrosine, as was confirmed by X-ray studies.

Other aspects of inhibitor binding, put forward prior to the X-ray studies, include the work of Neurath[19] on *d* and *l* amino acids. The former inhibit CPA at pH 7 while the latter only act at around pH 9.

Kinetic studies

The kinetic behaviour of carboxypeptidase with a range of substrates and inhibitors has been examined and shown to be very complex. These difficulties are often carried through from one metal to another, suggesting that they do not depend upon the metal ion. In fact they appear to be associated with dipeptide and analogous ester substrates or products having aromatic N-acyl groups.

A study of the rates of metal exchange in the presence of substrates shows that the metal is held more firmly in the presence of substrates, particularly by l,l dipeptides but also by d,l dipeptides but not by l,d dipeptides. This clearly shows the importance of the l configuration of the terminal peptide group in increased metal binding.

The kinetics of peptidase and esterase activity will not be discussed in detail but some mention will be made of pH dependence. The pH rate profile is different for peptidase and esterase activity. That for peptidase activity is bell shaped having inflexions at pH 6·7 and pH 8·5. If it is assumed that this behaviour reflects protonation equilibria of enzyme groups, (e.g. the increase in activity as the acidity is decreased being due to the increasing ionization of the carboxyl residue of Glu 270, while the final decrease in activity at higher pHs being due to the deprotonation of the tyrosyl residue 248) then it may be shown[20] that the two residues have pK_a values of 6·9 and 7·9, values that would not be expected for Glu 270 and Tyr 248. However, in view of the comments in Chapter 3 on the possibility of abnormal pK_a values for groups associated with the active site, it may be seen that these two pK_a values are rather difficult to interpret anyway. It should be noted in this context that the phenolic group Tyr 248, after conformational change, is near the positive residue Arg 145 and this could result in an increased acidity for Tyr 248.

The pH rate profile for esterase activity, apart from a class of short esters, is different from that for peptidase activity. A rise in rate from pH 5·5 to 7·0 is observed, followed by a plateau to pH 9·0 and a further rise to a maximum at pH 10·5. This suggests that the peptides and esters are hydrolysed by different mechanisms, but at present it is not possible to write mechanisms for ester hydrolysis with any confidence. It would obviously be of great interest to carry out X-ray studies on carboxy-peptidase-ester substrate complexes to see what information this would give on this problem.

Model building studies

Lipscomb and coworkers[3] have reported the results of model building experiments and have established two binding modes for substrates which may lead to inhibition. In one case the C-terminal side chain fits into the hydrophobic pocket; the aromatic N-acyl group has an interaction with Tyr 198; the second peptide carbonyl is bound to Arg 71, with the C-terminal carboxyl group bound to the zinc atom. In the second case the acyl group is in the pocket, the terminal carboxyl group is bound to

Arg 71 and the C-terminal side chain is near Tyr 198. Short substrates would bind less well in these positions in accord with their failure to show substrate inhibition. These ideas offer an explanation for certain of the kinetic anomalies in the reaction of CPA with dipeptide and ester substrates having aromatic N-acyl groups or large aliphatic groups, although, of course, other possibilities must be considered.

Lipscomb has also carried out model building experiments with ester substrates, assuming that the aromatic side chain is placed in the pocket and that the carbonyl group is again coordinated to the zinc. These suggest that binding of esters is essentially similar to peptides in some features but that it differs in that the carboxylate group may be bound to the zinc ion, in agreement with suggestions already made.[1] However, it appears unlikely that carboxylate–Arg 145 interaction is lost completely as the conformational change produced by this is necessary to bring Tyr 248 to the active site. A class of short ester substrates exhibit pH rate dependencies similar to those for peptides. It is reasonable that these should also bind like peptides.

Model building studies have also revealed that the 'Zinc-hydroxide' mechanism for carboxypeptidase activity involves serious steric interference in the binding stage. However, it is not possible to reject the mechanism purely on these grounds because of uncertainties in the positions of some of the relevant atoms in the X-ray structure determination.

Some conclusions from chemical and biochemical approaches to the study of carboxypeptidase

Despite the difficulties over the nature and number of the protein metal-binding groups, it is quite clear that a great deal of light had been thrown on the mechanism of carboxypeptidase activity by chemical and biochemical studies. The implication of the following groups in the active site had been clearly established:

(1) a group having a pK close to 6, involved in the catalytic step, possibly as a nucleophile;

(2) an acidic group, postulated as a proton donor, and suggested to be a tyrosyl residue, as implicated by chemical modification studies;

(3) an arginyl group, presumably to interact with the C-terminal carboxylate group of the substrate;

(4) the presence of the peptide bond near the zinc atom, and,

(5) the carboxyl group in inhibitor β-phenylpropionate at the zinc atom.

(a) *Zinc-carbonyl mechanism*

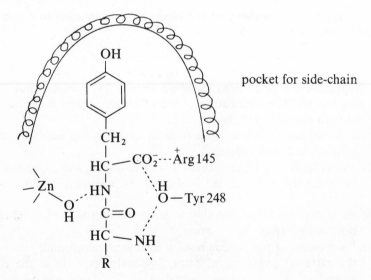

(b) *Zinc-hydroxide mechanism*

Figure 4.1 The binding of a peptide substrate to carboxypeptidase.

The mechanism of peptide hydrolysis

While the mechanism of ester hydrolysis is still unresolved it is now possible to write down with some confidence a peptide mechanism that accommodates the X-ray data and the chemical evidence previously mentioned. Certain other aspects of the chemical evidence, such as the role of the second tyrosyl group, are not yet understood.

Zinc-carbonyl mechanism

The feature of this mechanism is the polarization of the carbonyl group *via* coordination. Figure 4.1 schematically depicts substrate binding to carboxypeptidase.

A number of features call for further comment. The catalysis is initiated by the attack of the nucleophile upon the carbon atom of the polarized carbonyl group, as discussed in the section on hydrolysis in model compounds. The main unresolved question here is whether or not attack by this group (Glu 270) is direct, or whether it is *via* a water molecule (i.e. general base catalysis). In the former case an acyl intermediate would be formed. This, however, would be readily hydrolysed, so failure to show its presence is not meaningful. The X-ray results do not help here. The orientation and separation of the carbonyl group of Glu 270 is appropriate for nucleophilic attack on the hydrolysable bond of the peptide. But the necessary rearrangements that would have to occur for water attack to be feasible are quite possible. Possibly[18]O incorporation experiments could help resolve this problem. The actual timing of hydrogen ion transfer from the tyrosyl group is also not known. Presumably when hydrolysis is complete the reverse conformational changes occur in the enzyme to release the products.

Zinc-hydroxide mechanism

An alternative mechanism in accord with the X-ray data involves a mode of binding in which the susceptible carbonyl bond is directed away from the zinc, while the zinc is still linked to a water molecule or hydroxide ion. The coordinated OH^- group then attacks the carbon atom of the peptide bond. The OH of Tyr 248 would initially donate a hydrogen bond to one of the oxygen atoms of the terminal carboxyl group of the substrate. The scheme is shown in Figure 4.1. The results obtained from the studies of cobalt(III) catalysis of the base hydrolysis of glycine amides appear to show that the zinc-hydroxide mechanism might be a more realistic one. However, there are problems. Model building studies have revealed steric interferences in the binding stage. In addition NMR studies have shown that a water molecule or hydroxide ion is replaced on the binding of the inhibitor β-phenylpropionate to manganese carboxypeptidase. At present it appears, therefore that the zinc carbonyl mechanism is more feasible.

The schemes presented in Figure 4.1 are designed only to show the basic features of *enzyme-substrate* interaction. There are other finer points yet to be accommodated. The requirement for *l* configuration of the terminal residue is easily understood. The *d* species would not undergo the same interactions as the *l* form; as is demonstrated by the work of Neurath[19] on inhibition by *d* and *l* forms of amino acids.

It has also been noted that the activity of carboxypeptidase is affected by up to five residues, the second residue being specially important. Lipscomb's comments on the binding of longer substrates are important here. He has shown that maximum interaction of the aromatic groups of the substrate and the enzyme aromatic groups Tyr 198 and Phe 279 will occur if the phenolic OH group of Tyr 248 is also hydrogen bonded to the second NH group of the substrate. This will then allow the CO group of the third peptide link to be placed near residue Arg 71, so producing a further stabilizing interaction. This may in fact be the explanation for Vallee's results implicating two tyrosyl residues in catalytic activity.

The last point we will discuss is the question of the activity of the different metalloenzymes. The high activity of the cobalt enzyme has been discussed and may be readily understood. However, it is not certain that the other metals necessarily occupy the same binding site. Certainly Cu^{2+} will not favour the approximately tetrahedral site of cobalt and zinc, and so it is not surprising that the copper enzyme is inactive to both ester and peptide substrate.

Lipscomb[2] has reported on the structure of Hg^{2+} CPA. Here the metal ion is displaced by about 0.12 nm from the zinc position. This is due both to the necessity of providing longer metal-ligand bonds and the minimizing of Van der Waals repulsions which would be much greater for mercury than for zinc in the zinc binding site. The overall effect is to move the susceptible peptide bond away from Glu 270 in Hg CPA, so peptidase activity is lost.

CARBONIC ANHYDRASE

This was the first zinc metalloenzyme to be discovered, although in fact certain carbonic anhydrases in plants do not contain the metal ion. Both bovine and human carbonic anhydrase have been well studied and the experimental methods discussed under carboxypeptidase have also been applied[1,21,22] in this case. Of particular value have been the studies on metal substituted enzymes and their interaction with various inhibitors. Lindskog[21] has been able to prepare metalloenzymes where zinc has been replaced by manganese, cobalt, nickel, copper, cadmium and mercury with the new metal occupying the same site as the zinc. Lead, iron, beryllium and the alkaline earth enzymes have also been prepared. Again, the cobalt enzyme is active, the activity varying from 45% to 100% that of the zinc enzyme depending upon the conditions, while the nickel, manganese and iron enzymes are slightly active. The others are inactive.

Various aspects of the enzyme carbonic anhydrase have been reviewed in recent years. These reviews are summarized in a recent article by Maren[23] which in itself has a particular emphasis on the pharmacological and physiological aspects of the enzyme. A feature of its behaviour is the high specificity of a number of sulphonamides as inhibitors, while in addition to the hydrolysis of CO_2, it will catalyse the hydration[24,25] of molecules such as aldehydes and shows esterase activity towards several substrates[26] such as O-nitrophenylacetate.[27]

At physiological pHs CO_2 is converted to the bicarbonate ion HCO_3^-. This could either involve direct reaction of CO_2 with OH^-, or reaction with water and loss of a proton on ionization. In practice it appears that the first of these two steps is favoured, as carbonic anhydrase appears to bind HCO_3^- rather than H_2CO_3, while the ion itself is formed by attack of zinc bound OH^- on CO_2. The work of Wang on the infrared spectrum of enzyme bound carbon dioxide has contributed greatly to an understanding of these problems.

Infrared studies on carbonic anhydrase

Riepe and Wang[28] have measured the infrared spectrum of the CO_2 molecule bound in the cavity (already mentioned in the section on X-ray studies) at the active site of bovine carbonic anhydrase. The spectra were measured on 33% aqueous solutions of the enzyme at pH 5·5 under equilibrium CO_2 pressure against a reference cell containing the same enzyme solution without CO_2. The characteristic band of CO_2 observed at 2343 cm^{-1} for an aqueous solution, appears at 2341 cm^{-1} in the enzyme. That is, the spectrum is that of a normal CO_2 molecule.

Studies on the infrared of carbonic anhydrase solutions equilibrated with CO_2 and N_2O mixtures show that both compete for the active site suggesting that the CO_2 is not coordinated to the zinc. This is in accord with the fact that the d–d spectrum of the cobalt carbonic anhydrase is not affected by the binding of CO_2. Similar infrared studies on the enzyme in the presence of azide ion show that the carbonic anhydrase bound azide is in fact coordinated to zinc, as is reflected in the presence of an azide band at 2094 cm^{-1} compared with 2046 cm^{-1} for azide in inert protein solution. Bands at both 2094 and 2046 cm^{-1} are observed showing the presence of free and coordinated azide ion.

These workers have also studied the effect of azide ion on the enzyme/CO_2 system. Here the band at 2341 cm^{-1} due to bound CO_2 decreases as the 2095 cm^{-1} band due to coordinated azide increases. They deduced therefore that the hydrophobic cavity was right alongside the zinc ion, so that coordinated azide ion protrudes into the cavity, displacing CO_2. It was found also that bicarbonate ion displaced both CO_2 from the cavity and azide ion from the metal. This means that the bicarbonate ion must be coordinated to the zinc through its negative oxygen with the remainder of the molecule in the hydrophobic cavity. The implications of this are important. In the dehydration reaction proton transfer will accompany the C–O bond breaking, leaving CO_2 in the cavity and OH^- coordinated to zinc. In the hydration reaction, therefore, we can say at this point that the OH^- on the zinc must attack the bound CO_2 converting it to HCO_3^- (Figure 4.2).

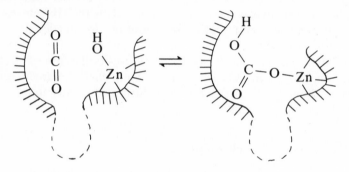

Figure 4.2   A scheme for the hydration of CO_2.

Metal-binding groups in carbonic anhydrase

X-ray studies have indicated that zinc is probably bound by three imidazole groups of histidine residues. It is clearly shown, however, that the cysteine group is too far away to be involved in metal binding. Chemical evidence was available for the implication of imidazole in the active site, but the situation regarding –SH groups was not clear cut. Thus titrations with Ag^+ had given conflicting results for bovine and human carbonic anhydrase, suggesting in the latter case only that –SH was a binding group, even though the d–d spectra of both cobalt substituted enzymes were similar.

Stability constant[22] measurements for metal substituted carbonic anhydrases and a comparison of log K and 3/2 log K_{en} suggest that zinc is bound by three nitrogen bases. The EPR spectrum of the copper(II) enzyme is similar to that of $Cu(dipyridyl)_2^{2+}$, while Δ_t, the crystal field splitting, of the cyanide complex with cobalt(II) carbonic anhydrase in which cobalt is in a tetrahedral environment, was found to be 5·3 kK, a value similar to that for $Co(benzamidazole)_4^{2+}$. In this particular case, therefore, chemical and physical evidence had given a satisfactory conclusion.

pH studies

These[21,29-32] have been of value in understanding the activity of carbonic anhydrase. Much has been done on the cobalt(II) enzyme. The $d-d$ spectrum is pH dependent, suggesting that there are two forms of the enzyme, related by a protonation equilibrium. The variation[21] of the $d-d$ spectrum with pH suggests that the group involved in this has a pK_a of 7·1. Figure 4.3 gives the $d-d$ spectrum over a range of pH values.

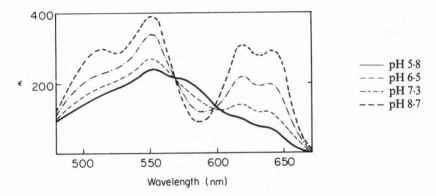

Figure 4.3 The effect of pH on the $d-d$ spectrum of cobalt carbonic anhydrase (Reproduced by permission from S. Lindskog, *J. Biol. Chem.*, 1963, **238**, 945).

This equilibrium between a protonated and non-protonated site is associated with the catalytic function of the enzyme. It appears that the non-protonated form is involved in hydration. Kernohan has shown, on the basis of pH-rate studies, that the pK_a of the group appears to be 7·1, in good agreement with the spectrophotometrically determined value.

Two alternative explanations have been offered for the pH effect on the cobalt carbonic anhydrase $d-d$ spectrum. One suggestion[1,21] is that the proton is released from a protein group which then coordinates to the metal ion. Support for this idea comes from a consideration of the $d-d$ spectrum of the alkaline form of the cobalt enzyme. While the acid form is that of a tetrahedral complex, the spectrum of the alkaline form is different from that of either tetrahedral or octahedral complexes, but may be interpreted in terms of a 5-coordinate complex. The second suggestion is that the ionization is that of a coordinated water molecule, generating a coordinated hydroxide group. This is in accord with a number of kinetic studies on the behaviour of carbonic anhydrase which implicate bound OH^-, i.e. E–Zn–OH. A fuller discussion of these two alternatives will be delayed until the results with inhibitors have been discussed, but it may be noted now that crystallographic evidence does not appear to support the former suggestion. The latter explanation has been criticized on the grounds that proton release from a coordinated water molecule is unlikely to occur at the low pH implicated by enzyme studies. However, once again, it is difficult to base too much on this type of argument as there is no guarantee that the dissociation of H_2O in a zinc complex bears any relation at all to the dissociation of water bound to zinc in carbonic anhydrase.

Inhibitors

Carbonic anhydrase is inhibited by a number of simple anions (e.g. azide, cyanide, cyanate, hydrosulphide) and by a range of sulphonamides $(X–SO_2–NH_2)$. The sulphonamides are efficient inhibitors. In general one mole is added to each mole of enzyme. The cobalt d–d spectrum is affected by the addition of inhibitors suggesting they are bound to the metal ion; in fact the shift in band position from inhibitor to inhibitor is in accord with the position of the ligand in the spectrochemical series. X-ray analysis of the iodide inhibitor complex with carbonic anhydrase shows that the iodide is very close to the zinc, although with a long Zn–I distance of about 0·4 nm. It is noteworthy that it had previously been suggested[32a] that a water molecule bridged I^- and the metal ion. The fact that the relative binding strengths of halides are in the order $F^- < Cl^- < Br < I^-$, i.e. the reverse of the order for zinc complexes, also suggests that other factors may be involved. The inhibitor acetazolamide does not bind to the apoenzyme.

The release of 1 mole of H^+ on addition of zinc to the apoenzyme between pH 6 and 9 does not occur in the presence of inhibitors, suggesting that the inhibitor competes with this ligand group for the metal.

The d–d spectra of the cobalt carbonic anhydrase complex with the inhibitors cyanide, cyanate, acetazolamide and hydrosulphide show that these complexes have an essentially tetrahedral structure (Figure 4.4).

Figure 4.4 also shows the spectrum of the alkaline form of the enzyme. It has been suggested that there is a trend towards this spectrum in the spectra of the complexes with inhibitors, and that this spectrum is probably that of a five coordinate species, brought about by the coordination of the deprotonated ligand group. The alternative of OH^- formation should not give rise to a new geometry whereas the position of the newly available

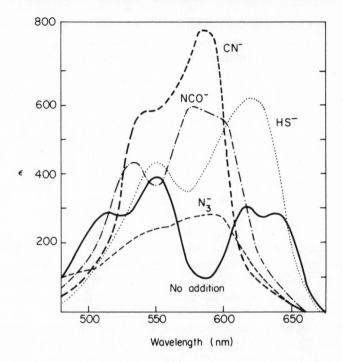

Figure 4.4 *d–d* spectra of inhibitor complexes with cobalt carbonic anhydrase (reproduced by permission from S. Lindskog, *J. Biol. Chem.*, 1963, **238**, 945).

protein group, one determined by the protein itself, could well force such a geometry upon the zinc.

A recent study[33] has involved the measurement of the rate of reaction between human carbonic anhydrase B and C and aromatic sulphonamides, using stopped flow techniques and measuring the changes in protein fluorescence associated with complex formation. In particular, rate constants k_1 and k_{-1} have been measured over the pH range 5·0–10·8 for a

$$E + S \underset{k_{-1}}{\overset{k_1}{\rightleftharpoons}} ES$$

number of sulphonamide inhibitors. There is a 240 fold variation in k_1 for the sulphonamides studied. There appears to be some evidence that presence of an *ortho* substituent in the aromatic ring ⟨⟨ ⟩⟩—SO_2NH_2 is important in determining the value of k_1.

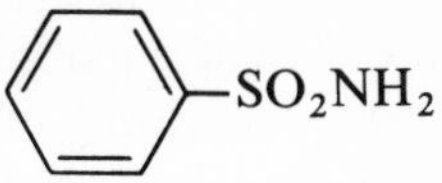

It is suggested that the inhibitor-enzyme complex is formed *via* the rapid pre-equilibrium formation of at least one less stable intermediate, and that the first intermediate could involve the hydrophobic interaction

$$E + S \rightleftharpoons (ES)_1 \rightleftharpoons (ES)_2$$

of the enzyme with the aromatic ring of the inhibitor. This then undergoes rearrangement to the more stable product involving metal ion coordination. The presence of an interaction between the aromatic nucleus of the inhibitor and the hydrophobic portion of the active site cleft has been demonstrated[33a] for bovine carbonic anhydrase by the spin label technique using the sulphonamide spin label 4-[(p-sulphonamido)benzoyloxy]-2,2,6,6-tetramethylpiperidine-1-oxyl). The presence of substituents in the aromatic ring could affect the relative position of the ring and the protein surface, and so the stability of the first intermediates.

The measured overall association rate constant is pH dependent, but the dissociation is pH independent. The results, which are essentially similar for cobalt and zinc enzymes, suggest that combination occurs between a neutral sulphonamide species and the alkaline form of carbonic anhydrase. Chemical modification (carboxymethylation) of a single histidine residue in human carbonic anhydrase markedly affects the sulphonamide binding kinetics, the pH independent dissociation rate is increased 45 times, while the pH dependence of association is shifted to higher pH. This same chemical modification results in a change in the pH dependence of the equilibrium between the coordination forms of the enzyme (i.e. *d–d* spectra of CoCa) which is in good agreement with the effect on association rate–pH profile just mentioned.

The pH-rate profile for the inhibitor-enzyme association is bell shaped, with pK's for the two functions of (for the inhibitor *p*-nitrobenzenesulphonamide) 6·60 and 9·30. The value of the second pK varies with the nature of the inhibitor and corresponds closely with the ionization constant of the sulphonamide group in the inhibitors. The lower pK does not correspond to any sulphonamide ionization, but does correspond to the pK derived from the spectral changes involved in the pH dependence of the cobalt carbonic anhydrase C (6·60), suggesting that the pK derived from the inhibitor studies is associated with the same function involved in the pH dependence of the enzyme, as is confirmed by the carboxymethylation results. There is a similar correlation between the association rate pK of zinc carbonic anhydrase B and the spectrophotometric pK of the cobalt enzyme B. Carboxymethylation can be prevented by the presence of sulphonamides or certain anions.

The spectrum of the Co(II) enzyme–sulphonamide complex is pH insensitive, consistent with the pH independence of the dissociation rate.

It seems therefore that the pH sensitive ligand has been replaced from the metal by the sulphonamide.

It appears that a sulphonamide proton is lost after the first hydrophobic interaction of the aromatic ring and the protein, as UV spectroscopy indicates the presence of a deprotonated group.[34]

Mechanism

Mention has already been made of two alternative schemes. The formation of a five coordinate complex by coordination of a newly deprotonated ligand group receives some support from the spectrum of the copper carbonic anhydrase enzyme. The spectrum of this species does not change with pH and can be interpreted in terms of a five coordinate structure. A comparison of formation constant data for copper(II) and cobalt(II), shows that copper(II) amine coordination is much more highly favoured than cobalt(II) amine coordination, so coordination of the fifth protein group could then take place at a lower pH in the copper case. The inactivity of the copper enzyme may be due to the fact that it binds this group too strongly.

The following scheme[33] is suggested; the parallel scheme for sulphonamide inhibitor complex formation is also given.

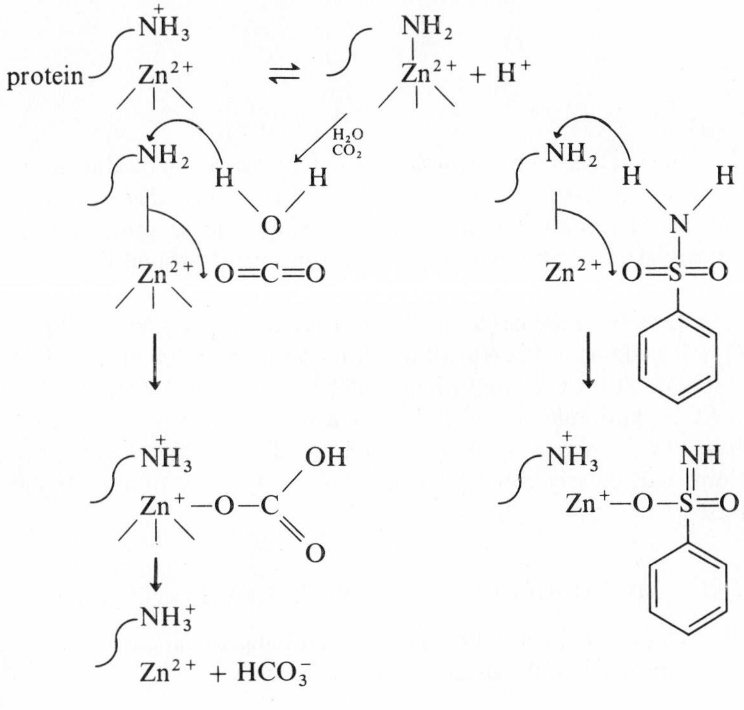

The alternative mechanism, involving the aquo-hydroxo interconversion has been supported by Coleman who has extensively[35,36] studied the interaction of a number of metallo carbonic anhydrases with the tritiated sulphonamide inhibitor ^{3}H-acetazolamide and anions over a pH range. These papers should be consulted for fuller details. Coleman has criticized the idea that the proton is liberated by a protein group, which subsequently coordinates, on the basis of stability constant measurements. He suggests that there should be a rapid increase in stability constant in the pH range 7–9 if this scheme holds. In fact there is not. On the other hand, change from coordinated H_2O to OH^- could not be expected to change the stability of the complex. Kinetic studies[37] support the following scheme.

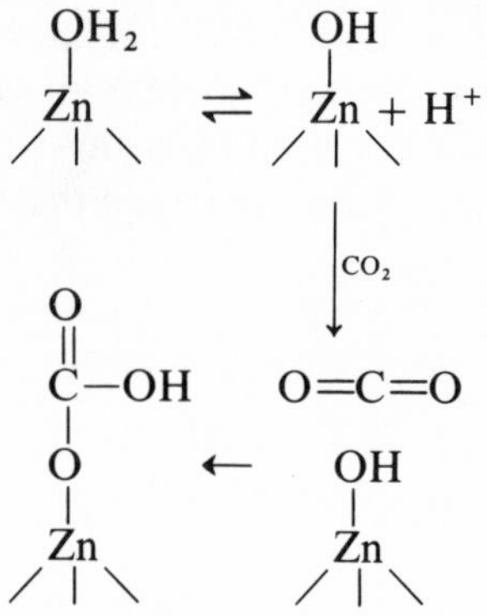

$$
\begin{array}{ccc}
\overset{\displaystyle OH_2}{\underset{\diagup\,|\,\diagdown}{|}\;Zn} & \rightleftharpoons & \overset{\displaystyle OH}{\underset{\diagup\,|\,\diagdown}{|}\;Zn} + H^+
\end{array}
$$

$$
\downarrow CO_2
$$

$$
\begin{array}{ccc}
\overset{\displaystyle O}{\underset{|}{\overset{\|}{C}-OH}} & & O=C=O \\
\overset{\displaystyle O}{\underset{\diagup\,|\,\diagdown}{|}\;Zn} & \leftarrow & \overset{\displaystyle OH}{\underset{\diagup\,|\,\diagdown}{|}\;Zn}
\end{array}
$$

There is some evidence for the implication of an imidazole residue in the activity of this enzyme, possibly in proton transfer. The change in d–d spectrum of the cobalt enzyme is then possibly due to some protein configurational effect altering the ligand arrangement around the metal rather than five coordination.

Both of these two schemes are probably oversimplified. The binding of the CO_2 molecule is unknown, although its position in the groove near the metal ion is known. Certainly there will be further interactions playing upon the CO_2 molecule, so it is difficult to assess the validity of the argument based upon models. However, the value of replacing zinc by other metal ions, particularly cobalt, has been clearly seen for both carbonic anhydrase and carboxypeptidase.

OTHER METAL-DEPENDENT HYDROLYTIC ENZYMES

The main classes of hydrolytic enzymes involving metal ions may be summarized in terms of the substrate hydrolysed.

Peptides

C-terminal amino acid residues are specifically hydrolysed by the carboxypeptidases. N-terminal residues are hydrolysed by the aminopeptidases, of which there are a number with varying degrees of specificity towards particular amino acid residues. Thus leucine aminopeptidases are particularly active towards leucine residues of L configuration. These enzymes are activated by Mg^{2+} and Mn^{2+} and are of value in amino acid sequence determination. A range of other enzymes contain zinc, sometimes with a second metal, usually calcium. An example of a dipeptidase is renal dipeptidase[38] which will catalyse the hydrolysis of a number of dipeptides and contains 1 mole of zinc per mole of enzyme (molecular weight 47,200). The apoenzyme may be prepared by dialysis against 1,10 phenanthroline, while enzymatic activity is restored by addition of the zinc. A whole range of microbial neutral proteases are known which appear to be metal ion dependent in that they are inhibited by added chelating agents. Often the metal ion has not been positively identified and analysed for. *B. subtilis* proteases are dependent upon zinc and calcium while *B. thermoproteolyticus* thermolysin, *B. megaterium* megateriopeptidase and *Streptomyces* naraensis neutral protease are all zinc dependent.

Esters

A wide range of enzymes catalyse the hydrolysis of a number of different types of organic and inorganic esters. Two examples of esterase activity have already been given, in which zinc is involved. Another zinc metalloenzyme, alkaline phosphatase, catalyses the hydrolysis of phosphate esters, while zinc is also found[39] in phospholipase C, which catalyses the hydrolysis of phospholipids. In this case the activity is inhibited by metal complexing agents and restored by the addition of zinc ions. Thiol groups are possibly involved in zinc binding. Magnesium ions are necessary for the activity of many enzymes involved in phosphate hydrolysis, such as ATP-ase (which catalyses the hydrolysis of ATP to ADP and orthophosphate), inorganic pyrophosphatase (which catalyses the conversion of pyrophosphate to orthophosphate) and other phosphatases. We will now consider the role of magnesium in these enzymes in more detail, as the work of Cohn has elucidated a number of problems here, and then briefly consider a few examples of other systems.

Magnesium ions

Most enzymes associated with phosphate group transfer or hydrolysis require Mg^{2+} as a cofactor, presumably to reduce charge on the phosphate group. There is now evidence to show that the metal ion is bound to the phosphate (as is the case for metal ion—free ATP interaction). Magnesium(II) may be replaced by manganese(II) in many cases with the activity maintained. This offers a means of studying metal ion interactions with enzyme and substrate; to tell, for example, whether the magnesium bridges substrate and enzyme. This work was pioneered by Cohn[40,41] who has applied both NMR and EPR techniques to the study of this problem. The latter technique is not so useful as the former as the EPR spectrum is sometimes difficult to obtain. However, the Mn^{2+} ion, with its high magnetic moment, has a large effect upon the proton relaxation rate of water, due to interaction with water in its coordination sphere. A concentration of 1M Mn^{2+} increases the relaxation rate by a factor of 10^4. The effect is dependent upon the environment of the metal ion, that is upon the type of complex formed. The displacement of the coordinated water by ligands should decrease the effectiveness of Mn^{2+} in increasing the relaxation rate of water. However, it is known that complexing Mn^{2+} with proteins can increase the relaxation rate as the relative rotational motion of the Mn^{2+} ion and water is hindered by the macromolecule. The resulting enhancement of the proton relaxation rate is thus a sensitive indicator of changes in configuration and the nature of the ligands occurring at the active site, provided the metal is at the active site.

About 100 known enzymes catalyse phosphoryl group transfer, for which we may write the general equation

$$ATP^{4-} + XH \underset{M^{2+}}{\rightleftharpoons} ADP^{3-} + XPO_3 + H^+$$

A number of kinase systems have been studied by Cohn[40] who has divided them into two groups on the basis of NMR and EPR data. For this the probe properties of Mn(II) have been utilized, together with spin labels. The two classes are

(I) where the substrate bridges metal and enzyme, M–S–E, and

(II) where the metal ion bridges enzyme and substrate, S–M–E.

The substrate will be ATP^{4-} or ADP^{3-}.

The existence of the first class has been demonstrated by EPR spectra. The Mn–ADP binary complex has an EPR spectrum which does not change on formation of the ternary complex with enzyme. This indicates that there is no change in the nature of the ligands bound to the metal ion when the ternary complex is formed. Clearly therefore the metal is bound to substrate only and cannot be bridging. Further support comes from kinetic studies,[42] using relaxation techniques, on the ADP–creatine phosphotransferase system using a range of metal ions as cofactors. Rate constants for the association and dissociation of M–S and enzyme are practically independent of the nature of the metal, implying that metal-ligand substitution is not occurring, and that therefore the metal ion is not bridging.

The existence of ternary complexes of the type S–M–E having bridging metal ion was shown by NMR studies on the proton relaxation rate of water bound to Mn^{2+}. A large enhancement of relaxation rate on formation of the metal-enzyme complex is followed by a de-enhancement upon binding of the substrate. This is consistent with substrate substitution into the coordination shell of Mn(II) with replacement of water, but it could be the result of a conformational change.

This division of the kinase into two groups is confirmed by studies on their activation by different metal ions. Certain kinases have a low metal ion specificity. These belong to Class I, M–S–E, where the requirement for the metal ion is mainly one of charge neutralization only. These that belong to Class II (E–M–S) have a greater selectivity, again reasonably, as here there are extra requirements in terms of the role of the metal ion as a bridge.

The use of spin labels has contributed to the elucidation of the geometry of the active site. An SH group is implicated in the active centre of creatine kinase. This can be spin labelled, with a nitroxide free radical and a second paramagnetic centre introduced in Mn^{2+}. These will interact and a study of the interaction will allow the distance between the —SH group and metal ion to be calculated. For E^-–Mn–ADP and E^-–Mn–ATP, this distance has been found to be in the range of 0·8–1·2 and 1·3–1·8 nm respectively.

Alkaline phosphatase[43–53]

This enzyme catalyses the hydrolysis of orthophosphate monoesters and shows maximum activity at pH 8 or above. It also shows phospho-transferase activity, catalysing phosphate transfer reactions. The dual specificity is a useful aid in the discussion of the activity of the enzyme.

The enzyme from *Escherichia Coli* has been most extensively studied. It contains four moles of zinc per mole of enzyme (molecular weight 89,000), and splits into two subunits on treatment with acid. The two units reassociate at neutral pH or on treatment with zinc ions. This suggests that zinc ions are necessary for linking the two subunits and that the protein dissociates in acid solution because protons compete successfully for the binding groups of the zinc.

On treatment with 1,10 phenanthroline, two zinc atoms are removed with loss of enzyme activity but without breakdown into two subunits. It appears therefore that two zinc atoms are involved in enzyme activity and two in structure maintenance. The enzyme is denatured more readily in the absence of Zn^{2+}.

The metalloenzyme is regenerated on addition of Zn^{2+} to the apoenzyme. Hg^{2+} and Co^{2+} also partially restore activity but other metals give an inactive species. A very high concentration of Mg^{2+} ions will restore the activity of the apoenzyme, but presumably by a different mechanism in which the metal ion reduces the negative charge of the phosphate group, so allowing the approach of water molecules.

Enzyme activity

The reactions catalysed by the enzyme are thought to involve an enzyme-phosphate intermediate. This may either react with water to give inorganic phosphate and hydrolysed substrate, ROH or with an acceptor giving the new phosphoester product. Kinetic data are consistent with the breakdown of the enzyme phosphate complex as the rate-controlling step. The phosphate group is esterified to the hydroxyl group of a serine residue.

Chemical modification

This has been carried out by Vallee and coworkers. The results on metal ion replacement have already been mentioned, but it should be noted that the cobalt enzyme, while retaining 10% of the hydrolase activity, has no phosphotransferase activity. Addition of 2 moles of cobalt ion to the apoenzyme gives a species having a d–d spectrum characteristic of cobalt octahedral complexes. This is inactive. Addition of a further two

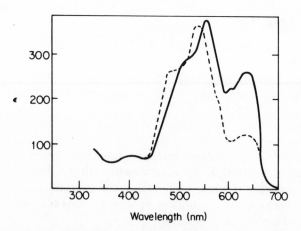

Figure 4.5 d–d spectra of cobalt(II) alkaline phosphatase (——) and its complex with phosphate (– – – –). (Reproduced by permission from M. L. Applebury and J. E. Coleman, *J. Biol. Chem.*, 1969, **244**, 709.)

moles of cobalt results in the generation of the active species. Now, however,[53] the $d-d$ spectrum (Figure 4.5) suggests a distorted structure, as in the case of cobalt carbonic anhydrase and carboxypeptidase. This is confirmed by circular dichroism studies. Interaction with the inhibitor inorganic phosphate gives a spectrum[53] (Figure 4.5) more characteristic of octahedral geometry. A pH study of the spectrum of the cobalt enzyme indicates that the distorted species is present in the pH range 6–8 with a pK_a of 7·0 for the appropriate enzyme group involved in this. Below pH 6 the spectrum is similar to that of the octahedral form. This pK_a of 7·0 is consistent with that obtained from the activity of the cobalt enzyme.

A number of organic modifications of the enzyme have also been carried out. For example, N-bromo succinimide, at pH 7·5, will oxidize a number of residues—tryptophan, tyrosine, methionine and cysteine. The reaction of the zinc enzyme with a 22 fold excess of NBS will double phosphotransferase activity and slightly increases hydrolase activity. However, a 50-fold excess of NBS, while increasing hydrolase activity in the cobalt enzyme, will also generate phosphotransferase activity, which is not present normally. Photooxidation of native alkaline phosphatase in the presence of methylene blue does not significantly affect activity, but similar treatment on the apoenzyme results in reduced zinc binding capacity. Histidine residues in the native and apoenzymes are destroyed as a function of the exposure but the rate is significantly faster for the apoenzyme. It appears that zinc protects three histidyl residues and is probably bound by them, as the treated apoenzyme is shown by amino acid analysis to have lost three groups which are not photooxidized in the zinc enzyme. Additional ligands are obviously required. The enzyme does not contain cysteine while the rate of oxidation of tryptophan is not affected by zinc.

While it is obviously premature to comment in detail on the activity of alkaline phosphatase, it is clear that the use of other metal ions has brought out a number of features which will provide extra checks on the correctness of any mechanism put forward.

Aminopeptidase

Certain microorganisms are able to function effectively at high temperature. Such a thermophilic species, *Bacillus stearothermophilus*, can maintain activity[54] at temperatures as high as 90° for short periods. This implies a remarkable stability of its enzyme systems towards denaturation. The aminopeptidase from this species is associated with the presence of very high metal ion concentration, so in this case the metal ion, in addition to its role in hydrolytic activity, is almost certainly providing a stabilizing influence on the protein structure. When isolated from 10^{-3}M cobalt(II) solutions, the enzyme contains eighteen moles of cobalt and two or three moles of zinc per mole of enzyme of M.Wt. 400,000. Treatment with EDTA at pH 5·3 converts it into the apoenzyme, which will recombine with a range of metal ions to give metalloenzymes of varying activity.

α-amylases

These are[55] calcium metalloenzymes containing one or more moles of calcium per moles of enzyme (M.Wt. = 50,000) that catalyse the hydrolysis of glucosides. The metal ion is necessary for biological activity and apparently also in maintaining the protein structure as the apoenzyme is less stable than the calcium enzyme.

The α-amylase of *B-subtilis* also contains $\frac{1}{2}$ mole of zinc per mole of enzyme apparently acting as a bridge, dimerizing two units.

REFERENCES

1. A. E. Dennard and R. J. P. Williams, '*Transition Metal Ions as Reagents in Metalloenzymes*' in *Transition Metal Chemistry*, Carlin (Ed.), Vol. 2, Arnold, 1966
2. *Brookhaven Symposia in Biology*, No. 21, 1968, pp. 1–119
3. *Phil. Trans. Roy. Soc. Lond.*, 1970, **B257**, 159
4. W. N. Lipscomb, *Acc. Chem. Res.*, 1970, **3**, 81
5. H. Neurath, R. A. Bradshaw, P. H. Petra and K. A. Walsh, *Ref. 3*, p. 159
6. B. L. Vallee and J. F. Riordan in *Ref. 2*, p. 91
   (a) B. L. Vallee, J. F. Riordan, D. S. Auld and S. A. Latt, *Ref. 3*, p. 215
7. W. N. Lipscomb, J. C. Coppola, J. A. Hartsuck, M. L. Ludwig, H. Muirhead, J. Searl and T. A. Stock, *J. Mol. Biol.*, 1966, **19**, 423
8. G. N. Reeke, J. A. Hartsuck, M. L. Ludwig, F. A. Quiocho, T. A. Stertz and W. N. Lipscomb, *Proc. Nat. Acad. Sci. U.S.*, 1967, **58**, 2220
9. W. N. Lipscomb, J. A. Hartsuck, G. N. Reeke, F. A. Quiocho, P. H. Bethge, M. L. Ludwig, T. A. Stertz, H. Muirhead and J. C. Çoppola, *Ref. 2*, page 24
10. D. E. Koshland, *Cold Spring Harbour Symp. Quant. Biol.*, 1963, **28**, 473
11. K. Fridborg, K. K. Kannan, A. Liljas, J. Lundin, B. Strandberg, R. Strandberg, B. Tilander and G. Wrien, *J. Mol. Biol.*, 1967, **25**, 505
12. B. Strandberg and A. Liljas, personal communication cited by S. Lindskog, *Structure and Bonding*, 1971, **8**, 153
13. (a) M. L. Bender and B. W. Turnquest, *J. Amer. Chem. Soc.*, 1957, **79**, 1889
    (b) M. D. Alexander and D. H. Busch, *J. Amer. Chem. Soc.*, 1966, **88**, 1130
    (c) D. A. Buckingham, D. M. Foster and A. M. Sargeson, *J. Amer. Chem. Soc.*, 1969, **91**, 4102
        D. A. Buckingham, D. M. Foster, L. G. Marzilli and A. M. Sargeson, *Inorg. Chem.*, 1970, **9**, 11
    (d) D. A. Buckingham, C. E. Davis, D. M. Foster and A. M. Sargeson, *J. Amer. Chem. Soc.*, 1970, **92**, 5571; D. A. Buckingham, D. M. Foster and A. M. Sargeson, *J. Amer. Chem. Soc.*, 1970, **92**, 6151

14. J. E. Prue, *J. Chem. Soc.*, 1952, 2331
    (a) A. E. Dennard and R. J. P. Williams, *J. Chem. Soc. (A)*, 1966, 812
    (b) M. Caplow, *J. Amer. Chem. Soc.*, 1971, **93**, 230
15. B. L. Vallee, R. J. P. Williams and J. E. Coleman, *Nature*, 1961, **190**, 633
16. B. L. Vallee, T. L. Coombs and F. L. Hoch, *J. Biol. Chem.*, 1960, **235**, PC 45
17. R. G. Shulman, G. Navon, B. J. Wyluda, D. C. Douglass and T. Yamane, *Proc. Nat. Acad. Sci. U.S.*, 1966, **56**, 39
18. H. M. Kagan, *Fedn. Proc.*, 1968, **27**, 455
19. H. Neurath, *The Enzymes*, Vol. 4, Boyer, Lardy, Myrback, (Eds), Academic Press, N. York, 1961, Ch. 2, p. 11
20. F. W. Carson and E. T. Kaiser, *J. Am. Chem. Soc.*, 1966, **88**, 1212
21. S. Lindskog, *J. Biol. Chem.*, 1963, **238**, 945
22. S. Lindskog and P. O. Nyman, *Biochim. Biophys. Acta*, 1964, **85**, 462
    (a) S. Lindskog, *Structure and Bonding*, 1971, **8**, 153
23. T. H. Maren, *Physiol. Rev.*, 1967, **47**, 595
24. Y. Pocker and J. E. Meany, *J. Am. Chem. Soc.*, 1965, **87**, 1809
25. Y. Pocker and J. E. Meany, *Biochemistry*, 1965, **4**, 2535; 1967, **6**, 239
26. T. A. Duff and J. E. Coleman, *Biochemistry*, 1966, **5**, 2009
27. Y. Pocker and J. T. Stone, *J. Am. Chem. Soc.*, 1965, **87**, 5497
28. R. Reipe and J. H. Wang, *J. Am. Chem. Soc.*, 1967, **89**, 4229
29. S. Lindskog, *Biochemistry*, 1966, **5**, 2641
30. J. C. Kernohan, *Biochim. Biophys. Acta*, 1964, **81**, 346
31. J. C. Kernohan, *Biochim. Biophys. Acta*, 1965, **96**, 304
32. J. E. Coleman, *J. Biol. Chem.*, 1967, **22**, 5212
    (a) Y. Pocker and D. G. Dickerson, *Biochemistry*, 1968, **7**, 1995
33. P. W. Taylor, R. W. King and A. S. V. Burgen, *Biochemistry*, 1970, **9**, 2638, 3894
    (a) J. F. Hower, R. W. Henkens and D. B. Chesnut, *J. Amer. Chem. Soc.*, 1971, **93**, 6665
34. R. W. King and A. S. V. Burgen, *Biochim. Biophys. Acta.*, 1970, **207**, 278
35. J. Coleman, *J. Biol. Chem.*, 1967, **242**, 5212
36. J. Coleman, *Nature*, 1967, **214**, 193
37. R. P. Davies, in *The Enzymes* (2nd Edit.) Vol. V, 545, (P. D. Boyer *et al* (Eds.)), Academic Press, New York, 1961.
38. B. J. Campbell, Y. C. Lin, R. V. Davis and E. Bellew, *Biochim. Biophys. Acta* 1966, **118**, 371
39. A. C. Ottolenghi, *Biochim. Biophys. Acta*, 1965, **106**, 510
40. M. Cohn, *Biochemistry*, 1963, **2**, 623; *Quart. Rev. Biophysics.*, 1970, **3**, 61
41. M. Cohn and J. Reuben, *Acc. Chem. Res.*, 1971, **4**, 214
42. G. G. Hammes and J. K. Hurst, *Biochemistry*, 1969, **8**, 1083
43. D. J. Plocke, C. Levinthal and B. L. Vallee, *Biochemistry*, 1962, **1**, 373
44. D. J. Plocke and B. L. Vallee, *Biochemistry*, 1962, **1**, 1039
45. F. Rothman and R. Byrne, *J. Mol. Biol.*, 1963, **6**, 330
46. I. B. Wilson, J. Dayan and K. Cyr, *J. Biol. Chem.*, 1964, **239**, 4182
47. S. R. Cohen and I. B. Wilson, *Biochemistry*, 1966, **5**, 904
48. G. H. Tait and B. L. Vallee, *Proc. Nat. Acad. Sci. U.S.*, 1966, **56**, 1243
49. R. T. Simpson and B. L. Vallee, *Biochemistry*, 1968, **7**, 4343
50. J. H. Schwartz and F. Lipmann, *Proc. Nat. Acad. Sci. U.S.*, 1961, **47**, 1996
51. M. J. Schlesinger and K. Barrett, *J. Biol. Chem.*, 1965, **240**, 4284
52. J. A. Reynolds and M. J. Schlesinger, *Biochemistry*, 1969, **8**, 588
53. M. L. Applebury and J. E. Coleman, *J. Biol. Chem.*, 1969, **244**, 308, 709

54. G. Roncari and H. Zuber, *8th Int. Cong. Biochem.* (Switzerland) 1970, 135
55. E. H. Fischer, W. N. Sumerwell, J. Junge and E. A. Stein, *Proc. 4th Int. Cong. Biochem.*, 1958, 124; B. L. Vallee, E. A. Stein, W. N. Sumerwell and E. H. Fischer, *J. Biol. Chem.*, 1959, **234**, 2901

CHAPTER FIVE

THE TRANSITION METALS IN BIOLOGICAL REDOX REACTIONS

Metal ions are involved in a number of different types of biological oxidations. The transition metals iron and copper are particularly well known in this context, but cobalt and molybdenum are also important. Here the redox change in the transition metal ion is associated with the catalytic activity of the metalloprotein. The non-transition elements zinc and magnesium are associated with certain redox systems but clearly have a different role, as outlined for zinc in Chapter 1.

There are a number of well-known transition-metalloproteins which act as storage and transport proteins rather than as enzymes. The oxygen carriers hemoglobin, hemerythrin and hemocyanin and the oxygen storer myoglobin will all be discussed in Chapter 7. The vanadium protein hemovanadin is also thought to act as an oxygen carrier (but is not an heme protein). Other proteins are involved in metal storage and transport. Ferritin[1] is a non-heme iron storage protein of molecular weight around 460,000, in which the ferric ion is present as an hydroxide phosphate complex. The apoferritin dissociates into about twenty-four subunits. The actual transport of iron (shown to be ferric by EPR studies) is carried out by transferrin and conalbumen[2] which are probably identical proteins. Transferrin has a molecular weight of 77,000 and possesses two iron binding sites. There is evidence to show that iron is bound to the protein by phenoxy groups of tyrosyl residues. Imidazole residues of histidine are possibly also implicated. The initial evidence for this was based upon pH studies and, in view of the complications noted in Chapter 3 for this approach, could only be regarded as leading to tentative conclusions. Thus, the effect on pH of adding Fe^{3+} to the apoprotein over a range of pH values had indicated that the pK_a of metal binding groups was near 11·2, suggestive of phenoxy groups. Titration studies on the iron- and apotransferrin also show that above pH 5·5 ten new groups in all are titratable in the apoprotein. It appears that six are the phenoxy groups mentioned above, while the fact that the pK_a of the remaining four groups is near 7 suggests that these are histidyl residues. As two iron atoms are bound per mole of protein it appears that each iron atom is bound by three tyrosyl and two histidyl residues. It is possible to replace iron by copper in these two proteins. EPR studies on the copper transferrin and conalbumen show ligand hyperfine structure, consistent with copper interaction with nitrogen donor groups. Accordingly, it has been suggested that copper has a square planar binding site with two nitrogen and two oxygen donor atoms.

Recently[2a] trivalent lanthanide ions have been used as fluorescent probes. There are two specific binding sites per mole for the ions Tb^{3+}, Eu^{3+}, Er^{3+} and Ho^{3+}, as for Fe^{3+}. There is only one site for Nd^{3+} and Pr^{3+}, possibly a size effect. Binding of Tb^{3+} to transferrin resulted in an increase in Tb^{3+} fluorescence intensity by a factor of 10^5. A fluorescence titration curve indicated that only Tb^{3+} ions bound to the specific metal binding sites of transferrin have enhanced emission, the fluorescence intensity levelling off after the Tb^{3+}/moles of transferrin ratio reaches a value of 2. Both UV and fluorescence measurements indicate that the metal ion is bound by the phenolic oxygen of two tyrosyl residues rather than three, as suggested above.

It has also been possible to estimate the distance between the two metal binding sites by a study of the quenching effect of Fe^{3+} on the Tb^{3+} fluorescence when they are both bound to the same protein molecule. The fluorescence lifetime of Tb^{3+} in this complex is about 5% less than the fluorescence lifetime of Tb^{3+} in the Tb^{3+} transferrin. If this is due to energy transfer it may be shown that the Fe^{3+}–Tb^{3+} distance is 4·3 nm. However, there may be other explanations for the shorter fluoroscence lifetime, in which case the separation between the two ions will be greater than 4·3 nm.

One other important and interesting metalloprotein, whose biological activity is not characterized with certainty at present is ceruloplasmin.[3] This is a glycoprotein of molecular weight 155,000 containing 8 atoms of copper per mole. It is known to catalyse the oxidation of ferrous iron to ferric ion. EPR and magnetic susceptibility measurements show that some 40–44% of the copper is present as copper(II).

Moving on to consider enzymatic oxidations, we will first present a summary of the different types of oxidation reactions usually observed. Most of the metalloenzymes involved are concerned with the oxidation of organic substrates, with oxygen, as the final electron acceptor, being reduced to water or hydrogen peroxide. This summary will be followed by a general survey of the principles involved in the study of the role of redox metal ions, including the use of small model complexes, concentrating particularly on iron and copper. The chapter will be concluded with an account of the main aspects of the biochemistry and associated inorganic chemistry of the metal ions mentioned in the introduction to this chapter.

TYPES OF BIOLOGICAL OXIDATION

Biological oxidations[4] may be brought about by one of the following processes (S = oxidized substrate, SH_2 = reduced substrate).

(a) electron transfer with oxygen (or another species) as the terminal electron acceptor,

(b) hydrogen atom abstraction, by (1) or (2)

$$2SH_2 + O_2 \rightarrow 2S + 2H_2O \tag{1}$$

$$SH_2 + O_2 \rightarrow S + H_2O_2 \tag{2}$$

(c) hydride group transfer,

(d) oxygen atom incorporation, and

(e) hydroxyl group incorporation.

The free energy liberated in these oxidative processes may be utilized in the production of ATP. Hydrolysis of ATP will then, in turn, provide the energy for a number of important processes, such as muscle contraction or synthesis.

Equations (1) and (2) represent an overall reaction. In practice, the reaction may proceed *via* one or more intermediate hydrogen carriers, with oxygen as the terminal hydrogen acceptor. Such a scheme is usually represented by (3), where A^n and A^nH_2 represent oxidized and reduced states of a series of acceptors. This scheme is very important as in effect we have an electron transport chain.

$$SH_2 \diagdown \diagup A^1H_2 \diagdown \diagup A^2H_2 \quad \cdots \quad A^nH_2 \diagdown \diagup H_2O_2$$
$$S \diagup \diagdown A^1 \quad A^2 \quad A^n \diagup \diagdown O_2 \tag{3}$$

Electron-transfer chains

It is possible to write schemes similar to (3) involving electron transfer or oxygen transfer in addition to hydrogen transfer. Some typical carriers that might be involved in such a chain of electron acceptors have already been referred to in Chapter 1. Examples include the flavin nucleotide coenzymes, the nicotinamide coenzymes, the cytochromes and the ferredoxins. The latter two examples are associated with a $Fe(II)–Fe(III)$ redox change. Oxygen is a very well known physiological electron acceptor, but often this (or other naturally occurring electron acceptors) may be replaced by artificial electron acceptors. Thus, if it is desired to examine an oxygen utilizing species under anaerobic conditions, it is usually possible to replace oxygen as the terminal electron acceptor by the dye methylene blue. Many other dyes have also been used. Significantly, these may usually be autoxidized readily.

One important example of electron transfer systems is that of linked reactions. Here one carrier intervenes between two separate dehydrogenases using two different substrates SH_2 and $S'H_2$ (Equation 4). Here

$$SH_2 \diagdown C_{red} \diagdown S'H_2$$
$$S \diagup C_{ox} \diagup S' \tag{4}$$

then, instead of the reduced carrier being reoxidized by a second carrier (and ultimately by a terminal electron acceptor), it is reoxidized by a molecule of a second substrate.

Many metalloprotein systems are often involved in these electron transfer chains, usually being arranged in sequence according to the values of their redox potentials. Thus, cytochromes, non-heme iron proteins and copper proteins may intervene between flavin coenzyme and oxygen. The result of this is that one end of the redox chain is of substantially different redox potential from the other end, so allowing the linking of quite different substrates. While the subject of the nature of the mechanism of electron transfer will be delayed until a later section, it is, however, appropriate to comment again on the unusual redox potentials and distorted metal site stereochemistries associated with some redox proteins.

The copper enzymes particularly involved in electron transfer reactions are the blue proteins where the copper is in the copper(II) state. The Franck–Condon principle, as outlined in Chapter 2, requires that for fast electron transfer reactions the geometries of the reactant and product complexes should easily be made equivalent. This then means that the energy involved in the promotion of the complex to an appropriately vibrationally excited state is small, otherwise there would have been a large additional contribution to the activation energy for electron transfer. As was discussed in detail in Chapter 2, copper(I) and copper(II) normally favour rather different stereochemistries. It may well be that the irregular structure imposed upon the copper in the copper blue proteins by the protein is one that lies between the favoured structures for the two oxidation states and therefore one that implies fast electron transfer.

The associated difficulty in ascertaining with certainty the oxidation state of the metal in certain copper proteins may also be a feature of a system ideally suited for fast electron transfer. This problem in the case of copper proteins may be demonstrated by noting references often found to 'oxygen-inert copper(I)' and 'EPR-inactive copper(II)'. The particular properties of these copper species of indeterminate valency may be broadly understood in terms of the nature of the metal-binding groups of the protein. Depending upon whether the ligands have appropriate empty or filled orbitals, metal electron density in the d^{10} case may be delocalized on to the ligand, or ligand electron density may be delocalized on to the metal in the d^9 case.

A similar but less well defined problem exists for iron. Both Fe(II) and Fe(III) complexes prefer an octahedral stereochemistry but bond lengths will be slightly different. In heme systems the rigid heme group provides an inflexible set of in-plane ligands: it should be noted in this context that X-ray studies on hemoglobin show the position of the iron atom to be raised slightly above the plane of the heme group. However, in general, the two axial ligands are rather more flexible in terms of the allowed metal-ligand bond length. Here again there is the possibility that fast electron transfer is favoured by the metal-ligand bond length lying in an intermediate position between the characteristic values of Fe(II) and Fe(III) bond lengths. Depending upon the nature of the axial ligands the cytochromes will show a range of redox potentials. Thus, an unsaturated ligand will prefer to bond to iron(II) and stabilize that state. Of greater importance, however, is the question of the spin state of the iron(III) system. It appears that there is an equilibrium between spin states in most examples of ferric hemeproteins. The redox potential of the Fe(III)–Fe(II) couple will depend upon the spin state. The spin state will, in turn, depend upon the axial ligands. Any effect which prevents the close approach of the axial ligand will result in a high spin complex, as also will the presence of weak ligands. It may be readily seen, therefore, why the redox potentials of the cytochrome vary with change of axial ligand.

The redox potentials (E'_0) of systems of biochemical importance are defined at pH 7. Some values are listed in Table 5.1, together with data for other couples for comparison. As has been noted, it may be seen that certain groups of metalloproteins show a wide range of redox potentials.

Dehydrogenases

There are three main groups, all catalysing dehydrogenations, but utilizing different acceptors.

(1) Copper enzymes

These are of high redox potential and therefore always use the oxygen molecule as the electron acceptor. It appears that oxygen oxidizes copper(I) to copper(II). This latter fact may also be associated with the high affinity of copper proteins for oxygen. Copper enzymes will therefore be involved

TABLE 5.1 Redox potentials for some biochemical and chemical couples

System	E_0' volts	System[a]	E_0' volts
Methemoglobin–hemoglobin	0·17	$Cu^{2+}-Cu^+$	0·167
		$Cu(imidazole)^{2+}-Cu(imidazole)^+$	0·255
Metmyoglobin–myoglobin	0·046	$Cu(imidazole)_2^{2+}-Cu(imidazole)_2^+$	0·345
NAD$^+$–NADH	−0·32	$Cu(NH_3)_2^{2+}-Cu(NH_3)_2^+$	0·340
Cytochrome c		$CuL^{2+}-CuL^+$	0·187
$Fe^{3+}-Fe^{2+}$	0·22	$CuL_2^{2+}-CuL_2^+$	0·243
Copper blue		$Cu(pyridine)^{2+}-Cu(pyridine)^+$	0·197
proteins		$Cu(pyridine)_2^{2+}-Cu(pyridine)_2^+$	0·270
		$Cu(1,10 \text{ phen})_2^{2+}-Cu(1,10 \text{ phen})_2^+$	0·174
$Cu^{2+}-Cu^+$	0·4	$Cu(2,Cl \ 1,10 \text{ phen})_2^{2+}-$	0·400
Riboflavin ox–red	−0·21	$Cu(2,Cl \ 1,10 \text{ phen})_2^+$	
Ferredoxin–			
$Fe^{3+}-Fe^{2+}$	−0·42	$Cu(2,9 \text{ Me}_2 \ 1,10 \text{ phen})_2^{2+}-$	0·594
Ceruloplasmin		$Cu(2,9 \text{ Me}_2 \ 1,10 \text{ phen})_2^+$	
$Cu^{2+}-Cu^+$	0·390	$Cu(L')_2^{2+}-Cu(L')_2^+$	0·400
Clostridial		$Cu(en)_2^{2+}-Cu(en)_2^+$	−0·38
ferredoxin	−0·420	$Cu(glycine)_2^{2+}-Cu(glycine)_2^+$	−0·160
Chromatium		$Cu(alanine)_2^{2+}-Cu(alanine)_2^+$	−0·130
ferredoxin	−0·490	$Fe^{3+}-Fe^{2+}$	0·77
Spinach		$Fe^{III}(phen)_3-Fe^{II}(phen)_3$	1·10
ferredoxin	−0·432	$Fe^{III}(dipyridyl)-Fe^{II}(dipyridyl)$	0·96
Adrenodoxin	+0·150 (?)	$Fe^{III}(oxalate)-Fe^{II}(oxalate)$	0·02
Rubredoxin	−0·057	$Fe(CN)_6^{3-}-Fe(CN)_6^{4-}$	0·22
Chromatium		$Fe^{III}(oxime)_3-Fe^{II}(oxime)_3$	−0·251
high potential N.H.I.P.	+0·350	Oxygen–water	0·815
Horseradish		Oxygen–H_2O_2	0·30
peroxidase		Methylene blue ox–red	0·011
$Fe^{3+}-Fe^{2+}$	−0·170	Acetaldehyde–ethanol	−0·20
Cytochrome b	0·050		
Laccase $Cu^{2+}-Cu^+$	0·415		
P. aeruginosa	0·328		
R. vericifica	0·415		
Plastocyanin	0·370		

[a] Aquogroups omitted.
  'met' = Fe(III) state
    L = 2 methylthioethylamine.
    L' = ethylenebisthioglycolic acid.
  phen = phenanthroline.
    en = ethylenediamine.

in the oxidation of substrates of high redox potential, such as the hydro-
quinones. They are often found as the terminal member of respiratory
chains.

(2) Flavoproteins
These are of lower redox potential than the copper proteins and accordingly catalyse the oxidation of substrates of intermediate redox potentials
and may interact with oxygen or with other electron acceptors.

(3) Nicotinamide coenzyme-dependent dehydrogenases
These are usually involved in the initial dehydrogenation of substrates
such as amines and alcohols, the coenzyme (to which initial transfer from
substrate occurs) having a redox potential of -0.32 V. They cannot be
linked directly to oxygen therefore, but are often associated with flavoproteins. They are well suited to act in the initial dehydrogenation of
many substrates.

Ferredoxins may remove hydrogen in addition to acting as electron
transfer agents.

The interrelationships between these three groups of dehydrogenases is
summarized in Figure 5.1.

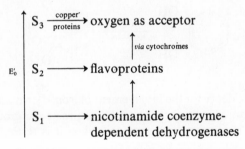

Figure 5.1   The relation between redox potential of substrate and electron transport.

It may be noted that in general these dehydrogenases contain a number
of metal atoms. Thus, ascorbic acid oxidase contains six atoms of copper
per mole of protein (M.Wt. 150,000), while xanthine oxidase contains six
atoms of iron and two of molybdenum per mole. It should be noted that
these are termed oxidases because O_2 is utilized as an electron acceptor.
They are dehydrogenases. In the first example, the product of hydrogen
atom abstraction will be the ascorbate free radical. The hydrogen
atom must then pass to the oxygen molecule which will be bound at
another site on ascorbic acid oxidase. However, we are dealing with a

non-complementary reaction. Oxygen is reduced to water (equation 1), a four electron change. A slow reduction of oxygen would generate free radical species of some life-time, allowing the possibility of direct reaction with the ascorbate radical. The presence of a number of metal centres may be associated with the necessity[4] of providing a rapid release of electrons to the oxygen molecule to prevent such a situation occurring. This would be followed by addition of H^+.

Enzymes catalysing oxygen atom incorporation

These include oxygenases (which catalyse the introduction of both atoms of molecular oxygen into a substrate) and hydroxylases, which catalyse the introduction of only one atom of a molecule of oxygen, and hence the formation of a monohydroxy derivative of the substrate. In both cases the introduction of oxygen may be demonstrated by the use of isotopically labelled oxygen.

The activity of oxygenases may be understood in terms of the inter-mediate formation of the dihydroxy derivative of the substrate. Thus, a typical reaction catalysed by an oxygenase is the cleavage of an aromatic ring (Reaction 5).

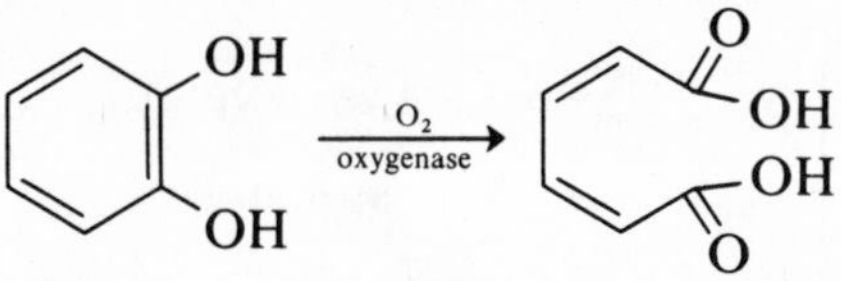

$$\tag{5}$$

There is good evidence for the involvement of iron(II) in many oxygenases. Thus tryptophan oxygenase (6) has a heme prosthetic group. Activity is lost when this is removed and regained when it is restored. In general,

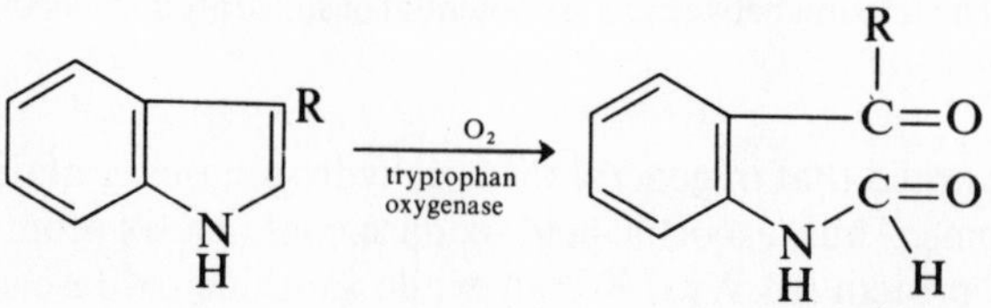

$$\tag{6}$$

the iron and copper enzymes active in catalysing the direct action of oxygen have an open structure or ligand groups that may easily be replaced, so favouring oxygen binding. Copper appears to be present as copper(I) and there is evidence for the reaction of copper(I) with oxygen. The action of hydroxylases may be illustrated by reaction (7), where it may be noted that oxygen-18 has appeared in the product water and that there is a requirement for a second oxidizable substrate, $S'H_2$.

$$SH_2 + O_2^{18} + S'H_2 \xrightarrow[\text{hydroxylase}]{} SH^{18}OH + S' + H_2O^{18} \qquad (7)$$

The hydroxylase is therefore acting as an hydroxylase and as a dehydrogenase. In many examples, there is a requirement for heme prosthetic groups and in some cases for copper. Monophenol oxidase and dopamine hydroxylase are examples of copper enzymes while peroxidase is a well known example of a heme enzyme that catalyses the hydroxylation of certain aromatic and heterocyclic compounds by hydrogen peroxide.

The following scheme has been suggested[5] for the catalysis of the hydroxylation of dopamine (SH_2) by dopamine hydroxylase. Here ascorbate is required as the second substrate ($S'H_2$).

$$Cu^{2+}.\text{protein} + S'H_2 \rightarrow Cu^+.\text{protein} + S'H + H^+$$

$$Cu^+.\text{protein} + O_2 \rightarrow [Cu.\text{protein}.O_2]$$

$$S'H + SH_2 \rightarrow S'H_2 + SH^{\cdot}$$

$$[Cu.\text{protein}.O_2] + SH^{\cdot} \rightarrow [Cu.\text{protein}.O_2.SH^{\cdot}]$$

$$[Cu.\text{protein}.O_2.SH^{\cdot}] + S'H_2 \rightarrow Cu^{2+}.\text{protein} + SOH + H_2O + S'$$

In an earlier section it was noted that fast electron transfer to oxygen was necessary to prevent the build up of oxygen free-radical species and their attack upon an organic radical. However, there are examples of oxidases where this does occur and where there is no binding of oxygen by the enzyme. In this case the redox metal will be in a higher oxidation state and will directly reduce the bound substrate (8).

$$M^{n+} + S^- \rightarrow M^{n+}S^- \rightleftharpoons M^{(n-1)+}S \qquad (8)$$

$$(M = Fe, \text{ possibly } Cu)$$

The free-radical substrate then adds on oxygen, eventually giving the appropriate products. Interactions of this type between redox metal ions and organic groups are particularly important in the flavoproteins, mentioned under dehydrogenases.

Normally, however, in oxygenase and hydroxylase activity the oxygen or hydrogen peroxide and oxidized substrate are both bound at the protein. A number of important and interesting features associated with this behaviour will be discussed later in the chapter.

REDOX POTENTIALS OF MODEL SYSTEMS

The value of the redox potential for any transition metal couple in a metalloenzyme system will determine the role it may play in electron transport or, for example, whether a metalloprotein will carry oxygen or be oxidized by it. The protein ligand groups will largely control the redox potential by providing a site symmetry and metal-binding groups that will favour the two oxidation states to different extents. In this section it is hoped to examine the way in which the redox potential of metals in model complexes respond to variation in ligand and then to apply this to enzyme systems.

A fair amount of information is available[6-8] for complexes of copper and iron, so allowing a study of the redox potentials for Cu(II)–Cu(I) and Fe(III)–Fe(II). Redox potentials are listed in Table 5.1 while formation constant data, contained in Table 3.3, tells us the favoured binding groups for these metals in the upper and lower oxidation state. We will not attempt an examination of other transition metal systems, partly due to a paucity of data and partly as the systems are not so convenient for study.

Copper

The formation constant data for copper may be readily understood in terms of the hard-soft formalism of Pearson, as outlined in Chapter 3. Copper(I) is much softer than copper(II), in accord with the fact that sulphur donor ligands are bound more strongly by copper(I) than copper(II), as are all unsaturated ligands such as o-phenanthroline and bipyridyl. Both nitrogen and sulphur donor ligands bind more strongly to copper(II) and copper(I) than do oxygen donors. For unidentate ligands, considerations of this type only are important. For polydentate ligands, however, a stereochemical effect will be superimposed upon this, as copper(I) and copper(II) have very different behaviour in terms of preferred coordination number and stereochemistry. Copper(I) is often linear, two-coordinate, but may increase its coordination number to four, in which case the geometry will be a tetrahedral one. Copper(II), in contrast, prefers to be six-coordinate, most complexes having a tetragonally distorted structure. For 1:2 stoicheiometries, therefore, with ligands such as bipyridyl which, in addition to being soft, cannot give a tetragonally distorted octahedral structure, it is quite clear that copper(I) complexes will be stable.

The data in Table 5.1 allow us to observe the effect on redox potential for Cu(II)–Cu(I) of successively replacing aquogroups by other ligands. Replacement of one and then two water molecules by nitrogen donor ligands results in a more positive potential. Increase in the number of nitrogen donor atoms beyond this will not further stabilize the copper(I) state, however, as this will involve the formation of a complex with a coordination number higher than the favoured one. The binding of groups such as the aliphatic α-amino acids and ethylenediamines results in negative potentials for the couple. They prefer to be bound to copper(II) and will, in fact, cause disproportionation of copper(I) to copper metal and copper(II). Cyanide and mercaptide ligands correspondingly favour copper(I) to such an extent that they will reduce copper(II) to copper(I).

James and Williams have commented[7] on the importance of steric hindrance in determining the values of redox potentials. 2,9 substituted phenanthrolines introduce a substantial degree of steric hindrance in four-coordinate square planar copper(II) complexes ML_2. Four-coordinate copper(I) complexes are tetrahedral and will not show steric hindrance. Thus, the high value of E'_0 in this case reflects the destabilization of the copper(II) state. Similarly the larger halide ions favour copper(I) rather than copper(II), reflecting the smaller size of the copper(II) ion.

Iron

The preferred stereochemistry for both Fe(II) and Fe(III) complexes is octahedral. We are not therefore concerned with major stereochemical problems as in the case of Cu(II)–Cu(I). However, it has been noted already that, in heme proteins, while there will be a preferred metal-axial ligand bond length for fast electron transfer, there is also another effect associated with the question of spin state of the ferric and ferrous ions.

Some redox potentials are listed in Table 5.1. As for copper there is a stabilization of the lower oxidation state by unsaturated ligands, with resulting high potentials. In the low spin state, interaction with the ligands will become greater, and so the effect of unsaturated ligands should be more pronounced. Increase in basicity of the ligand (i.e. σ electron donation) will obviously favour the ferric state, with a decrease in E°. Ultimately, however, this will cause a change from a high spin to a low spin state. But, the energy involved in this electronic rearrangement is smaller for Fe(II) than for Fe(III) and the rearrangement will therefore occur first in the former case. This means that we have to compare the effect of σ-donation on Fe(III) high spin with Fe(II) low spin. When both these oxidation states have become low spin (i.e. at a sufficiently high pK_a of ligand), then once again we are comparing similar spin states and ferric ion will be favoured by further increase of pK_a, with a corresponding fall in redox potential. It is difficult to assess the relative effect of increased basicity on low spin Fe(II) and high spin Fe(III) but Williams[4] has shown that for iron porphyrin complexes, where there is a spin-state change, increase of ligand pK_a results in a *rise* in redox potential, when Fe(II) only is low spin, followed by a fall when both oxidation states are low spin. The stabilization of Fe(II) when low spin can be understood in terms of ligand field stabilization energy changes.

The effect of protein upon redox potential

A feature of iron and copper proteins is that they may show high redox potentials compared to those usually observed for model compounds. That is, the lower oxidation state is stabilized with respect to the higher one. This may be understood quite readily for copper proteins. The copper(I) ion is much less specific in its steric requirements than the copper(II) species. The protein is therefore able to satisfy the steric requirements of the former species more readily, and probably with donor groups that bind more strongly to copper(I) than to copper(II). The steric effect is much more important for protein systems than for model complexes.

Of the protein groups available, it is probable then that the 'hard' groups, carboxylates, phenolates and amino groups will be specific for copper(II) while sulphur donors, RSSR, R_2S and particularly RS^-, will be bound to copper(I). However, under biological pH values, the proton competes effectively with the metal for the $-NH_2$, phenolate and sulphide groups and their availability will be reduced. In practice, very little is known about the nature of metal-binding groups in copper proteins.

Any protein system that involved copper bound to, say, sulphide groups (i.e. Copper(I) specific groups) would seem unlikely to be involved in fast electron transfer as the copper(II) would be generated in an environment characteristic of copper(I). The redox potential may be very high, in addition. It seems reasonable to suggest, therefore, that for copper proteins involved in electron transfer, at least some of the binding groups should be groups which are not particularly specific for either copper(I) or copper(II). It would also seem reasonable to predict that the copper in the oxygen carrier hemocyanin (copper(I)) *would* be bound by sulphide groups to prevent irreversible oxidation.

The varying potentials of the iron heme enzymes will depend upon the protein binding groups in the fifth and sixth positions. The redox potentials vary from -0.4 V to $+0.1$ V. The low potential could result from carboxylate being bound to iron, i.e. favouring Fe(III) as the harder species. High potentials would result from unsaturated soft donor groups favouring Fe(II), or also from a steric misfitting which would destabilize the iron(III) state.

Thus, in order to favour fast electron transfer in terms of the requirements of the Franck–Condon principle, the ferric-axial ligand bond length would be longer than that normally observed in ferric complexes. This would result in a greater tendency for high spin iron(III) than that seen in model complexes. This implies a high redox potential for the heme enzyme. Again, this fits in with the fact that electron transfer proteins have high redox potentials.

For non-heme iron proteins, there is also a range of redox potentials. Less may be said about these, but again it appears that this is largely associated with the binding of a range of ligand groups of varying hardness and softness.

'Valence-state' determination

The oxidation state of copper, in particular, in a protein is not always determined easily. EPR measurements should show the presence of copper(II), but the absence of a signal may be attributed to factors other than the absence of copper(II). Sometimes magnetic susceptibility measurements may help, but the possibility of the presence of organic free radicals is a complicating factor. Visible spectra should allow a distinction to be made between copper(I) and copper(II) proteins. However, in practice, charge-transfer bands occur at quite high wavelengths in copper(I) complexes and difficulty may be experienced in assigning the bands in the visible spectrum of a copper protein to charge-transfer or d–d transitions. The important guide-line here is the question of the intensity of the bands. If the intensity is low, they are probably d–d bands.

The information contained in the previous sections allows a chemical approach to the study of oxidation state. The extent to which added groups are bound by the enzyme at the metal ion will allow a distinction to be made between valence states. Obviously the very strongly binding 'valence specific' reagents are of little value as these will either cause disproportionation of copper(I) or reduction of copper(II) to give the favoured oxidation state anyway. The disproportionation of Cu^+ is represented below.

$$2\,Cu^+ \rightleftharpoons Cu^{2+} + Cu, \qquad E'_0 = 0\cdot167\,V$$

Clearly, normally copper(I) is very unstable. The equilibrium may be displaced in either direction. Thus the soft ligands CN^-, I^-, R_2S all reduce copper(II) to copper(I) while ethylenediamine, for example, will convert a copper(I) complex to a copper(II) ethylenediamine complex. An examination of the available data for a range of ligands shows that the equilibrium is dependent both upon the nature and the preferred stereochemistry of the ligand, as indicated earlier.

Provided the reduction of copper(II) by the protein binding groups is not possible, 2,2'-biquinoline is specific for copper(I), as also is the binding of oxygen or carbon monoxide. Chloride and bromide are bound more strongly by copper(II) than copper(I). This is probably due to a steric effect. Again, 2,2'-bipyridyl is specific for iron(II), although this is a kinetic phenomenon. Great care must be exercised in the use of this approach.

Spin-state determination

From the previous discussion it is clear that the matter of spin-state determination is also an important one. Iron(II) low spin systems will be diamagnetic, while high spin complexes would have high moments. Iron(III) systems may have either one or five unpaired electrons. In the latter case magnetic susceptibilities will clearly demonstrate in broad terms the composition of a complex in terms of its spin-state. It is much more difficult to say whether a complex is completely low or high spin. Thus the magnetic moment of a low spin complex may be raised above the spin only value either by orbital contributions or by the presence of a small amount of the high spin form and it will be difficult to rule out the latter explanation authoritatively. Another problem with measurement of magnetic moments is associated with the very high diamagnetic correction for heme groups.

EPR and Mossbauer spectroscopy should help in this problem, but the problems discussed elsewhere have always substantially lowered the value of these techniques. In Chapter 7 there is, however, an example of the application of Mossbauer spectroscopy, involving applied magnetic fields, which has led to useful conclusions.

It should also be noted that the absorption spectra of ferric heme proteins show correlations with spin properties.[9]

ELECTRON TRANSPORT IN REDOX METALLOPROTEINS

Electron transfer proceeds at a very fast rate in biological processes with a low activation energy estimated to be of the order of 7 kcal mole^{-1}. The inorganic reactions discussed in Chapter 2 usually proceed at a very much lower rate with activation energies of 15–20 kcal mole^{-1}. The highly organized arrangement of electron carriers in the mitochondrial membrane is evidently associated with this phenomenon, providing as it undoubtedly must, highly conjugated metal-ligand systems, and allowing maximum orbital overlap of the constituents of the electron transport chain.

For electron transport to occur between metal atoms in metalloproteins there must be electron transfer from one metal to a neighbouring atom of the protein binding group through overlap of orbitals. This electron must then be transported away before recombination with the positive 'hole' of the metal atom can occur. The importance of conjugated polarizable protein groups is plain in this connection and it is noteworthy that considerable attention has been given to the semiconducting properties of proteins in terms of band theory.

One general problem associated with electron transport is the nature of electron transport, whether it be by actual transfer of electrons or whether by group transfer, particularly that of hydrogen atoms. Electron transfer and group transfer are known for model systems, although sometimes it is not possible immediately to identify which is occurring. Thus the ferric–ferrous ion exchange, which one might expect to be by direct electron transfer, is catalysed by hydroxide ions. This suggests the following reaction pathway, in which electron transfer has really

$$[Fe^{III}(H_2O)_5(OH)]^{2+} + [Fe^{II}(H_2O)_6]^{2+}$$

$$\rightarrow [(H_2O)_5Fe^{2+}-OH\cdots H\cdots OHFe^{2+}(H_2O)_5]$$

$$\rightarrow [Fe^{II}(H_2O)_6]^{2+} + [Fe^{III}(H_2O)_5OH]^{2+}$$

involved H atom transfer. However, does this involve direct atom transfer or electron transfer followed by proton transfer? One conclusion is that it is important to check the dependence of redox potential upon pH.

Associated with this problem of electron transport is the question of the mechanism of the phosphorylation of ADP to ATP. This occurs at a number of stages in the electron transport chain and in general can be described in terms of two electrons transferred per phosphate bond formed. Any scheme for electron transfer must allow for this.

As a preliminary step before considering possible schemes for biological electron transport it is appropriate to summarize the various criteria for rapid electron transfer that we have discussed so far in this chapter and in Chapter 2. They are:

(1) that the requirements of the Franck–Condon principle be easily met,

(2) that electron spin be conserved, fastest electron transfer occurs when the net electron spin of the reactants is the same as that of the products,

(3) that the rates of outer sphere reactions are faster when polarizable ligands are bound to the metal,

(4) reactions tend to be faster when complementary redox species are involved.

Condition (4) would be illustrated by neighbouring members of a cytochrome chain where both reactants undergo one-electron changes. However, this does not always occur. Thus one problem arises in the interaction of oxygen with the terminal member of the respiratory chain. This may involve Fe(II) $\rightarrow$ Fe(III). However, oxygen is reduced to water, possibly *via* two two-electron changes. This has led to the suggestion that Fe(II) is oxidized to Fe(IV) in such cases. We have already discussed the way in which enzyme systems may overcome this problem—by allowing rapid electron release from a multienzyme system. A second problem is the way in which a two-electron step in oxidative phosphorylation utilizes one-electron changes in the members of the respiratory chain.

Models for electron transport

As models we may consider either reactions in solution or in the solid state. The problem with solution studies here is that condition(1) may not be readily met and measured activation energies will then also reflect the raising of molecules to vibrationally excited states. The protein system is more ordered than a solution system, being designed to overcome this problem. It may then be more appropriate to consider as models electron transfer in the solid state as in semiconductors and photo-conduction.

In the solid state, electron transfer in semiconductors is explained in terms of band theory; that is, the photoexcitation of electrons to upper bands and hence conduction. Williams[10] has pointed out that it is possible to consider instead a hopping mechanism in cases where there are two metal atoms in different valence states. This may be of some relevance to biological systems.

One model that has been used frequently is that of the metal phthalocyanines.[11] Metal-free phthalocyanine has a structure (Figure 5.2) similar to the porphyrins which are so important in biological redox systems. Thus they are both 18 π-electron aromatic systems. The main difference is the presence of aza rather than methine links. The understanding of the chemical and physical behaviour of the phthalocyanines is important therefore in the context of natural redox processes. These have been reviewed fairly recently.[12] Metal phthalocyanines show conductivity properties which are enhanced considerably by adding traces of an electron acceptor impurity such as o-chloranil, with a reduction of activation energy. In the undoped samples thermal energy is necessary to produce carriers. This is eliminated by the addition of an electron acceptor. Here the conductivity is associated with the mobile π-electrons of the phthalocyanine ring. The crystal structure of metal phthalocyanines is schematically represented in Figure 5.2. The structure involves parallel

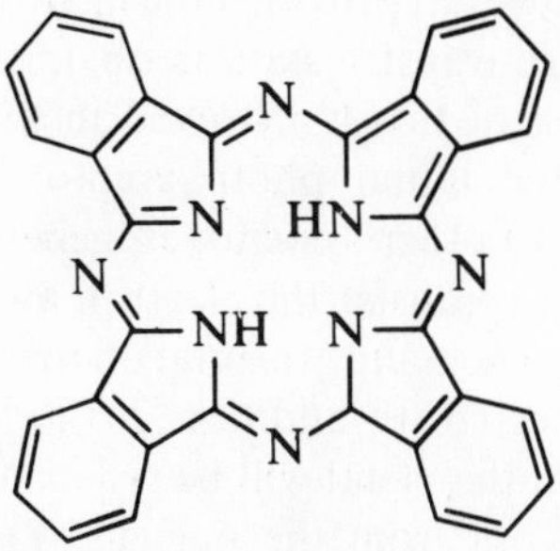

The structure of phthalocyanine.

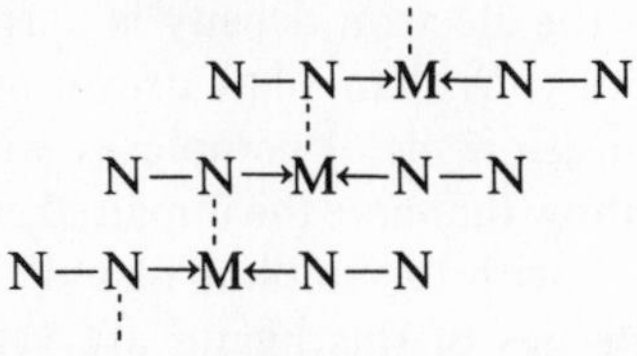

$$N-N{\rightarrow}M{\leftarrow}N-N$$
$$N-N{\rightarrow}M{\leftarrow}N-N$$
$$N-N{\rightarrow}M{\leftarrow}N-N$$

Schematic representation of the structure of metal phthalocyanines.

Figure 5.2

sheets with interaction between neighbouring sheets. Even so, the activation energy for conduction is still high and this certainly suggests that the mechanism involved in these cases could not adequately account for longer range electron transfer. The activation energy in this particular example could well be lowered by the presence of more polarizable ligands.

In solution redox reactions we have two general mechanisms; electron or group transfer *via* a bridge linking two metal centres (inner-sphere) or electron transfer *via* overlap of ligand orbitals (outer-sphere) in which case reaction is accelerated by the presence of polarizible ligands. It is interesting to consider the actual act of the electron transfer in the bridged system. Depending upon the timing of electron transfer to and from the bridging atom X (before group transfer), it is possible to have no electron concentration in this group or to generate a species, $M^{n+}.X^{-}.M^{n+}$. For either of these to be extended to polyatomic bridges it is essential to have a mobile conjugated electron system such as the porphyrin group.

Williams has discussed[10,13] the problems of biological electron transport on a number of occasions and has emphasized the importance of metal-ligand charge-transfer bands in this context. The position of the band will reflect the redox abilities of both ligand and metal. Now charge-transfer of this type will not, in itself, lead to electron transfer, as in the vast majority of cases the electron will fall back to the hole from which it was previously excited. However, provided the ligands have appropriate properties the excited charge-transfer state is obviously a step towards electron transfer. Something of this is involved in photosynthesis. The magnesium porphyrin is the initial photoacceptor while the excited electron is then transferred to other systems. In general terms, a mobile electron system is required to transfer the electron away and there must also be an appropriate low energy empty orbital on the ligand to which the electron may first be transferred. In addition, if the ligand is a good σ donor the positive charge on the metal will be reduced, so allowing the d electron density to expand out from the nucleus. This too will favour electron transfer to the ligand, particularly if the metal ion is in a low spin state, as then the electron density is particularly 'accessible' to the ligand. In this context Williams has drawn attention to EPR studies[14] on a range of transition metal complexes with the ligand maleonitril-edithiolate which show that here the unpaired electron density of the metal is largely associated with the sulphur ligand. The significance lies in the fact that the properties of this ligand are very much like those of the porphyrin group and demonstrate that ligands with the properties discussed above are able in effect to abstract an electron from the metal as a first step in electron transfer.

The outer sphere redox reaction between Fe(II) and Fe(III) involving H atom transfer is an important model in view of the possible significance of the H atom in general. In view of the associated phosphorylation reaction it seems unlikely that H atom transfer between OH and water could be a source of electron transfer; however, other RH groups could be involved in this. This is shown in Figure 5.3, where oxygen is used as the terminal electron acceptor and therefore two moles of substrate are oxidized. Williams[15,16] has suggested in particular that H-atom transfer takes place in mitochondria *via* phenol–quinone reactions.

Urry and Eyring have suggested an alternative scheme involving group transfer, the 'Imidazole Pump model of electron transport.' This scheme, on suitable modification, will also allow for the inclusion of oxidative phosphorylation by incorporating quinones. In particular, they have considered cytochrome c as a typical representative of the cytochromes involved in oxidative and photosynthetic phosphorylation. Cytochrome c will be discussed in a later section. It is sufficient to note now that it is a

$$Cu^+\text{---}R\text{---}Cu^{2+}$$

$$S\downarrow O_2$$

$$O_2\cdot Cu^+\text{---}R\text{---}Cu^{2+}S$$

$$O_2\cdot Cu^+\text{---}R\text{---}Cu^+S^{\cdot}$$

$$O_2\cdot Cu^+\text{---}RH\text{---}Cu^{2+} + S^{\cdot}$$

$$O_2^-\cdot Cu^{2+}\text{---}RH\text{---}Cu^{2+}$$

$$O_2^-\cdot Cu^+\text{---}R\text{---}Cu^{2+}$$

$$S\downarrow$$

$$O_2^{2-}Cu^+\text{---}R\text{---}Cu^{2+} + S^{\cdot}$$

Figure 5.3 A scheme for electron transport.

protein of molecular weight 12,000 having one iron atom in the prosthetic group hematoporphyrin. The prosthetic group is attached to the
protein *via* sulphur. Four-coordination positions of the iron will be filled
by the porphyrin nitrogen atoms and the fifth and sixth by an imidazole
group of histidine and a methionine residue.

Redox reactions of cytochrome *c* have been discussed in terms of
conduction through the π bonds of imidazole or through the plane
of the porphyrin ring system to the sulphur atom, although in the latter
case there is no continuous conjugated system.

Figure 5.4 shows the application of the model of Urry and Eyring.
A cross section through the iron porphyrin is presented to demonstrate
the function of the imidazole group more effectively. The imidazoles are
covalently attached to the iron atom to the left of the molecule (H^+ lost).
By shifting of electron pairs the imidazoles then become bound at the
right. It is then assumed that a donor group becomes covalently bound
to the iron porphyrin at the left and an acceptor is positioned at the other
end of the chain. The second set of electron pair shifts moves an electron
pair from the donor and delivers a pair to the acceptor, with regeneration
of the original arrangement. In other words, imidazole has pumped a pair
of electrons from donor to acceptor. The first of these steps will involve as
a transition state a situation in which the imidazole is symmetrically bound
between the two iron atoms. This is exactly analogous to the protonated
species.

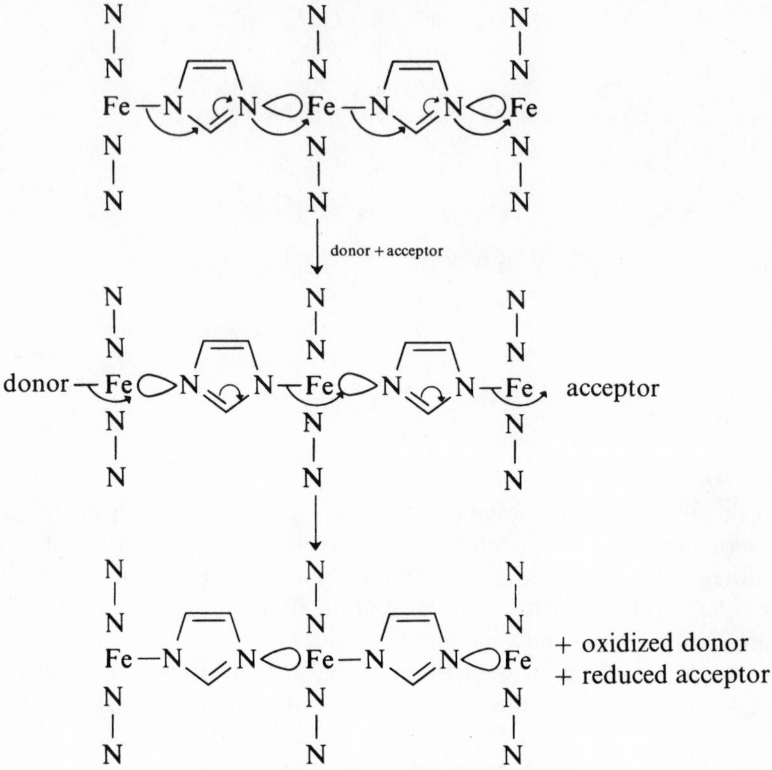

Figure 5.4   The imidazole pump model.

Normally the involvement of quinones in the respiratory chain is postulated,[18] (as a benzoquinone, coenzyme Q or a naphthoquinone, vitamin K) particularly in the context of phosphorylation.

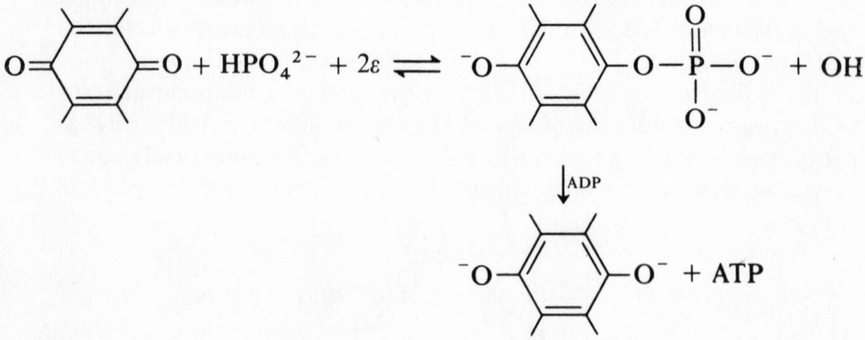

Eyring has modified the imidazole pump model (Figure 5.5) to include coupling by postulating the presence of a quinone between two imidazoles, each of which is attached to its porphyrin by a covalent bond. Electron transfer can occur as before. Orthophosphate may then attack the remaining quinol–imidazole bond, giving phosphoryl imidazole, quinone and OH⁻. The phosphoryl group will then react with ADP leaving imidazole with its pair of electrons to donate to the iron porphyrin so regenerating the initial arrangement. Support for this scheme lies in the fact that phosphoryl imidazole has been isolated from a phosphorylating mitochondrial preparation.[19]

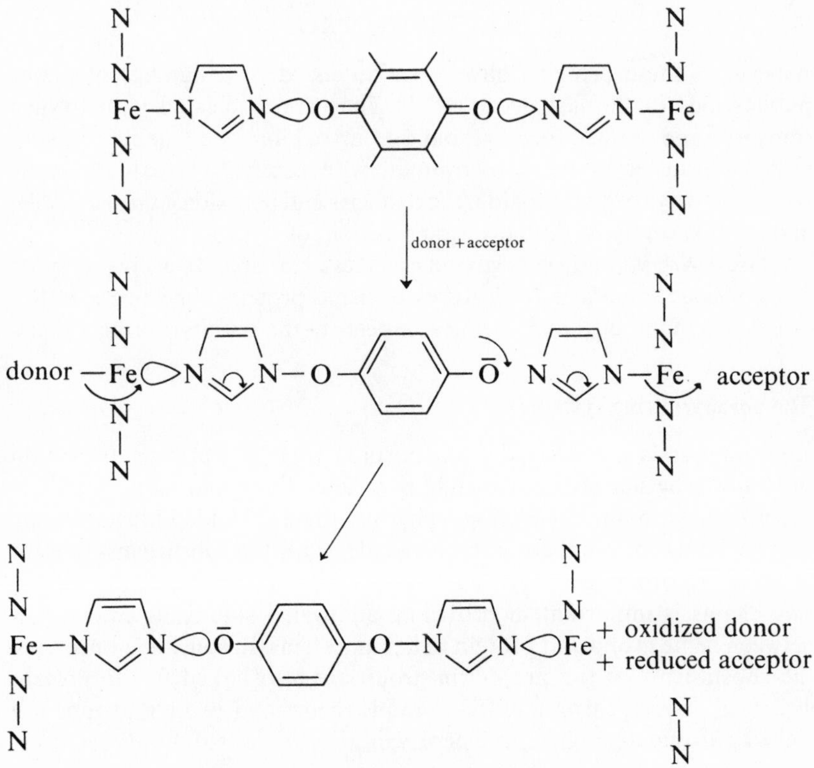

Figure 5.5 Modified imidazole pump model.

IRON METALLOPROTEINS

Iron metalloproteins may be divided into two general groups, the heme and non-heme iron proteins. The second classification involves a bringing together of a range of low molecular weight proteins, characterized in a negative way by the absence of the heme group. The first of these to receive prominence was ferredoxin, an important electron carrier associated with nitrogen fixation and other processes. Others fairly similar to ferredoxin and others rather different from ferredoxin have been discovered and classified together. A characteristic of these is the presence of metal-sulphur linkages. It should be noted that in the preceding section it was seen that certain sulphur groups could well show similar ligand properties to the porphyrin group. This certainly must be of deep significance.

THE HEME IRON PROTEINS

Aspects of these proteins have been discussed in a number of major publications in the last decade.[20–24] They are associated with oxygen transport and storage (hemoglobin and myoglobin, see Chapter 7), with electron transfer in the cytochromes, with catalysis of oxidations by oxygen and hydrogen peroxide (the oxidases and peroxidases respectively) and in the catalysis of the decomposition of hydrogen peroxide (the catalases). We will consider aspects of these reactions later, but first we will consider some general features of heme proteins, beginning with a discussion of physical and chemical aspects of the porphyrin ring system.

The porphyrin ring system

All compounds are derived from porphin (Figure 5.6) by substitution of H atoms by side chains. All eight pyrrole carbon atoms are completely substituted in natural metalloporphyrins. Figure 5.6 also illustrates the nomenclature for numbering the molecule, while the substituents present in some important porphyrins are listed in Table 5.2. The nature of the side chains is important in providing additional stabilizing interaction between heme group and protein in hemoproteins. In some hemoproteins the substituents of the porphyrin group are also linked to the protein structure. In cytochrome c, for example, heme and protein groups are linked *via* thioether groups. There will also, of course, be direct iron–protein linkages.

The porphyrins are highly conjugated molecules. Their electronic spectra show a number of bands, which will be characteristic of a particular porphyrin as the spectrum is dependent upon the nature of the substituents at the pyrrole carbon atoms. A very intense band, around 25 kK, is the Soret band (or γ-band) while there are two other bands of lower intensity, the α- and β-bands. These are close together in frequency. The presence

TABLE 5.2 Some naturally occurring porphyrins

Porphyrin	Substituents								E^0V^{20}			
	1	2	3	4	5	6	7	8	A	A'	B	C
Protoporphyrin	CH_3	$CH=CH_2$	CH_3	$CH=CH_2$	CH_3	CH_2CH_2COOH	CH_2CH_2COOH	CH_3	0·015	0·137	−0·033	−0·183
Hematoporphyrin	CH_3	$CHOHCH_3$	CH_3	$CHOHCH_3$	CH_3	CH_2CH_2COOH	CH_2CH_2COOH	CH_3	0·004	—	−0·099	−0·200
Etioporphyrin	CH_3	CH_2CH_3	CH_3	CH_2CH_3	CH_3	CH_2CH_3	CH_3	CH_2CH_3	−0·029	—	—	—
Deuteroporphyrin	CH_3	H	CH_3	H	CH_3	CH_2CH_2COOH	CH_2CH_2COOH	CH_3	—	—	—	—
Mesoporphyrin	CH_3	CH_2CH_3	CH_3	CH_2CH_3	CH_3	CH_2CH_2COOH	CH_2CH_2COOH	CH_3	−0·063	0·246	—	−0·229
Chlorocruoroporphyrin	CH_3	CHO	CH_3	$CH=CH_2$	CH_3	CH_2CH_2COOH	CH_2CH_2COOH	CH_3	—	—	−0·010	−0·113
Coproporphyrin	CH_3	CH_2CH_2COOH	CH_3	CH_2CH_2OH	CH_3	CH_2CH_2COOH	CH_2CH_2COOH	CH_3	−0·036	—	—	−0·247
Rhodoporphyrin	CH_3	CH_2CH_3	CH_3	CH_2CH_3	CH_3	COOH	CH_2CH_2COOH	CH_3	—	—	—	—

A, A': 5th and 6th ligands pyridine at pH 9·6 and pH 7.
B: 5th and 6th ligands α picoline at pH 9·6.
C: 5th and 6th ligands cyanide at pH 9·6.

of good σ- or π-donor axial ligands in metalloporphyrins will intensify the absorption of the α-band. These bands are due to $\pi \rightarrow \pi^+$ transitions. Porphyrin spectra may be explained satisfactorily in terms of molecular orbital theory.[25]

The two protons attached to nitrogen in a porphyrin are readily replaced by a metal ion. The structure of iron(II) protoporphyrin is shown in Figure 5.6. Here the metal is four-coordinate with a square planar environment.

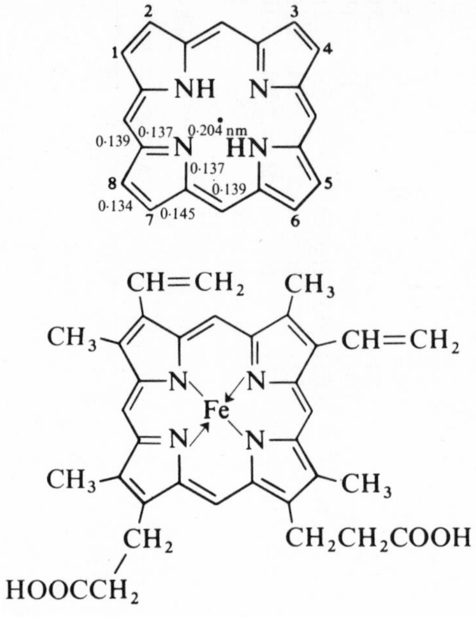

Figure 5.6   Porphin and iron(II) protoporphin.

Cytochrome c involves the protoporphyrin structure with the addition of protein cysteine side chains across the vinyl double bonds in positions 2 and 4. Hemoglobin, myoglobin, the peroxidases and the catalases also involve the protoporphyrin group.

The porphyrin molecule is essentially planar, but two pyrrole rings are tilted up and two are tilted down so that the nitrogen atoms are slightly out of plane. The molecule is not completely rigid and the presence of bulky substituents may cause puckering.[26] As indicated in Figure 5.6 the distance from the pyrrole nitrogen to the centre of the ring is 0·204 nm. The metal–nitrogen bond lengths in metalloporphyrins vary slightly (e.g. 0·210 nm for ferric porphyrins, 0·195 for nickel porphyrins), but not as much as a consideration of the ionic radii of the metal ions would suggest. This is a reflection of the inability of the porphyrin ring system to adjust itself to fit the metal ion in the metalloporphyrin.

Most metalloporphyrins involve four-coordinate metal ions. Others involve five-coordinate metal ions with square pyramidal structures, or six-coordinate metal ions with distorted octahedral geometry. These are formed by adding one or two axial ligands to the metal ion. Some five-coordinate metalloporphyrins are known with a tetragonal pyramidal stereochemistry. One example is provided by the high-spin iron(III) porphyrins, where the iron(III) is about 0·05 nm above the plane of the pyrrole nitrogen atoms. A second example is the vanadium in the vanadyl substituted etioporphyrin. In contrast the low-spin iron(III) porphyrin, bis(imidazole)iron(III) tetraphenylporphyrin, does have the iron lying in the plane of the nitrogen atoms.

Iron(II) porphyrins are readily autoxidized to the iron(III) form. It has been found[26] that an intermediate in this reaction is a five-coordinate species containing a bridging oxo group, the μ-oxobis(porphyrin iron(III)) dimer. Here the iron in the complex is 0·05 nm out of the plane of the nitrogen atoms towards the oxygen bridge.

The iron(II) heme proteins are also often readily autoxidized to the iron(III) state. An extra anion will be necessary. If it is a chloride or hydroxide the ferric heme is known as a chlorohemin or a hematin respectively. The remaining coordination positions in iron(II) and iron(III) hemes are filled by additional ligands, giving six-coordinate complexes in which there is a distorted octahedral geometry around the metal ion. Details of these additional ligands will be given when discussing each group of heme proteins. The iron(II) in oxygen carriers and storers will, of course, remain five-coordinate.

Substituent effects on porphyrin rings
These are of great importance. There are a great number of closely related iron hemes having slightly different redox properties. The presence of different substituents on the porphyrin ring could contribute to these finer points of difference. It does appear, though, that when major changes in property are required, other factors in the heme proteins such as the nature of the fifth and sixth ligand or the function of distant groups on the protein are important. Thus myoglobin, hemoglobin, peroxidase, catalase and certain cytochromes e.g. b_5, all have the protoporphyrin heme structure, but their physiological properties are quite different, as are their redox potentials.

It has already been noted that the bands in the spectra of the porphyrins are substituent sensitive. This will be carried through to the spectra of the iron hemes, and the substituents may well affect the d–d and porphyrin →

metal charge-transfer bands. The overall situation is complex. Spectroscopic changes caused by substituents are, however, the basis of the cytochrome classification scheme.

The presence of electron-withdrawing substituents will decrease the basicity of the porphyrin metal-binding groups. This is reflected in a decreased rate for metal ion incorporation into the porphyrin. The decrease in basicity of these groups will also be reflected in the binding of the fifth and sixth ligands to the metal. Extensive studies[27] have been carried out on the interaction of metal porphyrins and heme proteins with a variety of axial ligands.

The dimerization of porphyrins (which is solvent dependent and thought to occur *via* dipole–dipole interaction) is also dependent upon the nature of the substituents. In fact it does not occur unless electron-withdrawing groups such as vinyl groups are present. NMR studies have also shown differences in the reactivity of the various ring positions.

Redox properties of metalloporphyrins

Some redox potentials are given in Table 5.2. Due to dimerization and solubility problems these are often difficult to measure. In practice, it is often easier to do this for low spin complexes. Nevertheless, it is important that a study of these redox potentials be made as a foundation for the understanding of the redox potentials of the cytochromes and other electron-transfer catalysts. The presence of the porphyrin ligand will stabilize the higher oxidation state compared to the aquo couple. The presence of electron-withdrawing groups on the porphyrin will decrease the basicity of the nitrogen donors (i.e. σ donors). This will result in a destabilization of the iron(III) state.

The axial ligands will have a very significant effect upon the redox potential of the iron–porphyrin. The experimental study of iron protoporphyrin and other metalloporphyrins has confirmed the picture developed for simple model complexes. Strong σ donors stabilize the iron(III) state, but this may also result in a change from high spin to low spin, which will vary for iron(III) and iron(II). When high spin Fe(III) is converted to low spin Fe(II), the Fe(II) complex will be stabilized relative to Fe(III). On further increase of ligand σ donor strength, however, so that both redox states are low spin, the iron(III) complex will once more be stabilised. Spectral studies show that in iron(III) hemes there is a delicate balance between high and low spin types. Changes in the nature of the axial ligands in particular and also in the porphyrin substituents may exert a considerable effect.

Many heme systems involve the coordination of imidazole and water in the fifth and sixth positions. In view of this, it might be particularly useful to examine the behaviour of metalloporphyrins having one strong and one weak field additional ligand.

Metal ion incorporation into porphyrins

Enzyme systems have been identified[28] which catalyse the incorporation of iron into the porphyrin ring. These only utilize ferrous iron, and accordingly the insertion reaction is inhibited completely by the presence of oxygen.

Prosthetic groups of heme proteins

Some important porphyrins were listed in Table 5.2 along with the redox potential of the corresponding heme. The most important heme is undoutedly protoheme. It may occur free in tissues and is the prosthetic group of hemoglobin, myoglobin, erythrocruorin, peroxidases, catalases and the *b* cytochromes. Chlorocruoroporphyrin is the prosthetic group of chlorocruorin. However, recombination experiments have shown that the structure of the prosthetic group is relatively unimportant in determining the properties of certain heme proteins. Thus apohemoglobin may be recombined with a range of iron porphyrins with retention of oxygen-carrying properties, although, not unexpectedly, the details of the behaviour, such as the oxygen uptake curve, will be affected.

In the following sections a number of aspects of the behaviour of the iron heme proteins will be discussed. Inevitably this will be extremely selective.

The cytochromes

These are hematin compounds which are involved in electron transfer chains in the mitochondria, in which electron transfer is associated with the presence of the iron(III)–iron(II) redox couple. Terminal members of the cytochrome chain must also have the property of reacting directly with oxygen, so for these there will be extra requirements. Cytochromes are also involved in aspects of the nitrogen cycle and in enzymic reactions associated with photosynthesis.

Some fifty cytochromes have been characterized, but the most studied example is cytochrome *c*. The classification of the cytochromes is based upon the nature of the porphyrin ring system and in turn upon spectral data for the pyridine derivative. Type *a* have the α-band in the porphyrin spectra at longer wavelengths than 570 nm; type *b* have the α-band in the range 555–560 nm and have vinyl side chains on the porphyrin ring; while type *c* have the α band in the range 548–552 nm and have porphyrin rings in which the vinyl groups have been saturated, as in cytochrome *c* where these are associated with the thioether connection with the protein. In general the type 'a' cytochromes form a less cohesive group. The b and c types in general have two strong field donor groups in the fifth and sixth coordination position and are usually low spin complexes. These cannot normally combine with small molecules. Cytochrome a_3, on the other hand, the terminal member of the cytochrome chain, has water as one of the axial ligands (with imidazole as the other) and so binds oxygen.

The study of the cytochrome chain sequence involves its breakdown to single cytochromes or small complexes such as cytochrome c oxidase which contain a number of groups. The redox potential of single cytochromes may then be measured by standard experimental methods. It is important, however, to show that these will still react with the other components of the cytochrome chain as before and much effort has gone into attempts to reconstitute the cytochrome sequence.

At least two of the phosphorylation steps in the respiratory chain are associated with cytochromes. Any reconstruction must accommodate these reactions.

Cytochrome c *and cytochrome* c *oxidase*
These will be discussed in a little more detail. Cytochrome c oxidase contains copper as well as two type a cytochromes and will be discussed also in the section on copper. Cytochrome c has been reviewed recently.[22] The electron transport chain involving these is now known to involve the following sequence:

cytochrome c_1. cytochrome c. $\underbrace{\text{cytochrome } a. \text{ cytochrome } a_3}$ $- O_2$

$$\text{cytochrome } c \text{ oxidase}$$

Reduced cytochrome c is not readily reoxidized by molecular oxygen, even though the reaction is thermodynamically favoured. Presumably there is no appropriate kinetic pathway. Hence cytochrome c oxidase catalyses the reaction, reoxidizing reduced cytochrome c and in turn being oxidized by oxygen, which presumably is bound to the iron in cytochrome a_3.

Horse-heart cytochrome c has a molecular weight of around 12,400 and involves 104 amino acid residues in the protein chain. It contains four sulphur atoms per molecule, found in the two thioether (cysteine) bridges and in two methionine residues. The redox potentials of cytochrome c from various sources is around $+0.255 \rightarrow +0.285$ V. The axial ligands are imidazole and methionine residues.

Sano and Tanaka have been able[29] to reconstitute cytochrome c from its component parts and hence help confirm its structure. They combined the cytochrome c apoprotein with a reduced porphyrin species (protoporphyrinogen), autoxidized it to the porphyrin and then inserted the iron, giving a heme protein with the properties of cytochrome c. It is clear that the particular properties of cytochrome c must be due in part to the thioether links connecting protein and heme groups. Possibly this is important in controlling the nature of the coordinating residues in some way. ORD studies have been carried out[30] on cytochrome c. These show a marked effect associated with redox change, indicating the involvement of conformational changes in the heme with respect to the protein.

A number of other observations also indicate that conformational changes occur when cytochrome c is oxidized and reduced. Thus the NMR spectra differs from one state to another. For a full examination of this effect, and the observation of any possible significance it may have in terms of electron transport, high resolution X-ray diffraction studies on Fe(III) and Fe(II) cytochromes are necessary. Ferricytochrome c from two sources has been studied at 0·4 and 0·28 nm resolution. This has shown that the molecule is a prolate spheroid of overall dimensions $3 \times 3\cdot4 \times 3\cdot4$ nm, and has confirmed the nature of the fifth and sixth groups coordinated to the iron. The polypeptide chain is a continuous chain of 104 amino acid residues for both horse heart and bonito cytochrome c. The chain is wrapped around the heme in two halves, providing a crevice for the heme in which the heme group has one edge exposed to the solvent. As has been noted for earlier structures, the heme is surrounded by hydrophobic side chains, while the polar side chains lie on the outside of the particle. One propionic acid residue is deeply buried in the hydrophobic interior and is involved in a network of hydrogen bonds. One feature of the structure is the presence of two channels which lead from the surface to the hydrophobic interior and the heme. The main channel is associated with the presence of three aromatic residues (tryptophan 59, tyrosine 67 and tyrosine 74). This may give one pathway for the transfer of electrons from the iron to the surface. Lysine residues are gathered around the outlets of these channels and may represent the binding site. Clearly it is essential to compare the details of this structure with that of the reduced form, but it is attractive to speculate on electron transfer in terms of the above mentioned aromatic residues and the overlap of the aromatic π electron clouds. Tyrosine 74 and the six-membered ring of tryptophan 59 are parallel and 0·6 nm away from each other, while tyrosine 67 is 0·4 nm from tryptophan 59, 0·6 nm from tyrosine 74 and 0·48 nm from the heme plane.

A model for cytochrome c

Wang and Brinigar[31] have put forward a model for the oxidation of cytochrome *c*. An iron(III) protoporphyrin with the ligand 4,4-dipyridyl was reduced with dithionite and mixed with a buffer containing poly-L-lysine. A linear polymer was obtained in which the heme groups were bridged by the dipyridyl groups, while the polycation stabilized the polymer through interactions with the dissociated propionic acid side chains of the heme. This prevents the breakdown of the polymer on oxidation. It was found that the finely ground polymer, in suspension, could catalyse the oxidation of iron(II) cytochrome *c* by oxygen. The heme units in the catalyst are linked *via* a conjugated double bond system, so electron transfer between heme units should be facilitated. Wang suggested that this would allow the oxygen molecule bound at the terminal heme units in the Fe(II) polymer to be reduced to water *via* a four-electron transfer reaction. The Fe(III) polymer may then be reduced by cytochrome *c* and reoxidized by molecular oxygen in a cycle. This model certainly does demonstrate the value of conjugated double bond systems for rapid release of electrons to oxygen.

Peroxidase and catalase

Peroxidases are heme proteins that catalyse oxidations by hydrogen peroxide, while catalases catalyse the disproportionation of hydrogen peroxide (Equations 9, 10).

$$H_2O_2 + SH_2 \xrightarrow{\text{peroxidase}} 2H_2O + S \qquad (9)$$

$$H_2O_2 + H_2O_2 \xrightarrow{\text{catalase}} 2H_2O + O_2 \qquad (10)$$

Catalase activity may be regarded as a special example of peroxidase activity in which the substrate oxidized by hydrogen peroxide is another molecule of hydrogen peroxide. A large number of non-hemes show peroxidase and catalase activity to some extent, as do many metal ions and complexes.

There are a number of similarities in the behaviour of peroxidase and catalase. Thus, in both cases, complexes have been isolated in which a peroxide species is bound to the metal. The presence of iron in higher oxidation states, IV and V, has also been postulated in both instances.

Peroxidases

Heme peroxidases (in which the prosthetic group involves proto heme) and flavoprotein peroxidases are known. Horse radish peroxidase (HRPO) in particular has been studied. This has a low redox potential of -0.170 V, while spectral studies suggest the fifth and sixth ligands to be a nitrogen donor and a water molecule. Spectroscopically HRPO is much like myoglobin[10].

A green product (I) results if Fe(III)HRPO is treated with hydrogen peroxide. This has two oxidizing 'equivalents' above Fe(III). On slow decomposition of I, in the absence of a substrate, a red product (II) is formed, having one oxidizing equivalent above Fe(III). These observations, demonstrated by titration methods, may be explained either by postulating higher oxidation states of iron or by the association of oxidizing equivalents elsewhere in the system. Mossbauer spectroscopy has been applied[32] to this problem. It has been shown that there is a change in the Mossbauer spectra on going from HRPO to compounds I or II and it is suggested that the iron is Fe(IV) in both cases. The extra oxidizing equivalent in I must therefore be associated with some other group, possibly on the protein or porphyrin or possibly on some bound species derived from the peroxide. It is perhaps pertinent to add at this point that it is often quite difficult to assign formal oxidation states to compounds of this type. Thus hydrogen peroxide will react with nitrous acid giving the unstable species peroxonitrous acid, HOONO, which rapidly rearranges to nitric acid. The total redox capacity of 'HOONO' is then equivalent to that of HNO_3 where nitrogen is in the formal oxidation state of V. Quite obviously though the nitrogen atom in peroxonitrous acid is still in the formal oxidation state of III, with additional oxidative capacity associated with the peroxo component of the molecule.

Cytochrome c peroxidase has attracted attention recently.[33] This has a magnetic moment of 5.6 B.M., evidently largely high spin, while spectroscopic studies[10] suggest the presence of about 10% of the low spin form with the possibility of imidazole and water as axial ligands. As with HRPO the presence of a weakly bound axial ligand will allow its replacement by a peroxide species. The electronic spectrum of this peroxidase varies with temperature, there being a larger low spin component at lower temperatures.

The addition of a stoicheiometric amount of hydroperoxide results in the conversion of cytochrome c peroxidase to a red compound similar to HRPOII in terms of electronic spectra. The cytochrome c peroxidase adduct retains two oxidizing equivalents per heme. Yonetani has suggested that again the iron is now present as Fe(IV). EPR studies on the adduct show a very intense signal at $g = 2.004$, of a free radical type. It appears that one of the oxidizing equivalents is retained as a free radical, suggested to be that of 'an aromatic amino acid residue of the protein'.

The complex I of peroxidase (and the equivalent complex of catalase) have anomalous spectra. Smith and Williams[9] have pointed out, however, that it could be interpreted in terms of a simple Fe(III) peroxide in which the heme group was bent.

Catalases

These tend to be rather unlike other heme proteins, although as a group they tend to be very similar. The physiological role of catalase has been the subject of controversy. It has been suggested that it is a protective enzyme for preventing a build-up of hydrogen peroxide. More probably it is a two-electron oxidizer.

Catalases are heme proteins of molecular weight 240,000, having four heme groups. Unlike hemoglobin there is no interaction between them. The reactions of each hematin group are independent of the state of combination of the others. Histidyl and tyrosyl residues have been implicated by chemical modification techniques in the activity of the enzyme, but Williams and Smith[9] suggest that a carboxylate group is bound to iron.

Catalase is marked by a tendency to remain high spin. Even with cyanide as a ligand it is still about 50% in the high spin form. It is a very stable Fe(III) complex, thus dithionite is unable to reduce it to Fe(II).

Like peroxidase, but to a more marked extent, catalase binds hydrogen peroxide giving a product of unusual spectra. A marked feature of catalase activity is the formation of 'Compound I' with peroxide. Other peroxide compounds are also known. The existence of higher oxidation states of iron have been postulated for these compounds. Compound I may in fact be represented as $[Fe\,OOH]^{2+}$, a peroxide complex of Fe(III). Alternatively it has been suggested[34] that the peroxide has interacted with a heme group. It is known that the prosthetic group of catalase, protohemin, will combine with ethyl hydrogen peroxide.

The formation of compound I involves a change in magnetic susceptibility corresponding to a decrease from five to three unpaired electrons. Smith and Williams[9] have also pointed out that this too could be explained in terms of a distorted Fe(III) peroxide heme, as then the degeneracy of the e_g and t_{2g} orbitals would be lost, so allowing electron pairing.

The mechanism of action of catalase has also been the subject of controversy. It appears unlikely to operate through a Fe(III)–Fe(II) cycle. A reasonable alternative is that it serves to bring together substrate and oxidant. It is known that peroxide will coordinate readily (replacing carboxylate). It is assumed that the second peroxide molecule will displace a ligand at the fifth coordination position (or is bound to the porphyrin?) and that the ferric ion mediates the transfer of electrons between the two peroxide molecules.

Models for peroxidase and catalase

It is well known that iron and copper complexes in particular catalyse the decomposition of hydrogen peroxide. Much effort[35-37] has gone into producing suitable models, with some success. Sigel[35] has reviewed the use of copper complexes. This work has involved measuring the effect of a range of copper complexes upon the rates of decomposition and reaction of hydrogen peroxide. These complexes will only act as catalysts if they have donor sites available for the coordination of H_2O_2 or HOO^-, so tending to confirm the suggestion that catalysis occurs within the co-ordination sphere of the metal, with the metal mediating electron transfer from the oxidizable groups to peroxide. In fact the demonstration of catalytic activity can be used as a probe for the presence of coordinatively unsaturated complexes. A mechanism for catalytic activity of copper complexes is given below. A corresponding scheme for peroxidase activity may also be written. This will only occur if there is an oxidizable group in the coordination sphere of the metal. This has been demonstrated for pyrocatachol and o-phenylenediamine and other substrates.

THE NON-HEME IRON PROTEINS

Some examples of non-heme iron proteins are listed in Table 5.3, together with details of their function, molecular weight and moles of iron, cysteine and sulphide per mole of protein. It may be noted that certain of these proteins are involved in the transport and/or storage of iron and oxygen. Examples of the former class have been mentioned at the beginning of this chapter and examples of the latter class discussed in some detail in Chapter

TABLE 5.3 The non-heme iron proteins

Protein	Function	M.Wt.	Fe	Moles of Labile S	Cysteine
Ferredoxin (clostridial)	Electron transfer	6000	7	7–8	7
Ferredoxin (spinach)	Electron transfer	11,000	2	2	4–6
Ferredoxin (alfalfa)[a]	Electron transfer	11,500	2	2	6
Putida redoxin (*pseudomonas*)	Electron transfer	12,500	2	2	2
Adrenodoxin	Electron transfer	13,000	2	2	4
Rubredoxin	Electron transfer	5900	1	0	4
Azotobacter NHIP	Electron transfer	∼30,000(?)	1	?	?
High potential iron protein (chromatium)	Electron transfer	5600	3	4	3
Ferrichrome	Iron storage/ transport	—	—	—	—
Ferritin	Iron storage/ transport	480,000 (20 sub-units)	—	—	—
Transferrin	Iron storage/ transport	80,000	2	—	—
Hemerythrin	Oxygen transport	107,000[b]	2 per subunit	—	—
Pyrochatechase	Oxygenase	95,000	2	—	—
Meta-pyro-chatechase	Oxygenase	140,000	3	—	—
Xanthine oxidase	Electron transfer	300,000	8 (2 Mo)	—	—
Aldehyde oxidase	Electron transfer	300,000	4 (2 Mo)	—	—
Succinic dehydrogenase	Electron transfer	200,000	4	—	—
NADH de-hydrogenase	Electron transfer	80,000	4	—	—
Malate vitamin K reductase	Electron transfer	—	—	—	—

[a] S. Keresztes-Nagy and E. Margoliash, *J. Biol. Chem.* 1966, **241**, 5955.
[b] Eight subunits.

7. In this section, we will be particularly concerned with the electron transfer non-heme iron proteins of fairly low molecular weight (6000–13,000). We will, however, consider xanthine oxidase in a later section.

With the exception of rubredoxin the electron transfer non-heme iron proteins involve 'labile sulphur', that is H_2S is liberated on treatment with mineral acid. They are involved in a range of reactions. Thus one interesting development in the study of nitrogen fixation (see Chapter 6) was the demonstration that nitrogen fixation in cell-free extracts of *Clostridium Pasteurianum* involved the non-heme iron protein ferredoxin which operated at a more negative potential (-0.43 V) than any then known biological electron carrier. Non-heme iron proteins are also involved in electron transfer in photosynthesis, steroid hydroxylation and in oxidative phosphorylation in mitochondria.

A number of review articles[38–40] have been published fairly recently on aspects of the non-heme iron proteins. We will summarize a number of examples, and then group together some general comments, particularly on the valence state of the iron in reduced and oxidized forms of the proteins in the context of EPR measurements.

Rubredoxin

This species will be considered separately as it differs in many respects from the labile sulphur non-heme iron proteins. Rubredoxin was originally isolated from *C. pasteurianum*[41] but has since been found in a number of anaerobic microorganisms.[42] All have one iron atom and four cysteine residues per mole of protein, with molecular weights around 6000, there being some 55 amino acid residues. The redox potential for rubredoxin at pH 7 is -0.057 V.

It is of interest to note[43] that crystals of rubredoxin lose their red colour on treatment with sodium dithionite but recover it on reoxidation in air. This suggests that no major conformational change occurs during the redox cycle. The example of hemoglobin provides an illustration of the opposite case; here crystals of hemoglobin crumble on absorbing oxygen indicative of the major conformational changes occurring on uptake of oxygen.

The structure of oxidized rubredoxin (*C. pasteurianum*) has been determined to 0.25 nm resolution.[44] This has shown that the iron atom is bound to four cysteine sulphur atoms, forming an apparently tetrahedral complex. This is schematically represented in Figure 5.7. The sulphur–iron bond lengths are 0.221, 0.242, 0.237 and 0.222 nm. Comparison with the amino acid composition of the protein has allowed the tentative identification of certain of the amino acid residues. Many of the hydrophilic groups project

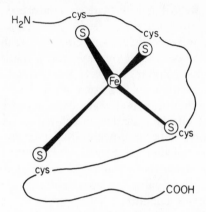

Figure 5.7   Schematic representation of the structure of rubredoxin[45].

from the surface of the molecule. A number of arguments suggest this tetrahedral stereochemistry is maintained in the reduced rubredoxin.

It has been suggested, prior to the X-ray study, on chemical evidence[43] that cysteine groups were involved in coordination. The apoenzyme may be prepared. The number of free SH groups has been found to be 2·7 by titration with *p*-mercuribenzoate. This is lower than the expected value of 4 probably due to the oxidation of the SH group during the isolation of the apoenzyme. However, the metalloprotein cannot be reconstituted by the addition of iron to a solution of the aporubredoxin at pH 7. This will only take place at higher pH in the presence of 2-mercaptoethanol. Again ^{59}Fe exchange with the iron in rubredoxin will only occur in the presence of 2-mercaptoethanol. All this is suggestive of a role for cysteine in metal binding.

A number of physical measurements have been carried out on both oxidized and reduced forms of rubredoxin. The two forms differ in ORD, optical and EPR spectra. The last mentioned technique showed an absorption at g = 4·3 for the oxidized form, but no absorption for the reduced form. A more recent report[45] quotes magnetic moments (by the NMR technique) of 5·85 and 5·05 B.M. for oxidized and reduced forms respectively, implying high spin Fe(III) and Fe(II). This has been confirmed by Mossbauer spectra, while the NMR results suggest a disordered tetrahedral array of ligands.

The near infrared circular dichroism spectrum of the Fe(II) rubredoxin has been reported.[46] This shows only one band in the range 9100–5550 cm^{-1} (at 6250 cm^{-1}). Circular dichroism studies reveal a very large anisotropy factor for this band and hence suggest it to be d–d in origin. This may be assigned as a $^5E \longrightarrow {}^5F$ transition for high spin Fe(II) in a tetrahedral field, so confirming the maintenance of this stereochemistry on reduction. This result also allows an estimation of $10\,Dq = 6250\,cm^{-1}$ for the SH group.

One recent result[46a] has also confirmed the tetrahedral structure. The laser Raman spectrum of oxidized *C. pasteurianum* rubredoxin in aqueous solution shows bands at 368 and 314 cm^{-1}. The spectrum of crystalline rubredoxin shows bands at 365 and 311 cm^{-1}. These may be assigned to v_{Fe-S} modes of the FeS_4 tetrahedron. Depolarization studies confirm that these bands are the v_3 and v_1 fundamentals of a tetrahedral complex. The results are also important in that they show that the environment around the metal ion is the same in the crystalline state and in solution.

Low molecular weight 'labile sulphur' electron transfer non-heme iron proteins. The ferredoxins

These can be classified into 'plant-type' and 'bacterial type' iron sulphur proteins. The former contain two moles each of iron and inorganic sulphur per mole of protein, while the latter group always contain more than this, usually eight moles each of iron and sulphide per mole of protein.

An example of a bacterial ferredoxin is that isolated from *C. pasteurianum*. This is a protein of molecular weight 6000. Analytical data has indicated the presence of 7–8 moles each of iron and inorganic, acid labile sulphide, while there are eight cysteine residues. Many studies have assumed the presence of seven moles of iron but it appears more probable now that there are in fact eight moles. X-ray studies have not been of great value as yet, partly because the method of isomorphous replacement cannot be applied to the ferredoxins.

Early studies[47] on ferredoxin gave an average magnetic moment of 2·0 B.M. per iron atom. These results were considered on the basis of the presence of seven iron atoms and it was suggested that these may be divided into two groups of five and two. It appears that it is uncertain whether this implies a division in terms of redox state or the environment. Blomstrom *et al*[47,48] have suggested the sulphur bridged structure given below, in which iron is tetrahedrally surrounded by four sulphur atoms, two cysteine and two inorganic. It involves a group of five iron atoms and a group of two terminal atoms with different ligand field environments, in accord with one view of the Mossbauer spectra. This structure would have to be modified to allow for the correct stoicheiometry in terms of iron and sulphide content.

An alternative structure based upon that of Blomstrom is given below. This accounts satisfactorily for the stoicheometry of this ferredoxin, but others may also be written.

A similar polymeric chain of iron atoms bridged by two sulphur atoms is found in the inorganic sulphide $KFeS_2$. In this compound the iron–iron distance[49] is 0.27 nm, a value which must imply some Fe–Fe interaction. If the model is a correct one then the implication is that the Fe–Fe distance in ferredoxin must also be of this order, with the further implication that this may be most important in understanding the electron transport properties of the protein and the difficulties associated with explaining the EPR data on oxidized and reduced forms of rubredoxin. It should immediately be noted, however, that X-ray diffraction studies on two ferredoxins[50,51] have shown the protein does not contain a linear array of seven iron atoms. On the other hand, the amino acid sequence of C. pasteurianum ferredoxin has been determined[52] and this has shown that the eight cysteine residues are contained in two groups of four each (with a small number of other residues between consecutive cysteines in each group). This fact certainly lends support[53] to the type of structure proposed by Blomstrom et al, as also does the fact that a number of other lines of evidence suggest that ferredoxin contains two different types of iron.[54] A further problem is associated with the previously mentioned difficulty in deciding whether these two groups differ in redox state or environment. It has been suggested[55,56] on the basis of a number of physical measurements that there are approximately equivalent amounts of Fe(II) and Fe(III) in reduced ferredoxin.

Apoferredoxin has been prepared. The ferredoxin may only be reconstituted though on addition of metal ion, sulphide and a mercaptan such as 2-mercaptoethanol. This again is consistent with the type of metal binding suggested for ferredoxin.

One non-heme iron protein obtained[57] from *Chromatium* differs from other bacterial type ferredoxins in that it has a redox potential of $+0.35$ V. This is termed 'high potential iron protein'. The molecular weight is about 5600; it contains three atoms of iron per mole together with three cysteine residues and inorganic sulphide. Mossbauer spectra indicate the possibility of two valence states as for clostridial ferredoxin.

A number of ferredoxins have been isolated from plant as opposed to bacterial sources and tend to show some differences from them. These are found in chloroplasts (from spinach, alfalfa, parsley and pea leaves) and those studied so far appear to be similar. Their role in photosynthesis has been discussed.[38] For illustration purposes we will consider spinach ferredoxin. The standard physical techniques have been applied[28] and the reduced form shows the typical $g = 1.94$.EPR signal. The molecule has a molecular weight of 11,000 and some 100 amino acid residues, including 4–6 cysteine residues. There are also two moles each of sulphide and ferric iron per moles of protein. On reduction of plant ferredoxin, one half of the iron is converted to ferrous.[53,59]

An example of a ferredoxin from an animal source is adrenodoxin, which is isolated from the adrenal cortex.[60] This is a component of the enzyme steroid-11-β-hydroxylase. The enzyme also contains a flavoprotein, and has the cytochrome P450 as oxygen acceptor. The following electron transfer scheme is suggested.[40]

$$\text{NADPH} \rightarrow \text{flavoprotein} \rightarrow \text{NHIP} \rightarrow \text{cytochrome P450} \rightarrow O_2$$

This protein is said to contain two ferric ions per mole of protein (Molecular weight 13,000) together with inorganic sulphide. Again on reduction the characteristic $g = 1.94$ EPR signal is seen, while only one half of the iron is reduced. Changes in ORD spectra have been observed on reduction, and by comparison of the spectra of apo- and metalloprotein, the d–d bands have been assigned.

The redox states of non-heme iron proteins

A number of difficulties in discussing the function of the non-heme iron proteins have already been noted. The oxidized form is diamagnetic and EPR inactive while the reduced form is paramagnetic and usually has a $g = 1.94$ EPR spectrum. EPR data are given in Table 5.4. It may be seen that rubredoxin differs from the other non-heme iron proteins in being paramagnetic in the oxidized form.

TABLE 5.4 'g' values for oxidized and reduced non-heme iron proteins

	Reduced		Oxidized
Clostridial ferredoxin	gx gy gz	1·89 1·96 2·00	—
Azotobacter NHIP	gx gy gz	1·93 1·94 2·00	—
Pseudomonas putida redoxin	g = 1·94		—
Spinach ferredoxin	gx = 1·89 gy = 1·96 gz = 2.00		—
Adrenodoxin	$\begin{cases} g_\perp = 1·94 \\ g_\parallel = 2·01 \end{cases}$		—
Rubredoxin			g = 4.43

It is not possible to interpret these magnetic properties on a simple basis, involving isolated Fe(II)–Fe(III) redox changes. The only diamagnetic form of iron is low spin, octahedral Fe(II), d^6, which implies that the reduced form of ferredoxin will be a low spin Fe(I), d^7 species.

However, the models we have considered will quite readily allow an explanation of the diamagnetic nature of the oxidized forms. Fe–Fe interaction will readily occur, i.e. with spin–spin interaction and an effective pairing of electrons.

It is known that ferredoxin undergoes a one-electron redox reaction. Now the source of the strongly temperature dependent g = 1·94 resonance in the reduced form has attracted some attention. It is not due straightforwardly to Fe(II). Brintzinger et al[62] have suggested low spin Fe(III) and low spin Fe(II) with a bitetrahedral structure. However, it is very unlikely that low spin tetrahedral iron species could be formed. None are known in conventional inorganic systems. An alternative suggestion[63] is that after reduction both iron atoms are still high spin but one is now formally Fe(II) and the other Fe(III); these will now interact as before but instead of resulting in a diamagnetic system this will result in a system with S = $\frac{1}{2}$; or, in other words, one electron has been added to a diamagnetic system. The g value, it is suggested, may then be explained quite readily on this basis. These models are of course in accord with physical data suggesting the presence of equal amounts of Fe(II) and Fe(III) in reduced ferredoxin.

A number of other results are of interest and should be noted. Thus the identification of the g = 1·94 band with the iron and sulphur was shown by feeding *azotobacter vinelandii* with sulphur-33 and observing the broadening of the EPR band. For ^{32}S, the nuclear spin $I = 0$, for ^{33}S, $I = \frac{3}{2}$ and hence interaction with the iron unpaired electron could occur. The hyperfine splitting[65] due to the nuclear spin of ^{57}Fe has also been observed associated with this signal at $g = 1\cdot94$.[56] Fe has no nuclear magnetic moment.

In conclusion it should be noted that when discussing structures such as

$$\diagdown \underset{\diagup}{Fe^{II}} \diagup \overset{S}{\underset{S}{\diagdown}} \diagdown \underset{\diagdown}{Fe^{III}} \diagup$$

it is going to be very difficult to assign formal oxidation

states. We will meet a corresponding problem when discussing the oxygenated oxygen carrier hemocyanin. The structure must be regarded as a unit in itself. Thus we describe the reduction of ferredoxin in terms of

$$\diagdown \underset{\diagup}{Fe^{III}} \diagup \overset{S}{\underset{S}{\diagdown}} \diagdown \underset{\diagdown}{Fe^{III}} \diagup$$

adding one electron to the unit. Ferredoxin may be

regarded as a delocalized complex of iron with sulphur ligands, one well suited for electron transfer. The earlier predictions, based upon models,[14] appear to have been relevant in this case. The unpaired electron in reduced ferredoxin can be regarded as being localized largely on the ligands utilizing the π acceptor properties of the sulphur donor atom.[66]

Rubredoxin is the exception in Table 5.4. The magnetic data[45] may be explained in terms of the reduction of high spin Fe(III) to high spin Fe(II). The differing magnetic behaviour may be readily understood as here there is only one mole of iron per mole of protein.

One other feature of the ferredoxin that has been the source of controversy is the question of the nature of the acid labile sulphur. It has been represented so far as sulphide S^{2-}. An earlier suggestion that this sulphur resulted from decomposition of the cysteine residues has now been disproved.[67,68] Thus it has been shown that the cysteine content of ferredoxin is unaffected by the removal of the iron and labile sulphur components. An alternative suggestion[69] is that the sulphur is present as a persulphide group (S_2^{2-}). This was presented on the basis of the study of the non-heme iron protein dehydroorotate dehydrogenase.

COPPER METALLOPROTEINS

Much less is known about copper metalloproteins than iron metalloproteins. The copper is very firmly bound by the protein and only in a few cases has it been possible to prepare the apoprotein. There is little direct evidence concerning the nature of the metal binding sites. As has been emphasized, these are usually very different from those for copper in model complexes. A number of workers have added copper ions to proteins or to apoproteins of other metalloproteins and in these cases the copper binding site is much as would be expected from model compounds.[70] EPR and visible spectra are similar to those of the models. It appears that in copper proteins, therefore, the metal takes up a very specific site, one which physical measurements have shown to be irregular in many cases.

Table 5.5 lists some of the copper proteins that have been studied. Aspects of their properties have been discussed in some recent publications.[71-73] A number of points emerge from a consideration of Table 5.5. Thus it may be noted that certain of these proteins have high molecular weights, and are in fact polymers made up of a number of subunits; secondly, their redox potentials are high: and thirdly, that the copper concentration varies from one atom per mole up to eight atoms per mole, usually with an even number of copper atoms. The latter observation may be significant in terms of its implication of pair-wise interactions.

One of the best studied copper proteins is the oxygen carrier, hemocyanin.[73] This involves pair-wise interactions of copper atoms, as will be discussed in Chapter 7. Certain of the tissue copper proteins[74] (so identified in Table 5.5) have no known enzymatic role, although they account for a large proportion of tissue copper. Other copper proteins in tissue are important enzymes, such as cytochrome oxidase and polyphenol oxidase.

We will consider the copper proteins under two general headings, those having one and many copper atoms per mole of protein, respectively. This will be followed by the consideration in more detail of some specific examples. In all this, EPR and visible spectra will be most important, although ligand hyperfine spectra have not so far been observed.

Copper proteins having one copper atom

The blue proteins form a distinctive group. These are copper(II), they have high redox potentials and are of low symmetry. In general they exhibit three absorption bands at about 450, 600 and 750 nm, with high extinction coefficients. The blue proteins also show a characteristic EPR spectrum, having two principal g values ($g\perp \sim 2\cdot1$; $g_{\parallel} = 2\cdot2$–$2\cdot30$), but showing very narrow hyperfine splitting.

The precise role of the blue proteins is not known with certainty, but it is probably that of electron transfer. Thus it has been suggested that *pseudomonas* blue protein may provide an alternative electron transport

TABLE 5.5 Copper proteins

Protein/enzyme	M.Wt	Cu per mole of protein	E^0V	Colour/function
Hemocyanin	Very large subunits 25,000–75,000	2/sub-unit	—	Oxygen carrier
Cytochrome oxidase	100,000	1	—	Cu(I)/Cu(II)
Transulfurase	35,000	1	—	—
Uricase	12,000	1	—	—
Phenol oxidase	34,000	1	—	—
D. Galactose oxidase	75,000	1	—	—
Monamine oxidase	170,000	1	—	Pink
Diamine oxidase (pea seedlings)	96,000	1	—	Pink
P. Aeruginosa blue protein	16,600	1	0·328	Blue
Rhus vernicifera blue protein (stellacyanin)	16,800	1	0·415	Blue
Mung bean blue protein	22,000	1	0·3–0·4	Blue
P. denitrificans	21,000	1	0·370	Blue
Diamine oxidase (porcine kidney)	190,000	2	—	Pink
Plastocyanin	21,000	2	—	Electron transfer in photo-synthesis
Erythrocuprein	35,000	2	—	Tissue copper protein
Cerebrocuprein	45,000	2	—	Tissue copper protein
Hepatocuprein	30,000	2	—	Tissue copper protein
Neonatal hepatic mitochondrocuprein	—	—	—	Tissue copper protein
Dopamine β-hydroxylase	290,000	2	—	—
Amine oxidase (Hog plasma)	195,000	3	—	—
Laccase (P. versicolor)	60,000	4	—	Blue
Laccase (rhus vernicifera)	120,000	6	0·45	Blue
Polyphenoloxidase (tyrosinase)	119,000	4	—	—
Ascorbic acid oxidase	150,000	8	—	Blue
Ceruloplasmin	150,000	8	0·390	Blue

system between cytochrome c and cytochrome oxidase. The redox potentials of this and the other blue proteins are consistent with their participating in electron transfer between the cytochromes.

The remaining one-copper proteins either appear to be pink copper(II) species or copper(I). In general they are associated with oxygen as an oxidizing agent and cycles involving Cu(I) and Cu(II) are suggested. In many cases the EPR spectra are more similar to those observed for copper models while the visible spectra of the pink species involve a broad band, more like the spectra of models. The extinction coefficients are still high (for amine oxidase, $\varepsilon_{500} \simeq 2500$) suggesting a possible involvement of charge-transfer.

Oxygen is always the final electron acceptor for copper(I). When there is only one copper atom, a ternary complex of SH_2, O_2 and copper enzyme is probably formed. Many of the substrates for galactose oxidase and other copper enzymes do in fact form strong complexes with copper in aqueous solution.

Much work has been done on optical rotation and circular dichroism spectra of many of the copper proteins. The irregular structure of the copper protein often results in the generation of optical activity characteristic for a particular enzyme.[75]

Copper proteins having more than one copper atom

Of these, plastocyanin has been suggested[76] to act as an electron carrier in photosynthesis. Laccases[77] are widespread in plants and fungi and have been extensively studied. Copper may be removed from *Rhus*-laccase by dialysis against cyanide, while the activity and spectral properties are partially restored on the addition of Cu^+ but not Cu^{2+}. Probably here, too, copper is reduced by the substrate and reoxidized by molecular oxygen.

The group of "tissue copper proteins" listed in Table 5.5 are probably not involved in redox reactions. Their redox potentials are low, while their EPR and electronic spectra are more like those of model compounds than the spectra of the blue proteins. The d–d bands have appropriately low extinction coefficients.

The blue proteins are generally similar to those already discussed in some respects. They show an additional band at 340 nm in their absorption spectra (as do oxygen binding copper proteins) and it has been suggested[78] that these result from charge-transfer between copper and copper or copper and oxygen.

EPR data have been very valuable in considering this class of copper proteins, particularly in the context of valence. In simple terms EPR data will distinguish copper(I) from copper(II). However, a diamagnetic complex may be the result of pairwise spin-spin interactions of copper(II) atoms. Thus EPR has indicated that only half of the copper in ceruloplasmin is copper(II). It was assumed[79] therefore that the remainder is copper(I), but later appreciated[80] that the result equally well implies pairs of copper(II).

On the basis of EPR evidence[81,82] copper atoms in proteins have been divided into three types: $Type-1\ Cu^{2+}$ typical of blue proteins with their characteristic absorption and electronic spectra; with high redox potential (0.7–0.8 V suggested for the type-1 Cu^{2+} in ceruloplasmin); $Type-2$ Cu^{2+}, the more 'normal' species discussed earlier; and, finally, diamagnetic copper.

It has been suggested that ascorbic acid oxidase, the laccases and ceruloplasmin contain all three of these forms of copper.[83,84] Thus the fungal laccase from $polyporus\ versicolor$, which catalyses the oxidation by molecular oxygen of diphenols and aromatic diamines, involves four copper atoms per mole. Two are diamagnetic and two are EPR active, one being a blue type-1 Cu^{2+} and the other being type-2 Cu^{2+}.

All these oxidases will reduce both atoms of molecular oxygen to water, and, as such, multielectron transfer is preferred rather than one-electron reduction. The blue proteins with only one copper atom are associated with electron transfer rather than with oxidase activity perhaps due to the lack of an 'open' site in the copper(I) form. In any case the reduced form is not autoxidizable.[83] In these blue oxidases it may well be then that one of the forms of copper other than type I is required for the reaction with oxygen. Kinetic studies[83,86] indicate that this reaction involves the diamagnetic pair. This means that oxygen will be reduced in two two-electron steps via the diamagnetic pair, which will then be reduced by appropriate cycling, and that the intermediate will be a peroxide species. This species would be bound by the protein, possibly by the type-2 Cu^{2+}. The binding of amonic inhibitors to the type-2 Cu^{2+} is known to interfere with the reoxidation reaction.

Cytochrome c oxidase[87,88]

The two heme components of this enzyme system, cytochromes a and a_3 have already been mentioned. Cytochrome a_3 is accepted to be the oxygen-interacting species. Thus it readily interacts with carbon monoxide. It is also accepted that the copper protein component of cytochrome c oxidase is associated with redox activity.

The copper protein contains one mole of copper per heme group (i.e. two per mole), but only about 40% of the copper gives an EPR detectable signal. This may infer the presence of Cu^+ or spin interaction between $Cu^+–Cu^{2+}$ or $Cu^{2+}–Cu^{2+}$ pairs. On denaturation all the copper is EPR detectable, thus possibly suggesting that the EPR inactive copper is present as a $Cu^{2+}–Cu^{2+}$ pair. There have, however, been conflicting reports on EPR data for cytochrome c oxidase, and Beinert[87] has suggested that problems of this nature may reflect differences in preparative and other experimental methods, which may result in adventitious copper, liberated on partial denaturation of the copper protein, being bound at non-active protein sites. Such copper will of course contribute to the EPR signal.

Titration experiments[89,90] support the idea that all copper is present as copper(II). Thus it has been shown that two equivalents per heme were taken up by the enzyme on anaerobic titration with NADH, using phenazine methosulphate as a mediator. It has been suggested that a band at 830 nm in the absorption spectrum of cytochrome c oxidase is associated with copper. A study of absorbance changes at this wavelength during the titration suggested that copper(II) was acting as an electron acceptor. This band has also been used for following $[Cu^{2+}]$ in a further study,[91] involving fast reaction techniques, which has confirmed that copper in cytochrome c oxidase undergoes redox reaction as fast as the cytochrome components.

On treatment[92] of the cytochrome c oxidase with substrate under anaerobic conditions in the presence of bathocuproine (a 'valence specific' copper(I) reagent) a copper deficient enzyme may be obtained. The activity can be restored on adding $Cu(I)$. Both this and the above mentioned result provide useful confirmatory evidence for redox cycling of copper in cytochrome c oxidase.

Polyphenoloxidase (tyrosinase)

This copper enzyme will catalyse two different aerobic oxidation reactions. These are (i) the dehydrogenation of catechols (i.e. 1,2-dihydroxybenzenes, Equation 11), which is termed 'catecholase activity' and (ii) the hydroxylation of monophenols ('cresolase activity').

$$\text{(11)}$$

At present there are two schools of thought regarding this aspect of polyphenoloxidase behaviour. First,[93] it is postulated that both reactions are catalysed by the enzyme, either on two different sites, or through the existence of two unlike subunits of the enzyme, each having different enzyme activity. The alternative suggestion[94] is that cresolase activity results from the hydroxylation of monophenols by the orthoquinoid product of catecholase activity and is therefore not due to the direct intervention of the enzyme molecule itself.

It is important to note in this context that the cresolase and catecholase functions of the enzyme are inhibited competitively by catechols and cresols respectively. This suggests that the idea of two active sites is a correct one.

The physiological function of the enzyme lies in the formation of the melanin pigments that, for example, give colour to the skin and eyes of animals. The enzymes are, however, widespread, but there is no satisfactory reproducible method for its isolation. It is possible to obtain enzyme preparations with high catecholase or high cresolase activity. It has been found[95] that there is little exchange of enzyme copper with radioactive ^{64}Cu while the enzyme is resting, but there is significant exchange of bound copper when both types of enzyme preparation are allowed to function catalytically on the two substrates. The 'high cresolase' preparation shows a significantly lower amount of exchange with both substrates. This led to the suggestion[93] that two types of active centres existed in the enzyme, the cresolase sites having 'non-exchangeable copper as compared with the catecholase sites'. The enzyme is known to have four copper atoms per mole of enzyme.

EPR studies[96] have shown the presence of o-benzosemiquinone when polyphenol oxidase catalyses the oxidation of catechols. These are produced by the product (o-benzoquinone) reacting with catechol giving two semiquinone molecules. There is evidence then for the reactivity of the product of catecholase activity. The quinone is also known to combine with the enzyme protein and in so doing may result in its deactivation. Quinones with 4 and 5 positions blocked will not do this.

Polyphenoloxidase is inhibited by carbon monoxide. This suggests the presence of copper(I). The apoenzyme may be prepared, but any copper(II) added is reduced to copper(I) in reconstitution.

THE COBALT B_{12} COENZYMES

The biological activity of cobalt is largely confined to its role in the B_{12} series of coenzymes. The vitamin B_{12} molecule has been studied by X-ray diffraction,[98] and its structure has been shown in Chapter 3. It involves a conjugated corrin ring (usually buckled), which has a porphyrin type structure apart from having one methine group less. The corrin system provides four nitrogen donor atoms, in a plane, for cobalt as found in the porphyrins. The fifth and sixth coordination positions are occupied by the benzimidazole of the nucleotide group and a cyanide ion. The presence of cyanide is an artifact of the isolation procedure and is of no biological significance. The cyanide complex is not active as a coenzyme. This complex without cyanide is termed cobalamin, so vitamin B_{12} is cyanocobalamin. A whole range of other substituents may take the place of cyanide.

On hydrolysis the benzimidazole group is displaced by water, giving the cobinamide series of complexes, while a variety of ligands may occupy the sixth position. The fifth position may be occupied by OH^- or CN^- instead of water. The cobinamides and cobalamins are represented schematically in Figure 3.7. Of particular interest, both chemically and biochemically, is the case when the sixth ligand is a carbanion, that is where there is a Co–C bond. Until the demonstration[99] of the direct cobalt–carbon σ bond linking an adenosyl group to cobalt in a B_{12} coenzyme, organo cobalt complexes were regarded as rare, unstable species. Now a series of very stable organo cobalt complexes have been prepared using model compounds based upon the B_{12} structure.

Some aspects of the biological function of the B_{12} coenzymes have already been given in Chapter 3 and will not be extended further. Some comments on model compounds were also made. The biochemical functions have been reviewed on a number of occasions.[100–102] In summary, it may be noted that these include methyl group transfer, skeletal rearrangements of molecules and various reduction reactions.

Models for cobalt B_{12} coenzymes

Much interesting inorganic chemistry has arisen from the work of the groups of Schrauzer[103] and Williams[104,105] in this context. The value as models of cobalt complexes of dimethylglyoxime, the 'cobaloximes', has already been noted in Chapter 3. The crystal structure[106] of a substituted alkyl cobaloxime shows that the Co–N (in-plane) and Co–C bond lengths are very similar to those found for the coenzyme.

It appears that the underlying reason for the good correlation between the properties of the coenzyme and models is the presence of a strong planar ligand field of the correct strength. Thus a number of other ligands[103] that satisfy this condition will also show model behaviour of varying appropriateness. One important difference between coenzyme and model compound, which probably contributes to the presence of a number of differences in behaviour, is the fact that there is a great deal of steric hindrance in the coenzyme which is not duplicated, for example, by the cobaloximes.

Williams and coworkers[104] have prepared cobalamins and cobinamides having a wide range of axial ligands including carbanions and have attempted to rationalize the behaviour of the molecule as a whole in terms of the properties of those ligands. This work has led to an increased understanding of the biochemical function of the coenzymes, and forms the basis of much of the following discussion.

The redox behaviour of the cobalt coenzymes

The role of the cobalt coenzymes in the catalysis of reduction and methyl group transfer reactions depends upon the presence of low oxidation states of cobalt. In the latter case, a low oxidation state methyl complex of cobalt will act as the transferring group, and cobalt (III) will act as the methyl-accepting group.[104,107]

The cobalt coenzymes contain cobalt in the formal oxidation state (III). The fact that cobalt may be reduced to cobalt(II) or cobalt(I) contributes to its biochemical activity. The cobalt(III) aquocobalamin is termed B_{12a} and this may be reduced to B_{12r} and B_{12s} forms in one-electron steps. The B_{12r} species involves[108,109] low spin cobalt(II); EPR studies have shown it to be paramagnetic, while the hyperfine structure due to interaction with the cobalt nucleus and the ligand hyperfine structure due to interaction with the benzimidazole nitrogen[110] have been observed. The B_{12s} species is Co(I). Similar redox behaviour is shown by the cobaloximes.

Studies on cobinamides and cobalamins with a range of axial ligands

Spectral studies on a range of complexes in which the nature of the axial ligand is changed show that there is dependence of the spectra on the ligand corresponding to the position of the ligand in the nephelauxetic series. The presence of carbanion ligands (particularly $C_2H_5^-$) with water as the other axial ligand, causes considerable changes in spectra. It has been shown[104] that this latter effect may be understood in terms of an equilibrium between 5- and 6-coordinate species. Weak axial ligands in the cobinamide series (with H_2O as the fifth ligand) only give the 6-coordinate species. Strong donors (e.g. $C_2H_5^-$) only give the 5-coordinate species (with water displaced), while $C_2H_3^-$ and CH_3^- give a mixture of the two forms. Similar behaviour has been observed for the cobaloximes,[111] and there is other supporting evidence.[104] Comparable changes have been seen for the cobalamin complexes, where in this case benzimidazole must be the leaving ligand for the cobalt complex to become 5-coordinate. Similarly, it has been shown that the coenzyme itself is an equilibrium mixture of 5- and 6-coordinate species with a fast rate of interconversion between these species.

The underlying chemical reason for the existence of the 5-coordinate species is the well known *trans* effect. It is particularly associated with carbanions and is clearly a bond breaking, σ effect. Evidence for the existence of the *trans* effect lies (1) in the long Co-ligand bond lengths found in certain cases; and (2) in the infrared spectra of a series of cyano-cobinamides,[112] where the effect of varying the second axial ligand on the ν C$\equiv$N frequency of the cyanide ion was observed. It was noted that as the electron-donor properties of the axial ligand were increased, so the ν C$\equiv$N frequency fell towards that observed for free cyanide ion, i.e. the Co–CN bond was being weakened.

However, while the cause is a well known one, the effect is most unusual for cobalt(III). Five-coordinates Co(III) complexes are very rare. Another anomalous feature is the fast rate of interconversion of 5- and 6-coordinate species. Reactions of cobalt(III) complexes are usually much slower. In general, however, the rate of ligand exchange at cobalt in vitamin B_{12} is extremely fast.

It is of great interest, though, that both these features are well known for cobalt(II). Such complexes are labile and stable five-coordinate species are known. Here again we have a situation ideally designed for electron transfer, for cobalt(III) to be converted to cobalt(II). It is significant that the ligands best able to cause this situation, the carbanions, are in themselves strong reducing agents.

The role of the carbanion group as a reducing agent

Many of the coenzymes contain cobalt–carbon bonds. This system is a flexible one in terms of redox reaction, (Equation 12).

$$Co(III)R^- \rightarrow Co(II)R^\cdot \rightarrow Co(I)R^+ \qquad (12)$$

In the preceding paragraph attention was drawn to the existence of five-coordinate coenzyme species, in the context of generating Co(III) in an environment that would favour Co(II) and hence electron transfer. This argument can be extended to Co(I). Cobalt(I) forms of the coenzyme and of the cobaloximes may be obtained. These are stable in alkali, decomposing in acid solution to give hydrogen and a cobalt(II) derivative. They are very reactive species. Now spinpaired cobalt(I) is isoelectronic with nickel(II). A characteristic feature of this configuration is the stability of square planar species. It is of great interest, therefore, to note that the axial ligand in five-coordinate cobaloximes is easily detached giving a square planar species. Furthermore, it appears that the binding of the fifth group in the coenzyme is weakened when the coenzyme is bound to the protein. There is, therefore, implied in this an extension of the 'entatic' state of the B_{12} coenzymes to cover the Co(I) state as well.

MOLYBDENUM ENZYMES

Molybdenum is associated with the electron transfer processes of the flavoenzymes, xanthine oxidase, aldehyde oxidase and nitrate reductase. In the first two examples non-heme iron is also involved. Molybdenum is also implicated in nitrogen fixation, but this topic is reserved for Chapter 6. Details of the above mentioned flavo enzymes are given in Table 5.6,

TABLE 5.6 Molybdenum metalloenzymes

Enzyme	Source	M.Wt.	Stoicheiometry	
			Metal	Coenzyme
Xanthine oxidase	Bovine milk	300,000	8Fe,[a] 2Mo, 2FAD	
Xanthine oxidase	Chicken liver	300,000	8Fe, 2Mo, 2FAD	
Xanthine oxidase	M. lactilyticus	250,000	8Fe, 2Mo, 2FAD	
Aldehyde oxidase	Hog liver	300,000	8Fe, 2Mo, 2FAD	
Nitrate reductase	N. crassa	?	Mo, FAD[b]	
Nitrate reductase	E. coli.	?	Mo, Fe, FAD[b]	

[a] Non-heme iron, with associated labile sulphur.
[b] Stoicheiometry unknown.

where it may be seen that flavin adenine dinucleotide (FAD) is a cofactor in each case. The presence of this particular coenzyme is a complicating factor in the study of the role of molybdenum, as this is largely based upon the electron paramagnetic resonance signals associated with Mo^V species. The generation of free radical semiquinones (and the presence of reduced non-heme iron) will result in more complex spectra.

Flavoproteins[113,114]
In general the flavoproteins catalyse a number of dehydrogenation reactions and are linked to a variety of electron acceptors, according to the nature of the substrate. Most flavoproteins do not involve meal ions. The flavins are of biochemical importance in that they may be involved in one- or two-electron reductions giving semiquinones and quinones respectively (Equation 13). Flavoproteins are often linked in pairs. This means that such a unit can be associated with redox reactions involving

from one to four electrons.

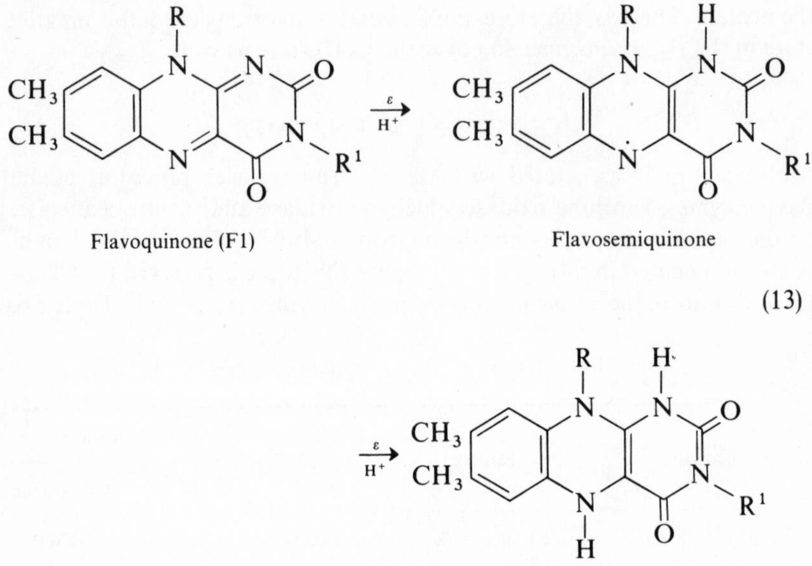

Flavoquinone (Fl)   Flavosemiquinone

$$\tag{13}$$

Flavohydroquinone (FlH$_2$)

It appears that the reduced species (FlH$_2$) interacts with Mo(VI) giving the flavoquinone and Mo(V) (Equation 14).

$$FlH_2 + 2Mo(VI) \rightarrow Fl + 2Mo(V) \tag{14}$$

Aspects of molybdenum chemistry[115]

Molybdenum is an element of the second transition series, having a $4d^5 5s^1$ configuration. In its compounds it shows a range of oxidation states and stereochemistries. The biologically important oxidation states are Mo(V) (d^1) and Mo(VI) (d^o), but the lower ones, particularly Mo(III), cannot be completely discounted. The preferred coordination numbers for Mo(V) and Mo(VI) are five and six, and four and six respectively.

Mo(V) species are paramagnetic and will have an EPR spectrum. For ^{95}Mo and ^{97}Mo isotopes, $I = \frac{5}{2}$ and hence six satellite peaks will be observed. For the other isotopes $I = 0$, and one broad band will occur. This means that for naturally occurring Mo(V), as the natural abundance of the $I = \frac{5}{2}$ species is only about 25%, it will be difficult to observe the hyperfine splitting. The synthesis of ^{95}Mo(V) enriched compounds is sometimes necessary.

Mo(V) complexes are usually anionic, having oxygen, halogen, thiocyanate or cyanide as ligands. Oxo species are particularly important and dimerization frequently occurs through oxo bridges. Under certain conditions this leads to diamagnetic species, through spin–spin interaction. Mo(V) compounds are usually prepared by reduction of molybdates or MoO_3 in acid solutions by chemical or electrochemical methods. Some oxo complexes with other ligands include $MoOCl_3(dipy)$, $Mo_2O_3Cl_4(dipy)_2$ and $\{MoO(oxalate)H_2O)_2O_2\}^{2-}$. The latter species is diamagnetic due to the presence of a Mo–Mo bond. An ethylxanthate complex $[(C_2H_5OCS_2)_2MoO]_2O$ has a structure[116] involving two distorted octahedrons linked by a Mo–O–Mo bond with *cis* double bounded oxygen atoms on each molybdenum. Most of these Mo(V) complexes are not particularly stable with respect to oxidation. Some others, including those with EDTA and with 8-hydroxyquinoline-5-sulphonic acid, are quite stable.

Mo(VI), $d°$, complexes are also usually anionic with oxygen as the preferred ligand. A variety of organic hydroxo complexes are known but these are not well defined. The tetrahedral oxo anion MO_4^{2-} is well known, but in acid solution polymerization occurs giving a range of polymolybdate ions of some complexity. Amines give complexes with molybdates which are usually ill characterized, but with diethylenetriamine MoO_3 . dien is formed. Mo(VI) also forms various binuclear oxo complexes.

It is to be expected, therefore, that, in biological systems, molybdenum will be bound through carboxy groups or hydroxyl groups of tyrosine and serine residues. It is clear, however, from studies with —SH bonding agents, that cysteine is a likely binding site of the metal ion. There is evidence to show that inhibitors (cyanide, etc.) are also bound to the molybdenum in the enzyme.

Models for molybdenum-flavin interaction

The interaction of molybdenum with the model compound lumiflavin-3-acetic acid has been studied[117] by EPR and polarography techniques. The model compound has similar redox properties to riboflavin and flavin mononucleotide (FMN) but is water soluble, light stable and has no complicating side chains. Its structure was given in Equation 13, for $R = CH_3$ and $R^1 = CH_2COO^-$. The flavohydroquinone is readily oxidized by Mo(VI) in acid solution, while the reverse reduction by Mo(V) occurs readily in alkaline solution. In the former reaction a deep red transient colour was observed which, under certain conditions, accounted for the major part of the total flavin concentration. This, it was suggested, was either a Mo(V) or Mo(VI) flavosemiquinone complex. EPR will distinguish between these two alternatives. Mo(V) will give an EPR signal but it is clear that normally under these conditions it will exist as an EPR inactive dimer. However, when first generated by reduction of Mo(VI) it will have a short life-time as a monomeric species (of high chemical activity). EPR measurements during the course of the reaction showed the presence of the free flavosemiquinone free radical ($g = 2.00$) and the Mo(V) monomer ($g = 1.95$). There was no signal due to the red intermediate compound. This must mean that it is Mo(V)-flavosemiquinone complex in which the d^1 and free radical spins are coupled as a result of the co-ordination of the radical N atom to the metal.

On reacting the flavosemiquinone with Mo (i.e. the reverse reaction), observation of the $g = 2.00$ signal indicates that a very much larger concentration of free radical semiquinone was formed. This may be attributed to the formation of a flavosemiquinone metal chelate involving the diamagnetic Mo(VI). These studies suggest that similar interactions occur in the metalloenzyme between molybdenum in both oxidation states and the flavin. The following overall scheme for these model reactions has been suggested.[117] The flavosemiquinone carries one proton that may be replaced by the metal ion on coordination.

$$Mo(VI) + FlH_2 \xrightarrow{slow} Mo(V)\dot{F}l \text{ (EPR undetectable)}$$

$$Mo(V)\dot{F}l \rightarrow Mo(V) + \dot{F}lH$$

$$2\dot{F}lH \rightleftharpoons FlH_2 + Fl$$

$$2Mo(V) \rightleftharpoons (Mo(V))_2$$

$$Mo(V) + Fl \rightarrow Mo(VI)\dot{F}l \text{ (EPR active)}$$

$$Mo(VI)\dot{F}l \rightarrow Mo(VI) + \dot{F}lH$$

The cations Zn^{2+} and Cd^{2+} have also been shown[118,119] to form complexes with flavosemiquinones under pH conditions of physiological importance, even though the oxidized and reduced flavin species have little affinity for these metal ions. It is clear then that metal ion coordination serves to stabilize the intermediate oxidation state of flavin.

Models for molybdenum–substrate interaction[120]

These have been discussed in Chapter 3.

Models for molybdenum–protein interaction

The interaction of molybdenum(V) with sulphur-containing model compounds, such as cysteine[121] and the peptide glutathione[122] has been studied. In the latter case a diamagnetic binuclear [Mo(V)]$_2$ glutathione complex was formed of stoicheiometry Na[Mo$_2$O$_4$(glutathione)(H$_2$O)]. 3H$_2$O, with the peptide serving as a pentadentate anionic ligand. EPR studies in phosphate buffer solution (pH 8–10) show an equilibrium between the diamagnetic binuclear species and a paramagnetic binuclear species, although the paramagnetic form never exceeds 4% of the total. The solution EPR spectrum of isotopically enriched [^{95}Mo(V)]$_2$ gluta-thione showed the presence of 11 hyperfine lines, resulting from an electron interaction with two magnetically equivalent molybdenum nuclei.

The electronic spectrum of the system at pH 8 shows a weak absorption at about 580 nm, similar to that observed in the Mo(V) cysteine complex. This suggests that in the glutathione complex, molybdenum is also bound by the cysteine residue. In addition the ligand will probably bind via one carboxylate group and one amino function per molybdenum.

The existence of the paramagnetic form appears to result from the dissociation of the oxo bridge, leaving the two molybdenum atoms linked through the bridging peptide. The existence of electron exchange coupling in the complex is clear from the EPR evidence.

Molybdenum flavoproteins, particularly xanthine oxidase

Xanthine oxidase and aldehyde oxidase are very similar in composition. It is of interest that two molybdenum atoms occur per mole. The catalytic activity is shown in Equation (15), where the substrate is xanthine, and the product uric acid.

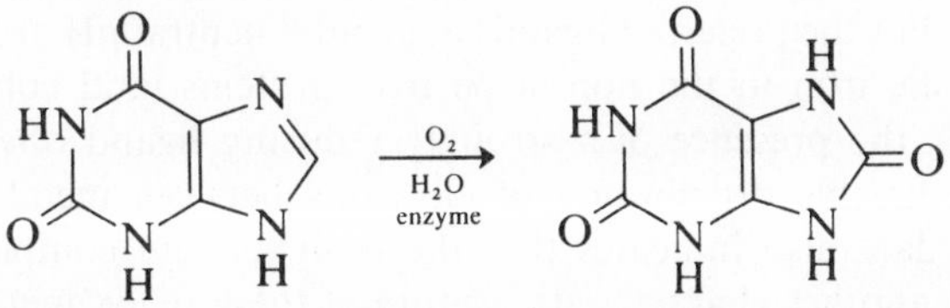

$$\tag{15}$$

The EPR spectrum[123] of reduced xanthine oxidase shows the typical non-heme iron protein signal at $g = 1.94$, and signals due to Mo(V) and FAD. The oxidized enzyme has no EPR spectrum, in accord with the presence of Mo(VI). A similar result has been obtained for aldehyde oxidase.[124]

Kinetic studies utilizing EPR and absorption spectra indicate that the electron transfer sequence for the xanthine oxidase catalysed oxidation of xanthine by oxygen (and for aldehyde oxidase acting upon appro-priate substrates) is given by:[125,126] xanthine $\rightarrow$ Mo(VI) $\rightarrow$ FADH

→ non-heme iron → oxygen. When xanthine is replaced[127] by salicylaldehyde, however, flavin reduction precedes Mo(VI) reduction in xanthine oxidase.

The results of model studies appear to be relevant to flavoproteins. Thus there is clear evidence for interaction between metal ions[48] including Mo(V)[129] and protein bound flavinsemiquinones. On the other hand, so far there seems to be no evidence[129] for Mo(VI) interaction with the semiquinones in flavoproteins.

The molybdenum EPR spectrum changes[130] when H_2O is replaced by D_2O as solvent. In water as solvent all bands have a doublet structure, implying that a proton $(I = \frac{1}{2})$ is involved in the molybdenum complex which gives the EPR signal. On changing to D_2O solvent the doublet bands coalesce into singlets, in accord with this suggestion. It is reasonable that the proton is derived from the enzyme FAD.

It has also been shown[130] that the Mo EPR signals never account for more than about a third of the molybdenum detected by chemical analysis. This suggests that either Mo(V) is not the main component or that the Mo(V) species are interacting pairwise. The postulate of a dimeric Mo(V) diamagnetic-paramagnetic equilibrium is, of course, consistent with model studies.[121,122]

The EPR data[131–133] for the Mo(V) enzyme indicates that sulphur ligands are bound to the metal. In accord with $I = \frac{5}{2}$ for the molybdenum nucleus, there are six satellite peaks in the EPR band. The separation between these bands is typical of molybdenum (V) bound to polarizable groups, such as RS^-. Molybdenum(V) complexes are more stable with RS^-, CN^-, Cl^- and aromatic nitrogen ligands than are molybdenum(VI) complexes but the preferred ligands at around neutral pH are still oxygen ligands. Like iron in the non-heme iron proteins (and cobalt with the carbanion), the presence of a strongly reducing ligand obviously lends itself to a flexible distribution of electrons between metal and ligand. The EPR data also indicates that the molybdenum complex is of low symmetry, another characteristic feature of these redox metalloenzymes.

Further support for the existence of a Mo–S chromophore results from the observation of an absorption band at 580 nm in the electronic spectrum of xanthine oxidase.[134] The Molybdenum(V) cysteine and glutathione complexes both show a band at this wavelength, as noted earlier.

Dennard and Williams[133] have suggested that it might be possible to replace molybdenum in these enzymes by niobium, tantalum and tungsten. The first two examples will not undergo redox reactions, but will give a stable M(V) complex, the study of which should be of interest.

It is clear that the use of model compounds has provided a profitable approach to the study of the action of xanthine oxidase. Light has been thrown on the possible binding of coenzyme, protein and substrate to the metal, and on the electronic factors that contribute to the behaviour of this redox enzyme.

REFERENCES

1. J. Bjork and W. W. Fish, *Biochemistry*, 1971, **10**, 2844
2. R. E. Feeney and S. K. Komatsu, *Structure and Bonding*, 1966, **1**, 149, Springer-Verlag, Berlin
2. (a) C. K. Luk, *Biochemistry*, 1971, **10**, 2838
3. I. H. Scheinberg in *The Biochemistry of Copper*, J. Peisach, P. Aisen and W. E. Blumberg, (Eds), 1966, p. 513. Academic Press, New York
4. R. J. P. Williams, *R.I.C. Reviews*, 1968, **1**, 13
5. M. Goldstein, E. Lauber, W. E. Blumberg and J. Peisach, *Fed. Proc.*, 1965, **24**, 604
6. C. J. Hawkins and D. D. Perrin, *J. Chem. Soc.*, 1962, 1351
7. J. C. Tomkinson and R. J. P. Williams, *J. Chem. Soc.*, 1958, 2010
8. B. R. James and R. J. P. Williams, *J. Chem. Soc.*, 1961, 2007
9. D. W. Smith and R. J. P. Williams, *Structure and Bonding*, 1970, **7**, 1, Springer Verlag, Berlin
10. R. J. P. Williams in *Non-Heme Iron Proteins. Role in Energy Conversion*, A. San Pietro (Ed), 1965, 7, Antioch Press, Yellow Springs, USA
11. P. Day and R. J. P. Williams, *J. Chem. Phys.*, 1962, **37**, 567
12. A. B. P. Lever, *Advances in Inorganic Chemistry and Radiochemistry*, Emeleus and Sharpe, (Eds), 1965, **7**, 27, Academic Press, New York
13. R. J. P. Williams in *The Biochemistry of Copper*, J. Peisach, P. Aisen and W. E. Blumberg, (Eds), 1966, 131, Academic Press, New York
14. A. H. Maki, N. Edelstein, A. Davison and R. H. Holm, *J. Amer. Chem. Soc.*, 1964, **86**, 4580
15. R. J. P. Williams, *J. Theoret. Biol.*, 1961, **1**, 1
16. R. J. P. Williams, *J. Theoret. Biol.*, 1962, **3**, 209
17. D. W. Urry and H. Eyring, *J. Theoret. Biol.*, 1965, **8**, 198, 214
18. G. E. W. Wolstenholme and C. M. O'Connor, *CIBA Foundation Symposium on Quinones in Electron Transport*, 1960
19. P. D. Boyer, D. E. Hultquist, J. B. Peter, G. Kriel, L. G. Butler, R. A. Mitchell, M. Deluca, J. W. Hinkson and R. W. Moyer, *Fed. Proc.* 1967, **22**, 1080
20. J. E. Falk, *Porphyrins and Metalloporphyrins*: 1964, Elsevier, Amsterdam.
21. B. Chance, R. W. Estabrook and T. Yonetani, (Eds), *Hemes and Hemoproteins*, 1966, Academic Press, New York
22. E. Margoliash and A. Scheztor, *Adv. Protein Chem.*, 1966, **21**, 113
23. J. E. Falk, R. Lemberg and R. K. Morton, (Eds), *Hematin Enzymes*, Pergamon, Oxford, 1961
24. P. D. Boyer, H. Lardy, K. Myrback (Eds), *The Enzymes*, Vol. 8. Academic Press, 1963
25. W. S. Caughey, R. M. Deal, C. Weiss and M. Gouterman, *J. Mol. Spect.*, 1965, **16**, 451; M. Gouterman, *J. Mol. Spect.*, 1961, **6**, 138; M. Gouterman, *J. Chem. Phys.*, 1959, **30**, 1139
26. E. B. Fleischer, *Acc. Chem. Res.*, 1970, **3**, 105

27. W. S. Caughey, J. L. York and P. K. Iber in *Electronic Aspects of Biochemistry*, B. Pullman, (Ed), 1966, 25, Academic Press, New York; S. J. Cole, G. C. Curthoys and E. A. Magnusson, *J. Amer. Chem. Soc.*, 1970, **92**, 2991

28. J. E. Falk and R. J. Porra, *Biochem. J.*, 1963, **70**, 66

29. S. Sano and K. Tanaka, *J. Biol. Chem.*, 1964, **239**, PC 3109

30. D. D. Ulmer, *Biochemistry*, 1965, **4**, 902

    (a) E. Margoliash, W. F. Fitch and R. E. Dickerson, *Brookhaven Symp. Biol.*, 1968, **21**, 259

    (b) R. E. Dickerson, T. Takano, D. Eisenberg, O. B. Kallai, L. Samson, A. Cooper and E. Margoliash, *J. Biol. Chem.*, 1971, **246**, 1511

31. J. H. Wang and W. S. Brinigar, *Proc. Nat. Acad. Sc. U.S.*, 1960, **45**, 958

32. T. H. Moss, A. Ehrenberg and A. J. Bearden, *Biochemistry*, 1969, **8**, 4159

33. T. Yonetani, H. Schleyer and A. Ehrenberg, in *Magnetic Resonance in Biological Systems*, A. Ehrenberg, B. G. Malmstrom and T. Vanngard (Eds), 1967, 154, Pergamon, Oxford; T. Yonetani and A. Ehrenberg, *Ibid*, p. 155

34. A. S. Brill and R. J. P. Williams, *Biochem. J.*, 1961, **78**, 246

35. J. H. Wang, *J. Amer. Chem. Soc.*, 1955, **77**, 822, 4715

36. R. C. Jarnagin and J. H. Wang, *J. Amer. Chem. Soc.*, 1958, **80**, 786, 6471.

37. H. Sigel, *Angew. Chem. Int. Edit.*, 1969, **8**, 167

38. A. San Pietro (Ed.) *The Non-Heme Iron Proteins. Role in Energy Conversion*, 1965, Antioch Press, Yellow Springs

39. R. Malkin and J. C. Rabinowitz, *Ann. Rev. Bioch.*, 1967, **36**, 113

40. T. Kimura, *Structure and Bonding*, 1969, **5**, 1, Springer-Verlag, Berlin

41. W. Lovenberg and B. E. Sobel, *Proc. Natl. Acad. Sci. U.S.*, 1965, **54**, 193

42. D. J. Newman and J. R. Postgate, *Europ. J. Biochem.*, 1968, **7**, 45

43. W. Lovenberg and W. M. Williams, *Biochemistry*, 1968, **8**, 141

44. J. R. Herriot, L. C. Sieker, L. H. Jensen and W. Lovenberg, *J. Mol. Biol.*, 1970, **50**, 391

45. W. D. Phillips, M. Poe, J. F. Weiker, C. C. McDonald and W. Lovenberg, *Nature*, 1970, **227**, 574

46. W. A. Eaton and W. Lovenberg, *J. Amer. Chem. Soc.*, 1970, 92, 7195

    (a) T. V. Long, T. M. Loehr, J. R. Allkins and W. N. Lovenberg, *J. Amer. Chem. Soc.*, 1971, 93, 1809

47. D. C. Blomstrom, E. Knight, W. D. Phillips and J. F. Weiker, *Proc. Natl. Acad. Sci. U.S.*, 1964, **51**, 1085

48. W. D. Phillips, E. Knight and D. C. Blomstrom, in *Non-Heme Iron Proteins, Role in Energy Conversion*, A. San Pietro (Ed), 1965, 69, Antioch Press, Yellow Springs

49. J. W. Boon and C. H. MacGillavry, *Rec. Trav. Chim*, 1942, **61**, 910

50. L. H. Jenson and L. C. Sieker, *Science*, 1965, **150**, 376

51. R. D. Gillard, E. D. McKenzie, R. Mason, S. G. Mayhew, J. L. Peel and J. E. Stangroom, *Nature*, 1965, **208**, 769

52. M. Tanaka, T. Nakashima, A. M. Benson, H. F. Mower and K. T. Yasunobu, *Biochem. Biophys. Res. Commun.*, 1964, **16**, 422

53. M. Tanaka, A. M. Benson, H. F. Mower and K. T. Yasunobu, in *Non-Heme Iron Proteins. Role in Energy Conversion*, A. San Pietro (Ed), 1965, 221. Antioch Press, Yellow Springs

54. R. D. Gillard, R. Mason, S. G. Mayhew, J. L. Peel and J. E. Stangroom, *Nature*, 1966, **211**, 769

55. G. Palmer, R. H. Sands and L. E. Mortenson, *Biochem. Biophys. Res. Commun.*, 1966, **23**, 357
56. B. E. Sobel and W. Lovenberg, *Biochemistry*, 1966, **5**, 6
57. R. G. Bartsch, in *Bacterial Photosynthesis*, 1963, 315. H. Gest, A. San Pietro and L. P. Vernon (Eds), Antioch Press, Yellow Springs
58. K. T. Fry, R. A. Lazzarini and A. San Pietro, *Proc. Natl. Acad. Sci. U.S.*, 1963, **50**, 652
59. G. Palmer and R. H. Sands, *J. Biol. Chem.*, 1966, **241**, 253
60. K. Suzuki and T. Kimura, *Biochem. Biophys. Res. Commun.*, 1963, **19**, 340
61. H. Watari and T. Kimura, *Biochem. Biophys. Res. Commun.*, 1966, **24**, 106
62. H. Brintzinger, G. Palmer and R. H. Sands, *Proc. Natl. Acad. Sci. U.S.*, 1966, **53**, 397
63. J. F. Gibson, D. O. Hall, J. H. M. Thornley and F. R. Whatley, *Proc. Natl. Acad. Sci. U.S.*, 1966, **56**, 987
64. T. C. Hollocher, F. Solomon and T. E. Ragland, *J. Biol. Chem.*, 1966, **241**, 3452
65. G. Palmer, *Biochem. Biophys. Res. Commun.*, 1967, **27**, 315
66. B. A. Goodman and J. B. Rayner, *J. Chem. Soc.*, (A) 1970, 2038
67. R. Malkin and J. C. Rabinowitz, *Biochemistry*, 1966, **5**, 1262
68. S. Keresztes-Nagy and E. Margoliash, *J. Biol. Chem.*, 1966, **241**, 5955
69. R. W. Miller and V. Massey, *J. Biol. Chem.*, 1965, **240**, 1453
70. F. R. N. Gurd and G. T. Bryce, in *The Biochemistry of Copper*, J. Peisach, P. Aisen and W. E. Blumberg (Eds), 1966, 115, Academic Press, New York
71. J. Peisach, P. Aisen and W. E. Blumberg (Eds), *The Biochemistry of Copper*, Academic Press, New York, 1966
72. A. S. Brill, R. B. Martin and R. J. P. Williams, Copper in Biological Systems, in *Electronic Aspects of Biochemistry*, B. Pullman (Ed), Academic Press, New York, 1966
73. F. Ghiretti (Ed), *Physiology and Biochemistry of Hemocyanins*, Academic Press, New York, 1968
74. H. Porter in *The Biochemistry of Copper*, J. Peisach, P. Aisen and W. E. Blumberg (Eds), 1966, 159, Academic Press, New York
75. B. L. Vallee and W. E. C. Wacker, in *The Proteins*, H. Neurath (Ed), (Second Edition) Volume V, *Metalloproteins*, 100, Academic Press, New York, 1970
76. S. Katoh and A. San Pietro, in *The Biochemistry of Copper*, J. Peisach, P. Aisen, and W. E. Blumberg (Eds), 1966, 407, Academic Press, New York
77. W. G. Levine, Ibid, p. 371
78. K. E. van Holde, *Biochemistry*, 1967, **6**, 93
79. L. Broman, B. G. Malmstrom, R. Aasa and T. Vanngard, *J. Mol. Biol.*, 1962, **5**, 301
80. R. Malkin and B. G. Malmstrom cited in L. E. Andreasson and T. Vanngard, *Biochim. Biophys. Acta*, 1970, **200**, 247
81. J. A. Fee, R. Malkin, B. G. Malmstrom and T. Vanngard, *J. Biol. Chem.*, 1969, **244**, 4200
82. T. Vanngard in *Magnetic Resonance in Biological Systems*, A. Ehrenberg, B. G. Malmstrom and T. Vanngard (Eds), 1967, p. 213, Pergamon, Oxford
83. B. G. Malmstrom, *Proc. VIII Int. Cong. of Biochem*, Switzerland, 1970, 124
84. B. G. Malmstrom, B. Reinhamnov and T. Vanngard, *Biochim. Biophys. Acta*, 1968, **156**, 967
85. L. E. Andreasson and T. Vanngard, *Biochim. Biophys. Acta*, 1970, **200**, 247

86. B. G. Malmstrom, A. Finazzi Agro and E. Antonini, *Eur. J. Biochem.*, 1969, **9**, 383
87. H. Beinert in *The Biochemistry of Copper*, J. Peisach, P. Aisen and W. E. Blumberg (Eds), 1966, 213, Academic Press, New York
88. T. E. King, H. S. Mason and M. Morrison (Eds), *Oxidases and related redox systems*, John Wiley, New York, 1965
89. B. F. van Gelder and E. C. Slater, *Biochim. Biophys. Acta*, 1963, **73**, 663
90. B. F. van Gelder and A. O. Muisjers, *Biochim. Biophys. Acta*, 1964, **81**, 405
91. D. C. Wharton and Q. H. Gibson, in *The Biochemistry of Copper*, J. Peisach, P. Aisen and W. E. Blumberg (Eds), 1966, 235, Academic Press, New York
92. P. M. Nair and H. S. Mason, *J. Biol. Chem.*, 1967, **242**, 1406
93. D. W. Brooks and C. R. Dawson, in *The Biochemistry of Copper*, J. Peisach, P. Aisen and W. E. Blumberg (Eds), 1966, 343, Academic Press, New York (and references cited)
94. M. W. Onslow and M. E. Robinson, *Biochem. J.*, 1928, **22**, 1327
95. H. Dressler and C. R. Dawson, *Biochim. Biophys. Acta*, 1960, **45**, 508, 515
96. H. S. Mason in *The Biochemistry of Copper*, J. Peisach, P. Aisen and W. E. Blumberg (Eds), 1966, 339, Academic Press, New York
97. D. Kertesz, ibid, 359
98. P. G. Lenhert and D. C. Hodgkin, *Nature*, 1961, **192**, 937
99. D. C. Hodgkin, *Proc. Roy. Soc.*, 1965, **A288**, 294
100. E. L. Smith, *Vitamin B_{12}*, 3rd Edition, Methuen and Co., London, 1965
101. R. Bonnett, *Chem. Rev.*, 1963, **63**, 573
102. Symposium on B_{12} Coenzymes, *Fed. Proc.*, 1966, **25**, 1623
103. G. N. Schrauzer, *Acc. Chem. Research*, 1968, **1**, 97
104. H. A. O. Hill, J. M. Pratt and R. J. P. Williams, *Chem. in Britain*, 1969, **5**, 156
105. R. J. P. Williams, *Roy. Inst. Chem. Reviews*, 1968, **1**, 13
106. P. G. Lenhert, *Chem. Comm.*, 1967, 980
107. A. Van den Bergen and B. O. West, *Chem. Comm.*, 1971, 52
108. H. P. C. Hogenkamp, H. A. Barker and H. S. Mason, *Arch. Biochim. Biophys.*, 1963, **100**, 353
109. P. K. Das, H. A. O. Hill, J. M. Pratt and R. J. P. Williams, *Biochim. Biophys. Acta*, 1967, **141**, 644
110. S. Cockle, Thesis, Oxford 1968, cited in Ref. 104.
111. G. N. Schrauzer and R. J. Windgassen, *J. Amer. Chem. Soc.*, 1966, **88**, 3738
112. R. A. Firth, H. A. O. Hill, J. M. Pratt, R. G. Thorp and R. J. P. Williams, *Chem. Comm.*, 1967, 400
113. H. R. Mahler and E. H. Cordes, *Biological Chemistry*, Harper and Row, 1968
114. E. C. Slater (Ed), *Flavins and flavoproteins*, Elsevier, Amsterdam, 1966
115. P. C. H. Mitchell, *Quart. Rev. Chem. Soc.*, 1966, **20**, 103; J. T. Spence and J. Y. Lee, *Inorg. Chem.*, 1965, **4**, 385
116. A. B. Blake, F. A. Cotton and J. S. Wood, *J. Amer. Chem. Soc.*, 1964, **86**, 3024
117. J. T. Spence, M. Heydanek and P. Hemmerich, in *Magnetic Resonance in Biological Systems*, A. Ehrenberg, B. G. Malmstrom and T. Vanngard (Eds), 1967, 269, Pergamon
118. P. Hemmerich, D. V. Der Vartanian, C. Veeger and J. P. W. Van Voorst, *Biochim. Biophys. Acta*, 1963, **77**, 507
119. F. Muller, A. Ehrenberg and L. E. G. Eriksson, in *Magnetic Resonance in Biological Systems*, A. Ehrenberg, B. G. Malmstrom and T. Vanngard (Eds), 1967, 281, Pergamon, Oxford

120. P. F. Knowles and H. Diebler, *Trans. Far. Soc.*, 1968, **64**, 977
121. T. J. Huang and G. P. Haight, Jr., *Chem. Commun.*, 1969, 985; *J. Amer. Chem. Soc.*, 1970, **92**, 2336
122. T. J. Huang and G. P. Haight, Jr., *J. Amer. Chem. Soc.*, 1971, **93**, 611
123. K. V. Rajagopalan, I. Fudovich and P. Handler, *J. Biol. Chem.*, 1962, **237**, 922
124. P. Handler, K. V. Rajagopalan and V. Aleman-Aleman, *Fed. Proc.*, 1964, **23**, 30
125. R. C. Bray, G. Palmer and H. Beinert, *J. Biol. Chem.*, 1964, **239**, 2667
126. H. Beinert and G. Palmer, *Advan. Enzymol*, 1965, **27**, 105
127. R. C. Bray and D. C. Watts, *Biochem. J.*, 1966, **98**, 142
128. H. Beinert and P. Hemmerich, *Biochem. Biophys. Res. Comm.*, 1965, **18**, 212
129. P. F. Knowles, in *Magnetic Resonance in Biological Systems*, A. Ehrenberg, B. G. Malmstrom and T. Vanngard (Eds), 1967, **265**, Pergamon, Oxford
130. R. C. Bray, P. F. Knowles and L. S. Meriwether, in *Magnetic Resonance in Biological Systems*, A. Ehrenberg, B. G. Malmstrom and T. Vanngard (Eds), 1967, **249**, Pergamon, Oxford
131. L. S. Meriwether, W. F. Marzluff and W. G. Hodgson, *Nature*, 1966, **212**, 465
132. R. C. Bray and P. F. Knowles, *Proc. Roy. Soc.*, 1968, **A302**, 351
133. A. E. Dennard and R. J. P. Williams, in *Transition Metal Chemistry*, R. L. Carlin (Ed), 1966, **2**, 158
134. K. Garbett, R. D. Gillard, P. F. Knowles and J. E. Stangroom, *Nature*, 1967, **215**, 824

CHAPTER SIX

NITROGEN FIXATION AND THE NITROGEN CYCLE

The element nitrogen is a constituent of many naturally occurring compounds. These complex molecules are built up from smaller organic molecules, while the original source of the element lies in simple inorganic compounds. The 'Nitrogen Cycle' represents the transformation of inorganic nitrogen to organic nitrogen, together with the reverse degradation process. In this cycle, soil ammonia is converted to nitrate, *via* a number of intermediates, by the action of microorganisms in the process of nitrification. Nitrate is assimilated by plants and built up into organic molecules. Higher species feeding on these plants convert such molecules into more complex compounds. Ultimately, on the death and decay of living matter, these compounds are broken down, leading to the reformation of ammonia. There is also the process of denitrification in which nitrates are reduced by microorganisms; in this some nitrogen is lost to the atmosphere, but in turn atmospheric nitrogen is 'fixed', that is converted to ammonia, by various bacteria.

The nitrogen cycle is of vital importance in agriculture in supplying soluble nitrates. Unless nitrates are extensively added as fertilizers the important step is that of nitrogen fixation. Nitrogen may be fixed industrially by a number of processes, of which the Haber process is most important, but it is obviously of great interest to investigate the mechanism of biological fixation of nitrogen partly in the hope of establishing new commercially viable processes and also for an enhanced understanding of the fundamental chemical and biochemical processes involved. For at present man can only do with great difficulty under extreme conditions what Nature does apparently quite readily under mild aqueous conditions with no extremes of pH and temperature. The particular aspects of the problem of interest to the inorganic chemist lie not only in the general difficulties associated with the reactivity of what has always been traditionally regarded as an inactive molecule, but also in the fact that the nitrogen-reducing species in nature, the enzyme nitrogenase, requires the presence of the transition metals molybdenum and iron. This has led investigators in recent years to postulate model systems for the reduction of nitrogen to ammonia in which the nitrogen molecule is coordinated to a transition metal ion. This has been stimulated by recent developments which have led to the preparation of dinitrogen complexes of a number of transition metals. It should be pointed out, however, that at present there is no biochemical evidence for the formation of metal complexes in the biological fixation of nitrogen.

While less dramatic than the area of nitrogen fixation, there are also a number of problems in the nitrogen cycle itself that should be of interest to inorganic chemists. A number of intermediates must lie between ammonia and nitrate, in both nitrification and denitrification pathways. Hydroxylamine and nitrite (with nitrogen in formal oxidation states $-I$ and $+III$ respectively) are now well established as intermediates, but the nature of the species with nitrogen in the oxidation state $+I$ is less certain (assuming that changes of two in oxidation states are occurring). The chemistry of possible intermediate compounds is not very well established, particularly under the pH conditions of biological processes, and much remains to be done. Some of the biochemical processes involved in these oxidation and reduction reactions are now being investigated at the molecular level. Thus a hydroxylamine oxidase has been isolated from *nitrosomonas*; this is an electron transfer particle containing cytochromes and a copper metalloenzyme. Again, in denitrification, the reduction of nitrate to nitrate is dependent on the presence of molybdenum, with the metal undergoing transition between the oxidation states VI and V. Here again is a potentially useful area of overlap between biochemistry and inorganic chemistry.

In this chapter, we will consider recent developments in the biochemistry and inorganic chemistry of both nitrogen fixation and, to a lesser extent, the nitrogen cycle. In each case a summary of the relevant biochemistry will be given first.

THE BIOCHEMISTRY OF NITROGEN FIXATION

Nitrogen-fixing microorganisms

These fall into two main groups.

(a) Free bacteria (asymbiotic). A well known example is *azotobacter*, but this is atypical in one sense in that this is an aerobic microbe, one requiring oxygen. A very much larger number of anaerobic species fix nitrogen in the absence of oxygen, including *Clostridium pasteurianum*, on which many of the earlier studies were carried out. A third type of free bacteria which fix nitrogen is the facultative group, those that have the property of being able to grow aerobically or anaerobically. These can only fix nitrogen when growing in the absence of oxygen as their biochemistry will depend upon the conditions. This general dependence upon the absence of oxygen seems intuitively reasonable as the process concerned is a reductive one.

(b) Symbiotic microorganisms. These fix nitrogen in association with plants, e.g. the bacterium *Rhizobium*, which is associated with the nodules on the roots of leguminous plants. In this case, the microorganism does not function until it has become a degenerate form inside the root nodule. Other examples include the lichens, where there is a combination of a fungus and nitrogen fixing blue-green algae. In general much less progress has been made with the study of nitrogen fixation by the symbiotic bacteria. One common feature of these root nodule systems is the presence of a hemoglobin-like protein which has been isolated and purified. This has been named leghemoglobin and contains one atom of iron per molecule. There appears to be a direct correlation between the presence of leg-hemoglobin and nitrogen-fixing properties.

TABLE 6.1 Some biological nitrogen-fixing systems

Asymbiotic	Symbiotic

(a) Anaerobic
    C. pasteurianum
    Chromatium
    Chlorobium (photosynthetic)
    Methanobacterium
    Desulphovibrio sp.
(b) Aerobic
    Azotobacter
    Anabena (photosynthetic)
    Azotomonas
    Beijerinckia
    Derxia
    Nocardia
    Pseudomonas
(c) Facultative
    B. polymyxa
    Klebsiella sp.
    Rhodospirillum (photosynthetic)
	(a) Root nodules
	    legumes, *Rhizobium* and
	    leguminous plant
	    non-legumes, microbe and tree
	    or shrub
	(b) Leaf nodules
	    bacteria and leaf of *Psychotria*
	    emetica
	(c) Lichens
	    Nostic, Anabena, Tolypothrix
	    and fungus
	(d) Mycorrhiza of pines
	    fungus and tree

Both classes of microorganisms appear to convert nitrogen directly to ammonia. There is no evidence for the intermediate formation of any other species, although diimine, $HN{=}NH$, has often been suggested. Ammonia is the first compound to be detected using nitrogen-15 techniques. Table 6.1 lists a number of nitrogen-fixing species.

Results obtained with whole cells

Until about 1960 little progress had been made towards the isolation of enzyme systems that could reproducibly fix nitrogen. This is at least partly reflected in the rate of progress in the understanding of nitrogen fixation that was being made up to that time. Even so, a number of important observations had been made using intact organisms. Thus the use of nitrogen-15 had shown the important role of ammonia as the first product of nitrogen fixation. Again, it had been shown that hydrogen was an inhibitor for nitrogen fixation, and that nitrogen-fixing systems contained hydrogenase. Inhibition by other molecules, nitrous oxide and carbon monoxide, had also been demonstrated, as also had the requirement for molybdenum and iron. Nitrogen-fixing organisms appear to contain other metals as well, but these are involved with associated enzyme systems and not directly with nitrogen fixation.

Fixation by cell-free extracts

The first substantial progress[2] in this direction involved the successful isolation of an extract from *Clostridium pasteurianum* which could reproducibly fix nitrogen, provided that oxygen was completely eliminated and that substantial quantities of sodium pyruvate were present. The extract itself could be divided into two fractions, one of which was the 'hydrogen donating system', abbreviated to HDS, involving the hydrogenase and the pyruvate metabolizing system, while the other was the nitrogen activating system (NAS). Both these fractions were air sensitive. The role of the pyruvate was two-fold; firstly in supplying the hydrogen for the stoicheiometric conversion of nitrogen to ammonia, and secondly in providing the necessary energy for the reaction, *via* ATP synthesis. It is interesting to note that the pyruvate-containing fraction, HDS, was shown to contain the non-heme iron protein ferredoxin, this being one of the first reports of its existence. In fact, ferredoxin is not directly involved in nitrogen fixation, but acts as an electron acceptor for pyruvate. It is still the only clearly demonstrated naturally occurring electron carrier operative in the process of nitrogen fixation. It may well be that in symbiotic bacteria the role of leghemoglobin is similar to that of ferredoxin in the whole cells of asymbiotic microorganisms.

In the example just cited, pyruvate supplies both cofactor requirements for the process. It is possible, however, to vary this. Thus, as an alternative to pyruvate, the reaction will also proceed in the presence of gaseous hydrogen, which will reduce ferredoxin, and the ATP generating systems of creatine phosphate, creatine phosphokinase and adenosine diphosphate ADP. It is usual now in studies on cell free extracts to provide sodium dithionite as the electron donor, in which case the ferredoxin pathway is by-passed. In blue-green algae there is a relationship between photosynthesis and nitrogen fixation; here the requirements of reducing power and ATP are met photochemically. A number of estimates of the ATP requirements for nitrogen fixation by various microorganisms have been made, and it appears that about two moles of ATP are required for each electron transferred. There is no real understanding at present of the details of the role of ATP, apart from the general energy requirement.

Further work[3] on the nitrogen-fixing system of C. pasteurianum has shown the presence of two metalloproteins, both oxygen sensitive, one containing both molybdenum and iron, and the other containing iron only. A similar result has been obtained[4] for the aerobic species Azotobacter Vinelandii. Two metalloproteins have again been isolated, one containing molybdenum and iron, and the other iron. These proteins are also oxygen sensitive and must be handled in the absence of air. They are isolated from particle components of cell free extracts which are in themselves stable in air. In the case of A. Vinelandii, the physiological reducing agent is not pyruvate. Usually in studies on this system too, dithionite is used as the reducing agent and ATP generated by the appropriate enzyme system. Neither of these two metalloproteins will fix nitrogen independently, nor will they carry out any of the related phenomena associated with nitrogen fixation. They will readily recombine, however, reforming the nitrogen-fixing system, and in fact, in the presence of dithionite and the ATP generating system, are most efficient fixers of nitrogen. A number of studies have been carried out on the composition of these proteins, but there have been conflicting reports on their molecular weights and on the metal ion content. The molybdenum-iron protein (fraction 1) of A. vinelandii nitrogenase has been isolated[4a] in crystalline form. It has a molecular weight between 270,000 and 300,000 and contains molybdenum, iron, cysteine and labile sulphide in the ratio 1:20:20:15. The second protein (fraction 2) has a molecular weight of about 40,000 and contains two iron atoms and two sulphide groups per molecule. The second protein is labile but can be stored at liquid nitrogen temperatures. It is particularly sensitive to oxygen. It is interesting that such a system should occur in an aerobic microorganism. Postgate[5] has discussed how Azotobacter could exclude oxygen but not nitrogen from this metalloprotein. He has suggested that the well-known extremely high energy requirements of these species provide a means of protecting the nitrogen-fixing site by utilizing the oxygen in other reactions.

It is possible to add fraction 1 obtained from one species to fraction 2 isolated from a second species and regenerate an active preparation, provided that the two species are not too dissimilar. It is not possible to combine fractions from aerobic and anaerobic species.

These cell-free preparations of nitrogenase are not specific for nitrogen reduction and can reduce a number of compounds that are analogous to nitrogen. These are listed below in Table 6.2. It is noteworthy that acetylene is reduced to ethylene and not ethane. Carbon monoxide is not reduced but it does inhibit the reduction of ammonia, indicating that it interacts with the system, although not reduced by it. The similarity between all these compounds suggests that they are all bound to the nitrogen-fixing site during enzymatic reduction.

TABLE 6.2 The reduction of various substrates by nitrogenase

Substrate	Product
N_2	NH_3
N_2O	$N_2 + H_2O$
N_3^-	$NH_3 + N_2$
CN^-	$CH_4 + NH_3$
$CH{\equiv}CH$	$CH_2{=}CH_2$
$CH_3C{\equiv}CH$	$CH_3CH{=}CH_2$
CH_3NC	$CH_3NH_2 + CH_4$

The result for the reduction of methyl isocyanide (CH_3NC) is interesting. It is usual in the chemical reduction of this species to obtain dimethylamine as product. This is not the product of enzymatic reduction. However, it has been shown[6] that the reduction of isocyanide coordinated to a transition metal ion does lead to the formation of methane as a major product. This suggests that a transition metal may be directly involved in the biological reduction.

All these reactions are accompanied by the evolution of hydrogen gas. In nitrogen fixation, this reaction competes with nitrogen reduction for electrons and this is reflected in a slowed rate of hydrogen evolution. Reference has been made to the fact that carbon monoxide inhibits nitrogen reduction; it does not, however, inhibit hydrogen evolution but chelating ligands inhibit both nitrogen fixation and hydrogen production. This fact has led to the suggestion that one metal is present as a hydride and that the other binds and activates the nitrogen molecule. The metal hydride may then react with the nitrogen complex giving an adduct that may be further reduced to ammonia.

$$-\overset{|}{\underset{|}{M}}-H + N{\equiv}N-\overset{|}{\underset{|}{M'}}- \longrightarrow -\overset{|}{\underset{|}{M}}-N{=}\overset{H}{\underset{|}{N}}-\overset{|}{\underset{|}{M}}-$$

The problem of the overall role of hydrogen in biological nitrogen fixation is complex. It has been shown that there are two hydrogen evolving systems, one ATP independent and the other ATP dependent. These results serve to emphasize the complexity of the biochemistry of nitrogen fixation. Little has been done towards investigating the metal-binding sites for these two metalloproteins and the effect of the protein on determining the electronic properties of the metal. One would anticipate that the iron in the iron protein is sulphur bonded.

THE CHEMISTRY AND STRUCTURE OF THE NITROGEN MOLECULE

A basic aspect of the problem of nitrogen fixation must be that of the structure of the molecule itself and the way in which it may be activated prior to chemical reaction. Until recently nitrogen has always been regarded as an inert molecule, its inertness being one of its characteristic features. The only reaction that it readily undergoes is the formation of nitrides, particularly with the most electropositive metals. Thus, it reacts directly with lithium at room temperature and under certain conditions with magnesium. In addition there are the catalytic processes such as the Haber process which involve the use of high temperature and pressure and appropriate catalysts. However, against this background of lack of reactivity, it is of great interest to note the increasing number of dinitrogen complexes that have been recently prepared.

The inertness of nitrogen has been discussed by Chatt and Leigh[7] and again by Chatt.[8] From some points of view it might be thought that its inertness is unexpected, for a number of apparently closely analogous triply bonded molecules are quite reactive. Thus acetylene can be reduced readily while carbon monoxide, isoelectronic with nitrogen, is well known for forming a vast number of carbonyl complexes. On the other hand, an obvious and important difference between the nitrogen molecule and carbon monoxide is that the former molecule is a non-polar species.

A study of the physical properties of the nitrogen molecule will indicate the feasibility of the different types of reaction that nitrogen could undergo. The high dissociation energy of 224 kcal mole^{-1} precludes the possibility of reaction *via* dissociation into nitrogen atoms, while the first ionization potential of 15·5 eV is appreciably higher than those observed for the analogous molecules already mentioned, so that oxidation would not be expected to occur under biological conditions. It should be mentioned, however, that both these processes, dissociation and oxidation (giving N_2^+), do occur under certain conditions. So-called 'active nitrogen', nitrogen which has been subjected to ionizing radiation or electric discharge at high temperatures, contains nitrogen atoms and, under certain conditions, N_2^+. Such 'active nitrogen' undergoes a number of reactions but this is usually due to the presence of atomic nitrogen and is not of great relevance to the chemical activation of molecular nitrogen.

The possibility of the reduction of nitrogen as a reaction pathway is clearly more feasible than these other two possibilities, as is evidenced by the reaction with lithium. Even so, such reactions as these plainly cannot be considered as a possible means of activating nitrogen under the aqueous conditions of nitrogen fixation, for such strong reducing agents as these would preferentially react with the water of the environment.

Despite this apparent lack of promise of reaction *via* routes involving dissociation, oxidation and reduction, a comparison with the chemistry of carbon monoxide suggests that it is not surprising that dinitrogen complexes have been formed. This certainly would offer a means of activating the nitrogen molecule, as will be discussed later. Carbon monoxide is a generally inert molecule, with poor donor properties in terms of Lewis Basicity. Yet carbonyl complexes are extremely well known. This can be rationalized in the case of carbon monoxide by the well-known synergic effect, in which the essentially weak σ bonding and the π bonding effects mutually reinforce each other. It is possible to apply a similar argument in the case of dinitrogen complexes. First we will consider the detailed molecular orbital picture of the nitrogen molecule. Here it is known that in addition to the formation of the π-molecular orbitals, there is overlap of the $2s$ atomic orbital of each nitrogen atom with the $2p_x$ orbital of the second nitrogen atom, x being the molecular axis, so giving two sets of bonding and antibonding σ molecular orbitals. This is shown in Figure 6.1. An alternative scheme would have involved the overlap of equivalent orbitals from the two atoms, i.e. $2s$ with $2s$ etc. This would have resulted in a

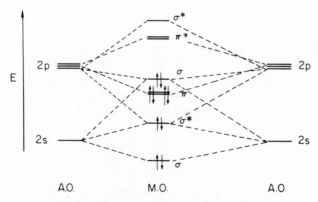

Figure 6.1 Energy levels in the nitrogen molecule.

different energy sequence for the molecular orbitals, with the highest filled molecular orbitals, as in carbon monoxide, being the degenerate π orbitals. That Scheme 6.1 does represent the nitrogen molecule is confirmed by the fact that it has been shown spectroscopically that the species N_2^+ has its unpaired electron in a σ orbital. In Figure 6.1 the $1s$ orbitals are ignored. The highest filled molecular orbital is now a bonding σ orbital which will lie deep in the molecule, shielded by the π-molecular orbitals, which are of low energy and strongly bonding. This means that the donor properties of the σ orbital will be limited and explains why the first ionization potential is high. It is then possible to write down a bonding scheme for the formation of the dinitrogen-transition metal complex, analogous to that for a carbonyl complex, in which the nitrogen molecule will be bonded end on to the metal with $N_2 \rightarrow$ metal σ bonding, and metal $\rightarrow N_2$ π-bonding *via* overlap of the metal d orbitals and the appropriate π^* M.O. of the nitrogen molecule (Figure 6.2). The overall result of this

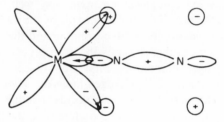

Figure 6.2   Bonding scheme for a transition metal–dinitrogen complex.

bonding is that electron density is removed from the bonding σ orbital of nitrogen and transferred *via* the metal ion to the antibonding π orbital of nitrogen. This means that there is a doubly weakening effect on the nitrogen–nitrogen bond; indeed, the metal–nitrogen bond is produced at the expense of the nitrogen–nitrogen bond. The stronger the former then the weaker the latter bond, and the more likely that it may be involved in chemical reaction. In addition, the result of the electronic effects will be seen in an increasing polarization of the nitrogen molecule, again possibly making it more susceptible to attack. All this is confirmed by spectroscopic studies.

Normally the nitrogen molecule has a Raman active nitrogen–nitrogen stretching frequency at 2331 cm^{-1}. This vibration is not active in the infrared as there is no associated change in dipole moment. However, the infrared spectra of metal-dinitrogen complexes usually show a strong band at around 2100 cm^{-1} assigned to $\nu_{N \equiv N}$, although in fact a fairly wide range of values has been noted. Two points are important. First, the activity of the vibration in the infrared confirms that the nitrogen molecule is now dipolar and secondly, the reduced frequency indicates a reduction in bond strength. It is also of interest to note that in the complexes $[Ru(NH_3)_5(N_2)]X_2$, $v_{N\equiv N}$ varies with the nature of the anion, ranging from 2105 (X = Cl) to $2167\,cm^{-1}$ (X = PF_6). This phenomenon also confirms the dipolar character of coordinated nitrogen. Values of $v_{N\equiv N}$ are a useful indication of the bond-strength of the nitrogen molecule.

The much greater ease of formation of carbonyl complexes can be understood in terms of the rather different molecular orbital scheme for carbon monoxide which results in a σ donor and π acceptor orbitals of more appropriate energy being available on the ligand.

A number of workers have isolated dinitrogen complexes of the transition elements and have attempted to reduce the coordinated nitrogen to ammonia, without success at present. Many more complexes have been observed in solution and a number of suggestions made regarding possible models for nitrogenase. It should be emphasized that prior to the isolation of these nitrogen complexes, it has been known that certain organometallic species can effect the reduction of nitrogen to ammonia in the presence of strong reducing agents, usually under anhydrous conditions. Recently evidence has been presented for the formation in benzene solution of a titanium(II) complex with molecular nitrogen, where the nitrogen can be converted to ammonia on reaction with strong reducing agents. In this case the nitrogen molecule is only weakly bound to the metal and is readily released.

TRANSITION METAL COMPLEXES OF DINITROGEN

In December 1965, Allen and Senoff[9] reported the preparation of the first well characterized molecular nitrogen–transition metal complexes, the species $[Ru(NH_3)_5(N_2)]X_2$, X = Cl, Br, I, BF_4, PF_6. The first complex was obtained by the action of hydrazine on ruthenium(III) chloride, in an attempt to prepare an ammine complex, although an ill-characterized complex had already been prepared[10] by zinc dust reduction of a THF solution of ruthenium(III) chloride under nitrogen. The presence of nitrogen in Allen and Senoff's species was confirmed by a number of reactions in which the nitrogen was displaced by other ligands (e.g. NH_3, Cl^-, pyridine), collected and analysed in a mass spectrometer. The assignment of the strong band at about $2100\,cm^{-1}$ in the infrared spectrum to $v_{N\equiv N}$ was confirmed by repeating the preparation with deuterated hydrazine and showing that the band was unchanged and not therefore due to a hydride species. Since this time a fair number of other complexes have been reported and doubtless many more will be prepared. The availability of a range of dinitrogen complexes with different types of ligand and metal ions will allow greater insight into their nature and structure.

Many of the early complexes were prepared accidently but there are now more clear-cut preparative routes. Thus, one route involves the reduction of a transition metal complex in the presence of nitrogen, giving a low valent metal dinitrogen complex. Others involve the decomposition of certain nitrogen-containing coordinated ligands. Dinitrogen complexes have been prepared with a range of metals, in a number of oxidation states (Ru, Os, Mo, Co, Rh, Ir, Fe, Re). Mono, bis (cis and trans) and bridging six-coordinate species have been prepared, with either phosphine or amine ligands. Usually the preparation of the complex is very dependent on the nature of the ligand, slight variations resulting in a failure to prepare them. However, there are now a number of cases in which a range of phosphines have been successfully used, e.g. $[CoH(N_2)(PR_3)_3]$[11] and $[OsX_2(N_2)(PR_3)_3]$[12] where X = Cl, Br. A number of crystal structures have been determined and will be discussed in detail in a later section.

None of the complexes prepared so far have given positive results for the attempted reduction of the nitrogen. An earlier report had suggested that the species $[Ru(NH_3)_5(N_2)]^{2+}$ could be reduced with sodium borohydride but it appears that this was due to the reduction of traces of a hydrazine complex carried through from the preparation as an impurity. Studies on a nitrogen-15 labelled dinitrogen complex gave no evidence for reduction of coordinated nitrogen. As has already been seen, the lower the value of $v_{N\equiv N}$ for the complex, then the more likely that a complex can be reduced. The lowest value of $v_{N\equiv N}$ observed so far is for the species $(PhMe_2P)_4ClRe-N\equiv N-MoCl_4$, where $v_{N\equiv N}$ is reduced down to 1680 cm^{-1}. In this case the frequency is of the same order as that observed for azo compounds which can be reduced chemically. However, no success was achieved in this instance. It is of interest to note, as will be discussed later, that a common feature of many suggested models for nitrogen fixation is the postulation of a nitrogen molecule bridging two metal ions.

No attempt will be made to review the dinitrogen complexes prepared so far, although a reasonably comprehensive list is given in Table 6.3 together with values of the nitrogen–nitrogen stretching frequency. It should be noted that dinitrogen complexes of molybdenum and iron are known.

Structural studies on transition metal complexes of dinitrogen

The structures of a number of dinitrogen complexes have been studied. In general, the results have confirmed the predictions made on the

TABLE 6.3 Dinitrogen complexes of the transition metals

Complex	Reference	$\nu_{N\equiv N}$ cm^{-1}	
[Ru(NH$_3$)$_5$(N$_2$)]X$_2$	9, 13, 14, 15	2105–2167	X = Cl, Br, I, BF$_4$, PF$_6$
[(NH$_3$)$_5$Ru–N$_2$–Ru(NH$_3$)$_5$](BF$_4$)$_4$	16	Raman 2100	—
(Ph$_3$P)$_4$RuH(N$_2$)	18	—	—
[Os(NH$_3$)$_5$(N$_2$)]$^{2+}$	19, 20	2033	—
[Os(NH$_3$)$_4$(N$_2$)$_2$]$^{2+}$X$_2$	21	2147	X = Cl, Br, I, cis
OsX$_2$(N$_2$)(PR$_3$)$_3$	12	2090–1998	X = Cl, Br, various PR$_3$
OsCl$_2$(N$_2$)(PBu$_2$Ph)$_3$	22	—	—
[Co(N$_2$)(PPh$_3$)$_3$]	23	2088	—
CoH(N$_2$)L$_3$	25, 26	2080	L = PPh$_3$, PEtPh$_2$
IrCl(N$_2$)(PPh$_3$)$_2$	27	2095	Also olefin adduct $\nu_{N\equiv N}$ 2190
RhCl(N$_2$)(PPh$_3$)$_2$	28	—	—
ReCl(N$_2$)(PR$_3$)$_4$	29	→1810	Various phosphines
Re(CO)$_3$(NH$_2$)$_2$(N$_2$)L	30	2220	L = PMe$_2$Ph
Re(CO)$_2$(NH$_2$)(N$_2$)L$_2$	30	2225	—
Mo(N$_2$)(PPh$_3$)$_2$·toluene	31	2005	—
Mo(diphos)$_2$(N$_2$)$_2$	32	2020w, 1970	trans
(PMe$_2$Ph)$_4$ClRe–N≡N–MeCl$_4$(PEtPh$_2$)	33	→1680	Variety of similar compounds Me = metal
FeH$_2$(N$_2$)L$_3$	a	2065–1989	L = PEtPh$_2$, PMePh$_2$, PBuPh$_2$
MoCl(N$_2$)(diphos)$_2$	b	1970	diphos = Ph$_2$PCH$_2$CH$_2$PPh$_2$
MoCl(N$_2$)(diphos)$_2$ + MoCl$_4$(THF)$_2$	b	1770, 1720	Probably dinitrogen bridged dinuclear complex
t[RuCl(N$_2$)(das)$_2$][PF$_6$]	c	2029	das = o-phenylene-bis(dimethyl arsine)
[C$_6$H$_6$Mo(PPh$_3$)$_2$]$_2$N$_2$	d	1910 (Raman)	
Ni(N$_2$)$_x$ (x probably 1 or 2)	e	2169, 2180	—
(Ph$_3$P)$_3$Co(N$_2$)$_2$Co(PPh$_3$)$_3$	f	No IR active band	Co(0) complex
NaCo(N$_2$)(PPh$_3$)$_3$	f	1840 cm^{-1}	Co(−1) complex

[a] M. Aresta, P. Giannoccaro, M. Rossi and A. Sacco, *Inorg. Chim. Acta*, 1971, **5**, 203.
[b] L. K. Atkinson, A. H. Mawby and D. C. Smith, *Chem. Comm.*, 1971, 157.
[c] P. G. Douglas, R. D. Feltham and H. G. Metzger, *J. Amer. Chem. Soc.*, 1971, **93**, 84.
[d] M. L. H. Green and W. E. Silverthorn, *Chem. Comm.*, 1971, 557.
[e] J. K. Burdett and J. J. Turner, *Chem. Comm.*, 1971, 885.
[f] M. Aresta, C. F. Nobile, M. Rossi and A. Sacco, *Chem. Comm.*, 1971, 781.

basis of bonding arguments. For both $[CoH(N_2)(PPh_3)_3]^{34}$ and $[Ru(NH_3)_5(N_2)]X_2^{35}$ it has been shown that the nitrogen molecules are bonded end-on and that the metal–nitrogen–nitrogen link is approximately linear. Ibers and his coworkers have shown that two modifications of the cobalt complex exist, in which the cobalt atoms have a trigonal bipyrimidal arrangement of ligands (Figure 6.3) but with slightly different bond lengths and angles in the two cases. The Co–N–N angle is 178°, while the N–N distances of 0·1123 nm and 0·1101 nm for the two complexes do not differ very much from that for free nitrogen, 0·10976 nm. This result is similar to that for carbonyl complexes where there is no significant lengthening of the carbon–oxygen bond on coordination.

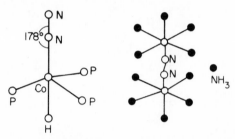

Figure 6.3 Structures of $Co(N_2)H(PPh_3)_3$ and $[(NH_3)_5Ru(N_2)Ru(NH_3)_5](BF_4)_4$.

The structure of the binuclear complex $[(NH_3)_5RuN_2Ru(NH_3)_5]^{4+}$ $(BF_4)_4$ has also been reported.[36] The Ru–N–N–Ru link (Figure 6.3) is again approximately linear with a Ru–N–N angle of 178.3°. The N–N distance is 0·1124 nm, similar to that observed for the cobalt complex.

General comments on transition metal complexes of molecular nitrogen

The ability of the nitrogen molecule to coordinate to a transition metal ion along the lines outlined earlier in this chapter is now well established. Despite the fact that there is no case where the nitrogen molecule can be reduced to ammonia using mild reducing agents, it is clear that the N–N bond is weakened by coordination as is indicated by the position of the nitrogen–nitrogen stretching frequency in the infrared spectrum. The fact that the metal–nitrogen bond is strengthened at the expense of the nitrogen–nitrogen bond is confirmed by a consideration of the stability of those nitrogen complexes which decompose by dissociative mechanisms. Thus the complex cation $[Os(NH_3)_4(N_2)_2]^{2+}$, with $\nu_{N\equiv N} = 2147 \, cm^{-1}$, decomposes at approximately 50°, while the cation $[Os(NH_3)_5(N_2)]^{2+}$ with $\nu_{N\equiv N} = 2033 \, cm^{-1}$ decomposes at over 300°.

Chatt, Richards, Sanders and Fergusson[31] have investigated the effect of adding strong acids to $CoH(N_2)(PPh_3)_3$, $[Ru(NH_3)_5(N_2)]X_2$, (X = Cl, BF_4), $Co(N_2)(PPh_3)_3$ and trans-$[IrCl(N_2)(PPh_3)_2]$ in the hope that reduction to ammonia might occur by protonation of the coordinated nitrogen with electron release from the metal, resulting in simultaneous oxidation of the metal and reduction of the nitrogen. In almost every case, however, the nitrogen was lost without reduction. The species $[Ru(NH_3)_5(N_2)]Cl_2$ did apparently give ammonia, but, as in the instance when reduction with borohydride was attempted, this arises from the presence of a hydrazine impurity. This is confirmed in the present case by the fact that use of $[Ru(NH_3)_5(^{15}N_2)]Cl_2$ resulted in the recovery of non-enriched ammonia. Obviously, this problem of reduction with mild reducing agents is a major obstacle in attempting to establish model systems for nitrogenase activity involving coordinated dinitrogen.

NITROGEN FIXATION VIA NITRIDE FORMATION

Over the last seven or eight years there have been a number of reports of successful procedures for the conversion of nitrogen to ammonia. In some cases processes have been developed for the regeneration of the reducing agent so resulting in the existence of an overall catalytic process for the production of ammonia via coordination, reduction and protonation of the nitrogen molecule. The chemistry of these processes is most interesting, but their relation to biological nitrogen fixation while possibly more realistic on first impression than the alternative approach involving stable dinitrogen complexes, is in fact a little uncertain as these reactions take place under anhydrous conditions with strong reducing agents and usually involve a hydrolysis step to liberate the ammonia. In the one case where there is evidence for the formation of a dinitrogen complex in solution (of Ti(II)), the nitrogen–nitrogen stretching frequency is not exceptionally low (1960 cm^{-1}), but the use of a much stronger reducing agent than would be allowed in aqueous solution allows its reduction.

The work began with the report of Vol'pin and Shur[39] of the reaction between transition metal halides and a reducing agent, such as a metal alkyl or hydride, in the presence of nitrogen (up to 150 atmospheres pressure) in an aprotic solvent, usually ether. The products on acid hydrolysis yielded ammonia, experiments[40] using $^{15}N_2$ confirming that this was derived from the nitrogen. It has since been shown[41] that the high pressure was unnecessary. Vol'pin and Shur have studied a wide range of reactants and have obtained maximum yields of 1·3 moles of ammonia per mole of transition metal ion. The more successful preparations were those involving the reduction of biscyclopentadienyltitanium dichloride with, for example, the Grignard reagent ethylmagnesium bromide or phenyllithium. The review article of Murray and Smith[1] discusses a

$$(Cp)_2TiCl_2 + EtMgBr \xrightarrow[ether]{N_2} Product \xrightarrow{acid} NH_3$$

number of aspects of this work.

The presence of hydride intermediates, for example $[(\pi C_5H_5)_2TiH_2]^-$, has been demonstrated by the use of EPR techniques but the detailed course of the reaction has not been clarified as yet. Brintzinger[43] has suggested a scheme for enzymatic reduction of nitrogen based upon nitrogen insertion into metal-hydride bonds.

Henrici–Olive and Olive[44] have also examined the reaction of metal halides with the strong reducing agent lithium naphthalide in THF. The reduced species will react with nitrogen leading, on electron transfer from the metal to nitrogen, to the formation of a complex nitride, which is converted to ammonia on hydrolysis. In their studies the best yield of ammonia

$$N_2 \xrightarrow{6\varepsilon} 2N^{3-} \xrightarrow{6H^+} 2NH_3$$

(2.0 moles per mole of metal) was obtained on the hydrolysis of the product obtained on treating vanadium(III) chloride with excess lithium naphthalide under a nitrogen pressure of 120 atmospheres for 30 minutes at 20°C.

van Tamelen and his coworkers have published a series of papers in which they have developed this type of reaction further. Initially[45] they carried out the reduction of dichlorodiisobutoxytitanium(IV) with potassium in diglyme or THF. On passing nitrogen through the solution ammonia was evolved. If, on the cessation of ammonia production, more potassium was added then the flow of ammonia recommenced. The source of hydrogen here is the solvent. More recently[46] they have replaced potassium by sodium naphthalide. Similar results have been reported[47]

$$\begin{array}{ccc} 6K & 3Ti^{II} & 2N^{3-} \\ & \times & \times \\ 6K^+ & 3Ti^{IV} & N_2 \end{array}$$

for a number of other metals (niobium, zirconium and tungsten). In a further development, it has been shown[48] that a titanium(II) dinitrogen complex is formed under certain conditions. An oxygen-free benzene solution of biscyclopentadienyl-titanium will pick up nitrogen at room temperature. Such solutions show an infrared absorption at about 1960

cm^{-1} attributable to $v_{N\equiv N}$. On flushing argon through the solution nitrogen is liberated, while the band at $1960\ cm^{-1}$ disappears. Molecular weight studies on the benzene solution indicate that the complex is dimeric. Treatment of a solution of this titanium(II) dinitrogen complex with sodium naphthalide and subsequent hydrolysis results in the production of ammonia.

One interesting aspect of this system is reflected in the development[49] of cyclic processes for the reduction of nitrogen to ammonia in which the reducing agent is regenerated electrolytically. Figure 6.4 represents one

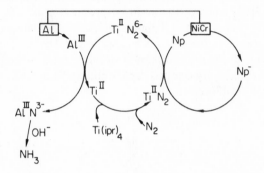

Figure 6.4   The reduction of nitrogen to ammonia *via* nitride formation.

example, and involves a cell fitted with an aluminium anode and a nichrome cathode. A typical electrolyte solution would contain the following: 1·68 mmol titanium tetraisopropoxide, 7·6 mmol naphthalene, 42 mmol aluminium isopropoxide, 8·6 mmol tetrabutylammonium chloride in 60 ml 1·2 dimethoxyethane. After electrolysis for eleven days at 40 V, treatment with 40 ml of 8M aqueous alkali and heating resulted in the formation of 10·2 mmol ammonia. It appears that naphthalene functions as an electron carrier, being reduced to naphthalide at the electrode and oxidized by the titanium nitrogen species. The aluminium isopropoxide, in addition to acting as an electrolyte, removes nitride from the titanium species, so allowing further interaction with nitrogen.

There has also been[50] a potentiometric study of titanium dinitrogen complexes. Based on the assumption that titanium(III) is generated from titanium(IV) under the conditions of the experiment, and that titanium (III) forms stable dinitrogen complexes, it has been shown that a 1:1 adduct $TiCl_3 . N_2$ is formed in solution in dimethyl sulphoxide, but that a 2:1 species with presumably bridging nitrogen is formed in propylene carbonate.

SOME INORGANIC STUDIES OF POSSIBLE RELEVANCE TO NITROGEN FIXATION

The first identifiable product of nitrogen fixation is ammonia, but much speculation[51] has centred over the reductive pathway, whether for example it proceeds through diimine (NH=NH) and hydrazine. The fact that these species have never been isolated from biological systems does not, of course, necessarily mean that they are not formed, diimine for example is an unstable compound, or it may remain bound to the enzyme for the duration of its lifetime. There have been a few reports[52-55] of complexes of species related to diimine. Thus the reaction between aqueous hydrazine and $(PPh_3)_2PtCl_2$ leads to the eventual production of a hydrido species $(PPh_3)_2PtHCl$, but an intermediate species is a dehydrodiimine whose structure has been confirmed by X-ray studies.[56] The reaction undergone by this species, that is the elimination of nitrogen, is not one that would suggest the parallel between this compound and nitrogenase is real, but the species is still of interest.

$$
\begin{array}{ccccc}
 & & H & & \\
 & & N & & \\
 & & \| & & \\
L & & N & & L \\
 & \diagdown \diagup & & \diagup & \\
 & Pt & & Pt & \\
 \diagup & & \diagdown \diagup & & \diagdown \\
L & & N & & L \\
 & & \| & & \\
 & & N & & \\
 & & H & &
\end{array}
$$

It has already been mentioned that a suggested model for nitrogenase involves two metal centres, one with nitrogen and the other with a hydride species. An inorganic analogue of nitrogenase has been suggested,[57] in which a benzenediazonium salt is a model for a metal nitrogen complex and a phosphine stabilized platinum hydride is a model for an iron or molybdenum hydride. It is found that p-fluorobenzenediazonium tetrafluoroborate and the complex $(PEt_3)_2PtHCl$ react together giving almost quantitative yields of a yellow crystalline p-fluorophenyldiimine complex.

$$
FC_6H_4N^+{\equiv}N \quad BF_4^- + H-\underset{\underset{L}{|}}{\overset{\overset{L}{|}}{Pt}}-Cl \rightarrow \left[FC_6H_4N{=}N-\underset{\underset{L}{|}}{\overset{\overset{H\ L}{|\ |}}{Pt}}-Cl \right] BF_4
$$

This reacts rapidly with hydrogen (or with other reducing agents) giving in about one hour the p-fluorophenylhydrazine complex

$[FC_6H_4NHNH_2PtL_2Cl]BF_4$. Further hydrogenation regenerates the platinum hydride complex and the tetrafluoroborate salt of the p-fluorophenylhydrazine, this latter compound may then be reduced to p-fluoroaniline and ammonia.

Other attempts to establish chemical models for nitrogenase have utilized the fact that nitrogenase will reduce acetylene to ethylene by using this reaction as a convenient marker for nitrogenase activity. It has been pointed out this is not valid as the sole criterion for the ability of any system to fix nitrogen.

Schrauzer and his group[58] have examined the reaction between acetylene (and other substrates of nitrogenase) and a variety of reducing agents in the presence of thiol complexes of transition metals (representing the active site of nitrogenase). The main feature of their results was the effectiveness of the molybdenum complex as a catalyst for the reduction of acetylene to ethylene, the relative activities of Mo, Ir, Ru, Pd, Rh, Pt, Fe being $100:15\cdot5:2\cdot9:1\cdot3:0\cdot4:0\cdot1:0$ respectively. Studies with molybdate, thioglycerol, borohydride and a trace of $FeSO_4.5H_2O$ cocatalyst indicated that the highest yield of NH_3 produced from nitrogen by borohydride reduction occurred when the molar ratio of molybdate to thiol is $2:1$, while replacement of the iron cocatalyst with salts of copper, palladium or nickel abolished the nitrogen-fixing ability of the system, so indicating a specific cocatalyst effect for iron. Schrauzer concludes that the molybdenum binding site in the Mo–Fe protein is important in nitrogenase for substrate binding and that this is reduced by electrons transferred by the iron protein portion of the enzyme.

Evidence has also been produced to indicate that the substrate binding site may be iron.[59] Thus, a number of nitrogenase substrates are reduced by a system comprising an iron(II) complex, reductant (borohydride or dithionite) and substrate. This exhibits true catalytic activity in that more than 1 mole of substrate was reduced per mole of iron, although only low levels of reduction of the nitrogen molecule were achieved. This result indicates that it is not really possible to use this sort of evidence to distinguish between molybdenum, iron, or a molybdenum–iron centre as the active site of nitrogenase.

It has been shown that the Mossbauer spectrum of nitrogenase is unaffected by the addition of substrate, suggesting that it may not be bound to the iron.

There are other cases in which it has been claimed that nitrogen is reduced to ammonia by treatment with reducing agent in the presence of assorted catalysts, as in the examples above. These claims have often not been substantiated by later workers. This may be due to the fact that the amounts of ammonia involved are very small and that impurities are present. The Nessler test for ammonia is not very specific in any case. Nitrogen-15 should really be used to allow the demonstration that N-15 ammonia is produced.

One other report of interest involves the demonstration of the catalytic reduction of nitrogen to hydrazine in protic solution.[60] It was found that nitrogen was reduced by Ti^{3+}, Cr^{2+}, V^{2+}, that is reducing agents that will reduce water to hydrogen. Reduction by Ti^{3+}, Cr^{2+} only occurs in the presence of molybdenum compounds such as MoO_4^{2-} and $MoOCl_3$. Mg^{2+} strongly catalyses the reaction, which only occurs above pH 7. A scheme is given below: it involves a bridging dinitrogen molecule. In fact, little is known of the mechanism of this type of catalytic reduction.

$$Mo^{III}\!\!-\!\!N\!\!\equiv\!\!N\!\!-\!\!Mo^{III} \qquad\qquad Mo^{IV}\!\!-\!\!NH\!\!-\!\!NH\!\!-\!\!Mo^{IV}$$

Complexes involving molybdenum bridged by the dinitrogen molecule are known (Table 6.3). It has been shown[33] that the bonding in the nitrogen molecule is considerably weakened when it bridges rhenium and molybdenum and suggested that this is à reasonable model for nitrogenase. The present example is rather different in that the molecule is bridging similar metal ions. The suggestion that nitrogenase has a dinuclear metal site is of long standing.[61] The main problem from the chemical point of view concerning these stable dinitrogen complexes is that the dinitrogen molecule has not been reduced. There is evidence, though, for the reduction of a dinitrogen titanium complex to ammonia (see p. 214). Here the complexes are only known in solution and the nitrogen is weakly bound and may readily be replaced. It may be that the degree of electrical asymmetry in the molecule is a critical factor. It may be necessary to prepare a wider range of stable dinitrogen complexes in the hope of finding one that may be reduced.

Schemes for nitrogen fixation

The following comprise a selection of the schemes that have been suggested in recent years. A number involve dinuclear metal sites with varying modes of bonding for nitrogen. In the following example[61] (Figure 6.5) ligand atoms facilitate transfer of hydrogen atoms to the nitrogen molecule, liberation of the ammonia regenerating the active site.

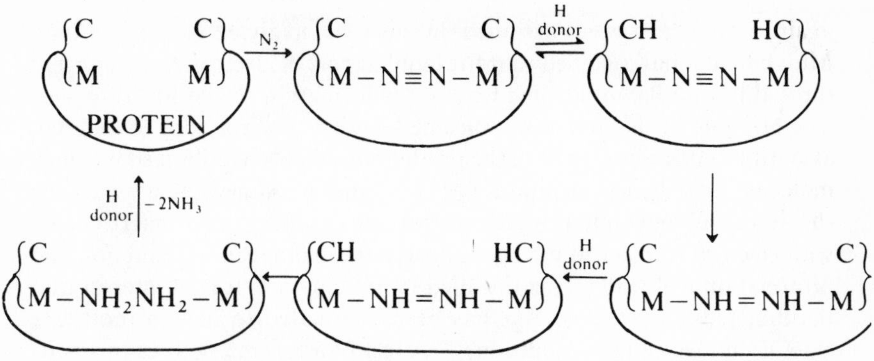

$$\left\{\begin{matrix} \text{C} & \text{C} \\ \text{M} & \text{M} \end{matrix}\right\} \xrightarrow{\text{N}_2} \left\{\begin{matrix} \text{C} & \text{C} \\ \text{M}-\text{N}\equiv\text{N}-\text{M} \end{matrix}\right\} \xrightarrow[\text{H donor}]{} \left\{\begin{matrix} \text{CH} & \text{HC} \\ \text{M}-\text{N}\equiv\text{N}-\text{M} \end{matrix}\right\}$$

PROTEIN

$$\left\{\begin{matrix} \text{C} & \text{C} \\ \text{M}-\text{NH}_2\text{NH}_2-\text{M} \end{matrix}\right\} \longleftarrow \left\{\begin{matrix} \text{CH} & \text{HC} \\ \text{M}-\text{NH}=\text{NH}-\text{M} \end{matrix}\right\} \xleftarrow[\text{H donor}]{} \left\{\begin{matrix} \text{C} & \text{C} \\ \text{M}-\text{NH}=\text{NH}-\text{M} \end{matrix}\right\}$$

H donor ↑ -2NH_3

Figure 6.5   A scheme for nitrogenase activity.

More recent schemes have attempted to explain the finer aspects of the problem. The successive involvement of bound diimine and hydrazine intermediates will involve increasing the nitrogen–nitrogen bond length. It has been suggested that this is brought about by changes in enzyme configuration, either through pH changes (with particular reference to symbiotic fixation[62]) or by the action of ATP[63] in free living species. The following scheme[63] (Figure 6.6) illustrates the latter suggestion. Once

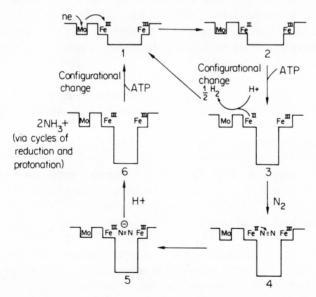

Figure 6.6   Scheme for nitrogenase activity with roles for iron and molybdenum ions.

again, a dinuclear site is assumed, in this case involving two iron atoms. Molybdenum has assigned to it the familiar role of electron transfer agent (probably with flavin) leading here to the formation of one Fe(II) species. The bridging of the nitrogen molecule between two iron ions in different oxidation states would lead to the production of a more polarized nitrogen molecule, and hence facilitate electron transfer, somewhat akin to the rhenium–molybdenum bridged species. Re-oxidation of iron(II) occurs with electron transfer to nitrogen. Protonation and further reduction and protonation will lead to the production of ammonia. It is of interest that the first reduced species, N_2^-, has been postulated to have a moderate stability under certain conditions.[64] A test of a scheme such as this is its effectiveness in explaining all the related phenomena associated with nitrogenase activity. In the absence of nitrogen, the model would allow the oxidation of Fe(II) by H^+, giving hydrogen and Fe(III), however other inhibitors will only be reduced if they can accept electrons. Thus carbon monoxide will inhibit nitrogen fixation but is not itself reduced. Parallels with nitrogenase are obvious. Another interesting scheme has been suggested by Hardy et al.[65]

Brintzinger[66] has postulated a scheme based upon the previously described studies on non-enzymatic nitrogen fixation involving titanium species. In this scheme he postulates the double insertion of a nitrogen molecule into two metal hydride bonds of a dimeric bridged hydrido complex of a metal in a reduced state. Again protonation is required for ammonia formation. In support of this scheme, it may be noted that both molybdenum and iron form stable hydride species, usually with π bonding ligands (e.g. Cp, CO, R_3P). The probable sulphide binding of iron in nitrogenase would give the appropriate π-bonding in this case. The occurrence of the ATP dependent hydrogen evolution in nitrogenase is evidence for the existence of some type of hydride. In the absence of nitrogen, hydrogen is evolved from nitrogenase. However, the acidic group that protonates the nitrogen species would also produce hydrogen if it reacts with a hydride complex.

It has already been mentioned that nitrogenase reduces acetylene to ethylene not ethane. If the reaction is carried out in D_2O, then cis $C_2H_2D_2$ is formed. This is consistent with the concept of the reduction as an insertion reaction as those characteristically lead to the formation of a cis product. In this example of nitrogenase activity, the full reducing power of the hydride system is not utilized. It would be of interest to see if hydrogen evolution were to occur alongside the reduction reaction, so confirming that only one of the two metal-hydrogen bonds was being consumed. In the reaction where nitrous oxide is converted to nitrogen and water, again not all the reducing power is utilized. Here hydrogen is released from nitrogenase at about half the normal rate, i.e. in the absence of an added substrate.

THE INORGANIC CHEMISTRY OF THE NITROGEN CYCLE

The ammonia formed by fixation of molecular nitrogen together with any ammonia resulting from decomposition of living matter is utilized by microorganisms which use the energy derived from the oxidation of reduced nitrogen compounds to reduce carbon dioxide, in order to obtain carbon for synthesis of new cellular material. These are autotrophic species. These oxidative reactions result in the eventual conversion of ammonia to nitrate in the process of nitrification. There are also reductive pathways for nitrate which lead to the production of ammonia, which may be incorporated into the cell, and also other gaseous products including nitrogen. These reduction pathways are more diverse than the oxidative reactions.

Nitrification

This is brought about mainly by *Nitrosomonas* species and *Nitrobacter* species that oxidize ammonia to nitrite and nitrite to nitrate respectively. It is now generally accepted that the oxidation to nitrite proceeds *via* two-electron changes, although the species with nitrogen in the formal oxidation state $+I$ has not been satisfactorily identified. Hydroxylamine appears to be the N^{-1} intermediate.

$$NH_3 \rightarrow NH_2OH \rightarrow \quad ? \quad \rightarrow NO_2^-$$
$$-III \qquad -I \qquad +I \qquad +III$$

It is not to be expected[67] that the reaction will involve appreciable build up of intermediate species as the rate of appearance of nitrite practically equals the rate of loss of ammonia. Whole cells of *nitrosomonas* will oxidize hydroxylamine nearly as rapidly as ammonia.

It has been shown that the use of different inhibitors affected the oxidation of ammonia to hydroxylamine and hydroxylamine to nitrite in different ways. In the presence of certain inhibitors an accumulation of hydroxylamine has been observed indicating that the two processes are independent. However, the presence of oxygen is essential if hydroxylamine is to be oxidized right through to nitrite. In the absence of oxygen, treatment of hydroxylamine with *nitrosomonas* leads to the production of nitrous (and nitric) oxides.

It has been shown[68] that trace amounts of *trans*-hyponitrite, $N_2O_2^{2-}$ are present in nitrification systems and this has led to the suggestion that this is the species with nitrogen in the formal oxidation state $+I$. However, added hyponitrite is not oxidized under these conditions, and it is most likely that it is formed as a result of a side reaction and does not lie on the main reaction pathway. An appropriate reaction leading to the formation of *cis*-hyponitrite species would be the dimerization of protonated nitroxyl ion, NOH. This would then be the N^{+1} intermediate.

The oxidation of nitrite to nitrate by *Nitrobacter* is inhibited by cyanide, and from spectroscopic studies it has been suggested that a cytochrome was involved.

Studies with cell free extracts

It is only relatively[69] recently that studies have been carried out on some of the enzyme systems involved in these reactions. While whole cells of *Nitrosomonas Europaea* quantitatively oxidize both ammonia and hydroxylamine to nitrite, cell free extracts can only oxidize hydroxylamine, although with reduced efficiency. Provided molecular oxygen is present nitrite is formed. The oxidation is complex and involves the transfer of electrons to an electron acceptor, usually a cytochrome. The enzyme itself has been termed hydroxylamine oxidase[70] as it is the initial acceptor of electrons from hydroxylamine. It lacks terminal oxidase activity as the naturally occurring terminal electron acceptor is not physically associated with hydroxylamine oxidase and is separated from it in the purification procedure. In cell free studies on the enzyme an artificial terminal electron-accepting compound has to be added.

The molecular weight of hydroxylamine oxidase has been estimated[69] as being near to 196,000 while electron microscopy studies have shown it to be an essentially spherical particle with a diameter of about 16 nm, containing at least one *b* and one *c* type cytochrome. The terminal oxidase system has a molecular weight of 128,000. The enzymes responsible for the initial oxidation of ammonia are thought to be near the membranes of the cell and are hence lost on destruction of the cell.

The normal yield of nitrite is about 30–40%. Incomplete recovery is due to the interaction of a soluble *nitrosomonas* cytochrome with the hydroxylamine oxidation system, resulting in hydroxylamine being oxidized to products other than nitrite.

An EPR study has been carried out[71] on hydroxylamine oxidase and indicates the presence of a copper(II) species. On addition of hydroxylamine the copper signal undergoes changes which indicate that the metal is involved in the oxidase system. The presence of an iron species is also observed.

Particulate species have been separated[72] from *Nitrobacter*. The enzyme was inhibited by cyanide while the addition of ferrous or ferric ions enhanced activity.

Nitrate reduction

A number of nitrate-reducing pathways are known. Certain microorganisms will reduce nitrate *via* nitrite to ammonia, which ends up as cell nitrogen. This is the process of assimilation. There is also the process of denitrification which leads to the production of gaseous nitrogen, while certain microorganisms use nitrate as an hydrogen acceptor, in which case the nitrate is not incorporated into cell nitrogen (the dissimilation of nitrate).

The first stage in the reduction of nitrate in microorganisms and higher plants involves the formation of nitrite. Nitrate reductase from *Neurospora Crassa*[73] has been shown to contain molybdenum and may be separated from it, in which case activity is lost. The addition of metal chelating agents also causes inhibition. Activity is restored on the addition of molybdenum. Studies have shown that the metal links nitrate with flavin dinucleotide (FAD) in the electron-transfer sequence, and that it is likely that two molybdenum atoms are required to effect the two-electron transfer for the reduction of nitrate. There is also a phosphate requirement

$$FAD \diagdown \quad Mo^{VI} \diagdown \quad NO_3^-$$
$$FADH_2 \diagup \quad Mo^V \diagup \quad NO_2^-$$

at the molybdenum site. It is suggested that the phosphomolybdate complex facilitates oxidation-reduction between flavin and nitrate, while preventing overreduction of the molybdenum in the enzyme.

The dissimilatory nitrate reductase such as in *pseudomonas aeruginosa* also has a molybdenum requirement, together with an iron requirement (as a cytochrome). Again the formation of molybdenum(V) is involved. Certain organisms can contain assimilatory and dissimilatory nitrate reductases depending upon the conditions. The presence of oxygen favours assimilation.

The appearance of a molybdenum enzyme system appears to be a general feature of nitrate reduction systems.

The pattern of behaviour for reduction beyond nitrite will vary with the type of system. The following are examples of nitrate assimilating species, all having reductases which reduce nitrite to ammonia: *azotobacter agile*, *Bacillus pumilis*, *Escherichia coli*, *Neurospora crassa*.

In the case of *N. crassa* a nitrite reductase has been isolated and shown to contain iron and copper. It has been suggested that the copper is coupled to the electron transfer sequence, copper(I) reducing nitrite chemically, while the copper(II) so formed is enzymatically converted back to copper(I). The pathway between nitrite and ammonia is not well defined, however hyponitrous acid and hydroxylamine have been isolated, and *N. crassa* contains an Mn^{2+} enzyme dependent system that will reduce hydroxylamine to ammonia. There has also been a report of a hyponitrite reductase.[75] Overall, therefore, the evidence seems to suggest a series of two-electron steps.

$$NO_3^- \rightarrow NO_2^- \rightarrow (N_2O_2^{2-}) \rightarrow NH_2OH \rightarrow NH_3$$

Reduction pathways in denitrification are no better defined.[76,77] A number of denitrifying species have enzyme systems that convert nitrite to nitric oxide. Thus *P. aeruginosa* contains iron and copper atoms and converts nitrite to nitric oxide, probably *via* reduction through copper(I). Again, *P. denitrificans* has a nitrite reductase, a copper protein of molecular weight 130,000, that converts nitrite to nitric oxide. The use of whole cells leads to the production of nitrous oxide and nitrogen. It appears that the reaction sequence is as follows:

$$NO_2^- \rightarrow NO \rightarrow N_2O \rightarrow N_2$$

with nitrogen as the eventual final product, although usually nitrous oxide and nitrogen are both produced.

Some general comments on the chemistry of the nitrogen cycle

Many of the intermediates involved in the nitrogen cycle are well defined. Figure 6.7 represents the current state of affairs. The greatest uncertainties arise over the nature of the species with nitrogen in the formal oxidation state I. Under biological pHs hyponitrite would exist partly as $H_2N_2O_2$ and partly as $HN_2O_2^-$

$$H_2N_2O_2 \underset{pK_1 = 7}{\rightleftharpoons} HN_2O_2^- + H^+ \underset{pK_2 = 11}{\rightleftharpoons} N_2O_2^{2-} + 2H^+$$

It has been shown[78] that at room temperature, hyponitrous acid and the hyponitrite anion $N_2O_2^{2-}$ are stable and that the hydrogen hyponitrite anion $HN_2O_2^-$ is unstable, 'hyponitrite' decomposing by the following scheme

$$H_2N_2O_2 \rightleftharpoons H^+ + HN_2O_2^- \quad \text{Fast}$$
$$HN_2O_2^- \rightarrow N_2O + OH^- \quad \text{Slow}$$

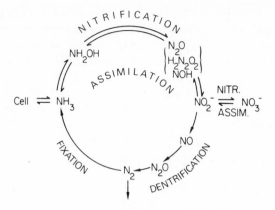

Figure 6.7   The nitrogen cycle.

At pH 7 and 8·8 the half lives for the decomposition are about 60 and 16 minutes respectively. There is also the possibility[79] that a free-radical pathway could result in much more rapid decomposition. The instability of hyponitrite presents a problem in carrying out studies using enzyme systems. On the other hand, however, the alternative nitrogen(I) species, nitroxyl NOH, is much more unstable, its dimerization leading to the formation of highly unstable *cis*-hyponitrous acid and hence nitrous oxide. The existence of NO^- in solution has been disputed in the past, but there is now evidence for its existence in alkaline solution.[80] However at pH 7, NO^- is almost certainly protonated, and appropriately more unstable as the tendency for two neutral species to dimerize would be greater than that for two negative ones. The instability of NOH appears to rule out the possibility of its existence in solution long enough for involvement in enzymatic reduction. It may well be, however, that nitrous oxide itself is the appropriate species. A number of other compounds have been suggested as intermediates in the nitrogen cycle, such as α-oxohyponitrite $N_2O_3^{2-}$, nitramide HN_2NO_2, dihydroxyammonia $NH(OH)_2$ but there is no evidence for their formation.

One of the major problems in this type of work, however, is that while it may be shown that enzyme systems will utilize certain added substrates, at a fast enough rate, there is no guarantee that this substrate lies on the normal reaction pathway.

It is not possible to make definite comments at present about the detailed mechanism of enzyme action in these systems, but it is of interest to note a parallel between nitrification and the chemical autoxidation of hydroxylamine in alkaline solution. This latter reaction[81] leads to the formation of nitrite in a copper(II) catalysed reaction. At pH 11, with a trace of metal ion, up to 80 % of the hydroxylamine is converted to nitrite. It is suggested that the nitroxyl ion, NO^-, is the first product of interaction of hydroxylamine and oxygen. In the absence of oxygen this would dimerize giving nitrous oxide, but in the presence of more oxygen, further reaction occurs giving the species peroxonitrite ONO_2^-. Peroxonitrite then reacts with hydroxylamine *via* copper coordination, giving nitrite and regenerating nitroxyl ion, from which peroxonitrite ion may once more be formed. This means that a trace of peroxonitrite will catalyse the autoxidation of hydroxylamine directly to nitrite.

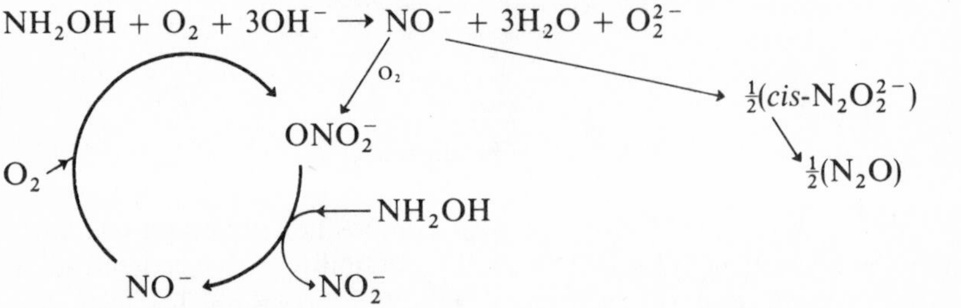

$$NH_2OH + O_2 + 3OH^- \rightarrow NO^- + 3H_2O + O_2^{2-}$$

The parallel with the behaviour of the copper protein hydroxylamine oxidase is marked.[82] However, under biological pH conditions, the autoxidation reaction does not proceed as above, but rather results in the production of nitrous oxide (due to enhanced dimerization of NOH over NO^-). However, in the cell there is always the possibility of localized higher pH values, or of abnormal pK_a values or alternatively, if the hydroxylamine were attached to the active site of the enzyme, it is possible that the species NOH would also remain bound long enough to allow oxygen addition to compete with dimerization. The oxygenated species ONO_2^- would then react with hydroxylamine *via* copper coordination giving nitrite. Nitric oxide could be formed by a one-electron oxidation of NO^- in the absence of oxygen.

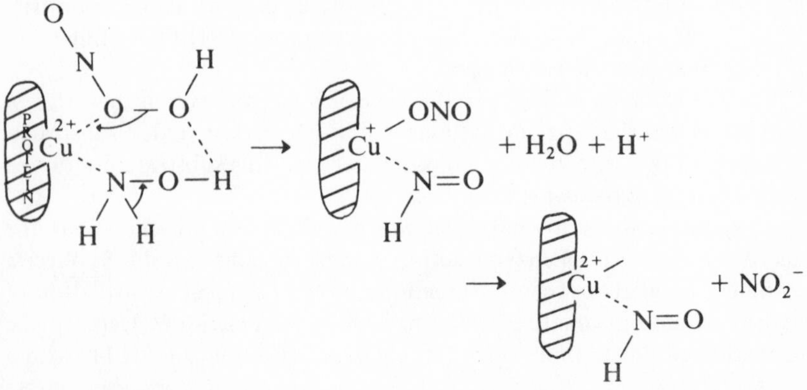

REFERENCES

1. The material contained in this chapter has been the subject of a number of recent articles, selected examples are as follows:
   A Discussion on Nitrogen Fixation, *Proc. Roy. Soc.* (*B*), 1969, **172**, 317; J. R. Postgate, *Nature*, 1970, **226**, 25; J. R. Postgate, *Science Journal*, 1968, 69; R. Murray and D. C. Smith, *Coord. Chem. Revs.*, 1968, **3**, 429; A. D. Allen and F. Bottomley, *Acc. Chem. Res.*, 1969, **1**, 360; D. J. D. Nicholas, *Biol. Rev.*, 1963, **38**, 530; J. Chatt and G. J. Leigh, *Chem. Soc. Rev.*, 1972, **1**, 121

2. J. E. Carnahan, L. E. Mortensen, H. F. Mower and J. E. Castle, *Biochim. Biophys. Acta*, 1960, **44**, 520

3. L. E. Mortensen, *Biochim. Biophys. Acta*, 1966, **127**, 18

4. W. A. Bulen, and J. R. Lecomte, *Proc. Nat. Acad. Sci. U.S.*, 1965, **53**, 532
   (a) R. C. Burns, R. D. Holsten and R. W. F. Hardy, *Biochem. Biophys. Res. Comm.* 1970, **39**, 90

5. J. R. Postgate, *Proc. Roy. Soc.* (*B*), 1969, **172**, 355

6. Cited by J. R. Postgate, *Science Journal*, 1968, 69

7. J. Chatt and G. J. Leigh, *Nitrogen Metabolism*, V. Cutting (Ed), Academic Press, London, 1968

8. J. Chatt, *Proc. Roy. Soc.* (*B*), 1969, **172**, 327

9. A. D. Allen and C. V. Senoff, *Chem. Comm.*, 1965, 621

10. A. E. Shilov, A. K. Shilova and Yu. G. Borod'ko, *Kinetika i Kataliz*, 1966, **7**, 768

11. A. Sacco and M. Rossi, *Chem. Comm.*, 1967, 316

12. J. Chatt, G. J. Leigh and R. L. Richards, *Chem. Comm.*, 1969, 513; *J. Chem. Soc.* (*A*) 1970, 2243

13. A. D. Allen, F. Bottomley, R. O. Harris, V.. P. Reinsalu and C. V. Senoff, *J. Amer. Chem. Soc.*, 1967, **89**, 5595

14. A. A. Diamantis and G. J. Sparrow, *Chem. Comm.*, 1969, 469

15. D. E. Harrison and H. Taube, *J. Amer. Chem. Soc.*, 1967, **89**, 5706

16. D. E. Harrison, E. Weissberger and H. Taube, *Science*, 1968, **159**, 320

17. J. Chatt, A. B. Nikolsky, R. L. Richards and J. R. Sanders, *Chem. Comm.* 1969, 154

18. A. Yamamoto, S. Kitazume and S. Ikeda, *J. Amer. Chem. Soc.*, 1968, **90**, 1089

19. Yu. G. Borod'ko, V. S. Bukneev, G. I. Kozub, M. L. Khudekel and A. E. Shilov, *Zh. Strukt. Khim.*, 1967, **8**, 542

20. A. D. Allen and J. R. Stevens, *Chem. Comm.*, 1967, 1147

21. H. A. Scheidegger, J. N. Armor and H. Taube, *J. Amer. Chem. Soc.*, 1968, **90**, 3263

22. J. Chatt, G. J. Leigh and D. M. D. Mingos, *Chem. and Ind.*, 1969, 109

23. A. Yamamoto, S. Kitazume, L. S. Pu and S. Ikeda, *Chem. Comm.*, 1967, 79; *J. Amer. Chem. Soc.*, 1971, **93**, 371

24. A. Yamamoto, S. Kitazuma, L. S. Pu and S. Ikeda, *J. Amer. Chem. Soc.*, 1967, **89**, 3071

25. A. Sacco and M. Rossi, *Chem. Comm.*, 1967, 316

26. A. Misono, Y. Uchida and T. Saito, *Bull. Chem. Soc. Japan*, 1967, **40**, 700

27. J. P. Collman and J. W. Kang, *J. Amer. Chem. Soc.*, 1966, **85**, 3459; J. P. Collman, M. Kuota, F. D. Vastine, J. Y. Sun and J. W. Kang, *J. Amer. Chem. Soc.*, 1968, **90**, 5430

28. L. Yu. Ukhin, Yu. A. Shvetsov and M. L. Khidekel, *Bull. Acad. Sci. USSR, Div. Chem. Sci.*, 1967, 934

29. J. Chatt, J. R. Dilworth and G. J. Leigh, *Chem. Comm.*, 1969, 687
30. J. T. Moelwyn-Hughes and A. W. B. Garner, *Chem. Comm.*, 1969, 1309
31. M. Hidai, K. Tominari, Y. Uchida and A. Misono, *Chem. Comm.*, 1969, 814
32. M. Hidai, K. Tominari, Y. Uchida and A. Misono, *Chem. Comm.*, 1969, 1392
33. J. Chatt, J. R. Dilworth, R. L. Richards and J. R. Sanders, *Nature*, 1969, **224**, 1201
34. J. H. Enemark, B. R. Davis, J. A. McGinnety and J. A. Ibers, *Chem. Comm.*, 1968, 96; B. R. Davis, N. C. Payne and J. A. Ibers, *J. Amer. Chem. Soc.*, 1969, **91**, 1240
35. F. Bottomley and S. C. Nyburg, *Chem. Comm.*, 1966, 897; F. Bottomley and S. C. Nyberg, *Acta. Cryst.*, 1968, **24B**, 1289
36. I. M. Treitel, M. T. Flood, R. E. Marsh and H. B. Gray, *J. Amer. Chem. Soc.*, 1969, **91**, 6512
37. J. Chatt, R. L. Richards, J. R. Sanders and J. E. Fergusson, *Nature*, 1969, **221**, 551
38. J. Chatt, J. E. Fergusson, R. L. Richards, J. L. Love, *Chem. Comm.*, 1968, 1522
39. M. E. Volpin, V. B. Shur, *Nature*, 1966, **209**, 1236; *Dokl. Chem. Proc. Acad. Sci. USSR*, 1964, **156**, 1102
40. M. E. Volpin, V. B. Shur and L. P. Bichin, *Bull. Acad. Sci USSR, Div. Chem. Sci.*, 1966, 698
41. R. Maskill and J. M. Pratt, *Chem. Comm.*, 1967, 950
42. H. Brintzinger, *J. Amer. Chem. Soc.*, 1966, **88**, 4305, 4307; 1967, **89**, 6871
43. H. Brintzinger, *Biochemistry*, 1966, **7**, 3947
44. G. Henrici-Olive and S. Olive, *Angew. Chem. Intern. Ed.*, 1967, **6**, 873
45. E. E. van Tamelen, G. Boche, S. W. Ela and R. B. Fechter, *J. Amer. Chem. Soc.*, 1967, **89**, 5707
46. E. E. van Tamelen, G. Boche and R. Greeley, *J. Amer. Chem. Soc.*, 1968, **90**, 167
47. D. R. Gray and C. H. Brubaker, *Chem. Comm.*, 1969, 1239
48. E. E. van Tamelen, R. B. Fechter, S. W. Schneller, G. Boche, R. H. Greeley and B. Akarmark, *J. Amer. Chem. Soc.*, 1969, **91**, 1551
49. E. E. van Tamelen and J. D. A. Seeley, *J. Amer. Chem. Soc.*, 1969, **91**, 5194
50. T. C. Franklin and R. C. Byrd, *Inorg. Chem.*, 1970, **9**, 986
51. E. K. Jackson, G. W. Parshall and R. W. F. Hardy, *J. Biol. Chem.*, 1968, **243**, 4952
52. W. G. Hanstein, J. B. Lett, C. E. McKenna and T. G. Traylor, *Proc. Nat. Acad. Sci. U.S.*, 1967, **58**, 1314
53. Y. G. Borod'ko and A. E. Shilov, *Russ. Chem. Rev.*, 1969, **38**, 355
54. H. A. Itano, *Proc. Nat. Acad. Sci. U.S.*, 1970, **67**, 485
55. P. C. Huang and E. M. Kosower, *Biochim. Biophys. Acta*, 1968, **165**, 483
56. G. C. Dobinson, R. Mason, G. B. Robertson, R. Ugo, F. Corti, D. Morelli, S. Cenini and F. Bonati, *Chem. Comm.*, 1968, 274
57. G. W. Parshall, *J. Amer. Chem. Soc.*, 1967, **89**, 1822
58. G. N. Schrauzer and G. Schlesinger, *J. Amer. Chem. Soc.*, 1970, **92**, 1808; G. N. Schrauzer and P. A. Doemeny, *J. Amer. Chem. Soc.*, 1971, **93**, 1608; G. N. Schrauzer, G. Schlesinger and P. A. Doemeny, *J. Amer. Chem. Soc.*, 1971, **93**, 1803
59. W. E. Newton, J. L. Corbin, P. W. Schneider and W. A. Bulen, *J. Amer. Chem. Soc.*, 1971, **93**, 268
60. A. Shilov, N. Denisov, O. Efimov, N. Shuvalov, N. Shuvalova, and A. Shilova, *Nature*, 1971, **231**, 460

61. M. E. Winfield, *Rev. Pure Appl. Chem.*, 1955, **5**, 217
62. K. Abel, *Phytochemistry*, 1963, **2**, 429
63. W. A. Bulen, J. R. Lecomte, R. C. Burns and J. Hinkson, in *Non-Heme iron proteins: Role in Energy Conversion*, A. San Pietro (Ed), Antioch Press, Yellow Springs, 1965, 261
64. N. Bauer, *J. Phys. Chem.*, 1960, **64**, 833
65. R. W. F. Hardy, E. Knight and A. J. D. Eustachio, *Biochem. Biophys. Res. Comm.*, 1965, **20**, 539
66. H. Brintzinger, *Biochemistry*, 1966, **5**, 3947
67. T. Hofmann and H. Lees, *Biochem. J.*, 1954, **54**, 579
68. A. S. Corbet, *Biochem. J.*, 1928, **28**, 1575
69. M. K. Rees, *Biochemistry*, 1968, **7**, 353, 366
70. A. B. Falcone, A. L. Shug and D. J. D. Nicholas, *Biochim. Biophys. Acta*, 1963, **77**, 199
71. D. J. D. Nicholas, P. W. Wilson, W. Hernen, G. Palmer, and H. Beinert, *Nature*, 1963, **196**, 433
72. M. I. H. Allen and A. Nason, *Proc. Nat. Acad. Sci. U.S.*, 1960, **46**, 763
73. D. J. D. Nicholas, A. Nason, and W. D. McElroy, *J. Biol. Chem.*, 1954, **207**, 341
74. D. J. D. Nicholas and H. M. Stevens, *Inorganic Nitrogen Metabolism*, Ed. W. D. McElroy and B. Glass, Johns Hopkins University Press, 1958
75. A. Medina and D. J. Nicholas, *Nature*, 1957, **179**, 553, *Biochim. Biophys. Acta*, 1957, **25**, 138
76. M. Miyata, T. Matsubara and T. Mori, *J. Biochem. (Tokyo)*, 1969, **66**, 759 and earlier papers
77. G. C. Walker and D. J. D. Nicholas, *Biochim. Biophys. Acta*, 1961, **49**, 350
78. M. N. Hughes and G. Stedman, *J. Chem. Soc.*, 1963, 1239
79. J. R. Buccholz and R. E. Powell, *J. Amer. Chem. Soc.*, 1963, **85**, 509
80. M. N. Hughes, *Quart. Rev. Chem. Soc.*, 1969, 1
81. M. N. Hughes and H. G. Nicklin, *J. Chem. Soc.*, (A), 1971, 164
82. M. N. Hughes and H. G. Nicklin, *Biochim. Biophys. Acta*, 1970, **222**, 660

CHAPTER SEVEN

OXYGEN CARRIERS

The transport and storage of oxygen is an extremely important physiological function. This is carried out by a number of well-known iron and copper species, which occur in the blood of animals. The vanadium species, hemovanadin, is also an oxygen carrier. These are listed in Table 7.1, where it may be seen that both heme and non-heme iron proteins are involved.

TABLE 7.1 Oxygen carrier and storage proteins

	Metal	Function
Hemoglobin	Iron	Carrier
Hemerythrin	Iron	Carrier
Myoglobin	Iron	Storage
Hemocyanin	Copper	Carrier
Hemovanadin	Vanadium	Carrier

The heme oxygen carriers, which give the red colour to human blood, absorb one mole of oxygen per mole of carrier, while hemocyanin (responsible for the blue colour of fresh crab blood) absorbs one mole of oxygen for every two moles of metal ion. Certain marine worms (e.g. *golfingia elongata*) have a violet colour which may be attributed to the presence of the non-heme iron protein hemerythrin.

Vanadium is present in the blood cells of certain ascidians. These are the vanadocytes and are of interest in that they involve a pH of 2·5 due to the presence of sulphuric acid. The vanadium protein may be isolated quite readily. It has a molecular weight of 27,500, with vanadium present as V(III).

All these carriers provide sites at the transition metal ion for the reversible attachment of oxygen (Equation 1). Normally iron(II) for example, is readily oxidized by molecular oxygen. In hemoglobin iron(II) will bind oxygen and yet not be oxidized by it.

$$\text{Carrier} + O_2 \rightleftharpoons \text{Carrier} (O_2) \tag{1}$$

The oxygen is then carried *via* catalysts that are oxidized by oxygen, with a resulting concluding oxidation reaction. Some mention has already been made of these oxygen carriers. Thus, for example, the five-coordination of the Fe(II) in hemoglobin (and myoglobin) has been noted. This provides an open site for oxygen addition and is, in fact, the S_N1 transition state structure observed for ligand substitution reactions. This implies a decrease in the activation energy for oxygen addition compared to the activation energy for other ligand substitution reactions. The addition of oxygen is associated with a change in the magnetic properties of hemoglobin from paramagnetism to diamagnetism, while X-ray studies show that there is a change in the position of the iron atom with respect to the plane of the nitrogen donors.

A large number of model compounds, usually coordination complexes of cobalt, will reversibly carry oxygen. Much useful information has been obtained from the study of these systems and is presented in a number of recent review articles.[1-6] In theory, oxygen may be held in a number of ways by the metal complex. First, therefore, we will discuss the electronic properties of the oxygen molecule and relate these to its oxidizing properties. Most importantly, we will attempt to assess the factors that determine whether or not these model systems will carry oxygen. We will then consider current theories on the biological oxygen carriers, concentrating particularly on recent X-ray structural studies and the configurational and conformational changes resulting from oxygenation of hemoglobin.

THE ELECTRONIC STRUCTURE OF THE OXYGEN MOLECULE

In the ground state the oxygen molecule has two unpaired electrons. Its biradical nature is readily explained by simple molecular orbital theory. This is schematically represented in Figure 7.1, together with the molecular orbital schemes for the possible reduction products, superoxide and peroxide.

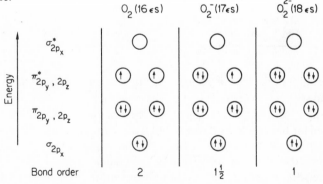

Figure 7.1 Molecular orbital scheme for oxygen and its reduction products.

The degeneracy of the π orbitals naturally leads to the biradical character of oxygen. As the oxygen molecule is reduced, electrons are fed into the antibonding π molecular orbitals with a decrease in bond order, so that the peroxide species contains a single bond only. This is reflected in the increasing oxygen–oxygen bond length from 0·121 to 0·128 and 0·149 nm for oxygen, superoxide and peroxide respectively.

Four electrons in all may be added to the oxygen molecule. In this case there will be no overall bonding effect as the effect of the electrons in the bonding molecular orbitals will now be completely cancelled out. The product will therefore be oxide (Equation 2).

$$O_2 \xrightarrow{e} O_2^- \xrightarrow{e} O_2^{2-} \xrightarrow{2e} 2O^{2-} \qquad (2)$$

In addition to the scheme presented in Figure 7.1 for oxygen, it is also possible to write down excited state configurations either $\uparrow\downarrow\,\bigcirc$; or $\uparrow\,\downarrow$, 23·4 and 37·5 K cal. above the ground state respectively.

THE REACTIONS OF OXYGEN

Many reactions of oxygen are complex, chain reactions. Its reactions with transition metal ions are often much simpler. It is obviously of interest to compare redox potentials for the various reduction reactions of oxygen with other redox potentials[7] (Table 7.2). The potential for reaction (5)

TABLE 7.2 Some redox potentials

Couple	E^0 (volts)	
$O_2 + \varepsilon \rightleftharpoons O_2^-$	$-0\cdot4$	(3)
$O_2 + 2H^+ + 2\varepsilon \rightleftharpoons H_2O_2$	$+0\cdot68$	(4)
$O_2 + 4H^+ + 4\varepsilon \rightleftharpoons 2H_2O$	$+1\cdot23$	(5)
$O_2 + H^+ + \varepsilon \rightleftharpoons HO_2$	$-0\cdot1$	(6)
$Fe^{3+} + \varepsilon \rightleftharpoons Fe^{2+}$	$+0\cdot77$	(7)

indicates that molecular oxygen is a strong oxidizing agent. The full redox potential will only be realized when the oxidation carried out involves an essentially synchronous four-electron reaction. Such reactions have been suggested for biological oxidations, but if this cannot occur, then reduction will take place either by one- or two-electron steps (Equations (3) and (4)). The geometry of the π antibonding molecular orbitals of the oxygen molecule will, of course, impose limitations on the feasibility of different redox reactions occurring. Reduction will involve the transfer of electrons from the valence shell of the metal ion to the oxygen molecule and there will therefore be a symmetry requirement on the orbital of the metal ion. The implications of the Franck–Condon principle, discussed in Chapter 2, will also hold here.

In a number of reactions of oxygen with transition metal species in acid solution (such as iron(II)) it has been shown that the rate law involves a square dependence on metal ion concentration. This can be understood in terms of the redox potentials in Table 7.2. This data shows that the reduction of oxygen to superoxide (O_2^-) by ferrous ion is not favoured. However, as the stoicheiometry involves two moles of ferrous ion it may be seen that under these conditions the oxygen will now be reduced to peroxide O_2^{2-}, a more favoured reaction, particularly as the peroxide ion product is probably stabilized by coordination to the Fe(III) produced. The presence of ligands other than water in the coordination shell of the ferrous ion has a marked effect on the rate of reaction. Oxygen donor ligands will particularly affect the rate of reaction as these bases have π-donor properties and hence will enhance the transfer of electron density from the appropriate metal d orbital into the oxygen molecule. σ-donor ligands (such as N donors) *trans* to the oxygen molecule will, by contrast, have little influence.

In general, in the following discussion of model compounds, we will consider transition metal complex systems in which the oxygen molecule is present as O_2, O_2^- and O_2^{2-} and may or may not be carried as such. There is, however, no evidence for further reduction of bound oxygen in model compounds. We will discuss the binding of the oxygen molecule to the metal in a later section.

MODEL COMPOUNDS FOR OXYGEN CARRIERS

The air oxidation of cobalt(II) complexes

It has been known since the time of Werner[8] that cobalt(III) complexes may be prepared by assembling the ligands and cobalt(II) (which is labile and therefore the complex is rapidly formed), and then air oxidizing to give the cobalt(III) complex, which is inert. In particular it was known that ammoniacal solutions of cobalt(II) complexes were weak oxygen carriers but that they were ultimately air oxidized to give brown salts containing the diamagnetic cation $[(H_3N)_5CoO_2Co(NH_3)_5]^{4+}$. Treatment with hydrogen peroxide gives the green paramagnetic $[(H_3N)_5CoO_2Co(NH_3)_5]^{5+}$. Much work[9] has been done on these diamagnetic and paramagnetic species, which are now fairly common. Thus air oxidation of cobalt(II) cyanide leads to the formation of brown, diamagnetic $[(NC)_5CoO_2Co(CN)_5]^{6-}$

while further treatment with bromine in alkali gives the red paramagnetic species $[(NC)_5CoO_2Co(CN)_5]^{5-}$. The diamagnetic species involve two cobalt(III) ions bridged by a peroxide group: $Co^{III}-O_2^{2-}-Co^{III}$. Werner formulated the paramagnetic complexes in terms of a peroxide group bridging two cobalt ions in different oxidation states $Co^{IV}-O_2^{2-}-Co^{III}$, but EPR studies indicate that both cobalt atoms are equivalent, a conclusion that is intuitively reasonable as the two cobalt atoms have identical environments. It appears that both cobalt atoms are cobalt(III), with a bridging superoxide ion O_2^-. Molecular orbital theory predicts that the odd electron is in an orbital which extends over both cobalt atoms and both oxygen atoms. If this orbital were heavily concentrated on the oxygen atoms it is consistent with the superoxide picture. Oxygen–oxygen bond length will also indicate the nature of the oxygen species. An X-ray study[10] has shown the oxygen–oxygen bond length to be 0.131 nm, a value consistent with a superoxide species. It appears that the structure (Figure 7.2) involves σ-bonding of the oxygen to the cobalt atoms. The structure of $[(NH_3)_5CoO_2Co(NH_3)_5]^{4+}$ has also been determined[11] confirming the longer O–O bond length expected for a peroxo group.

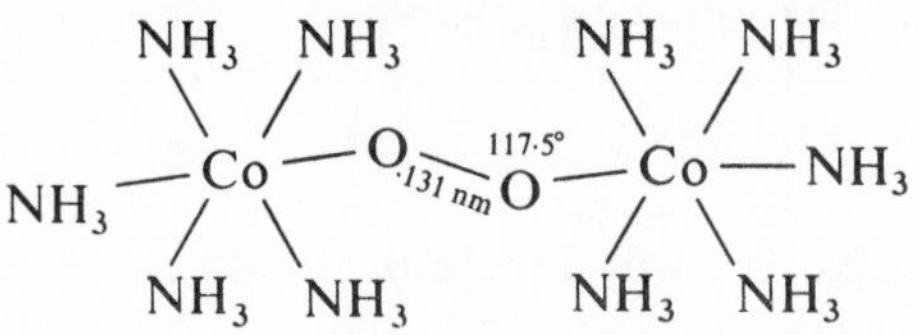

Figure 7.2   The structure of $[(NH_3)_5CoO_2Co(NH_3)_5]^{5+}$.

Cobalt(II) complexes with bis(salicylaldehyde)ethylenediimine[1,12] and related ligands

Certain of these chelates are very effective oxygen carriers both in the solid state and in solution in certain solvents. The complexes I and II shown in Figure 7.3 are particularly well known as the parent complexes of these carriers. Carriers related to Complex I have been most studied,[12-17] i.e. those with ligands related to bis(salicylaldehyde)-ethylenediimine ('Salen'), formed by the Schiff base condensation of two moles of salicylaldehyde with ethylenediamine. Complex II involves a Schiff base formed between salicylaldehyde and a triamine.

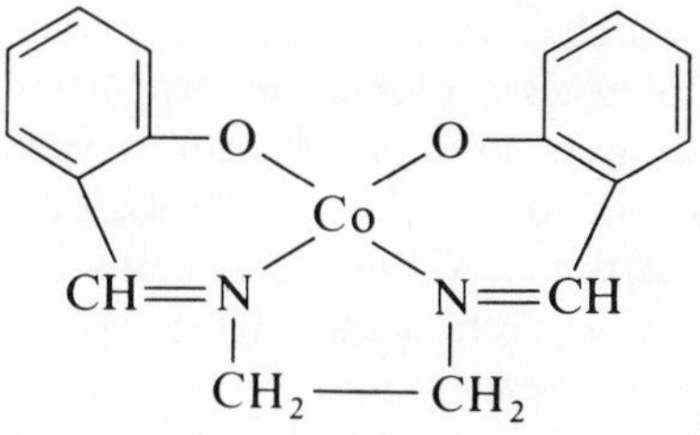

Figure 7.3 Some synthetic oxygen carriers.

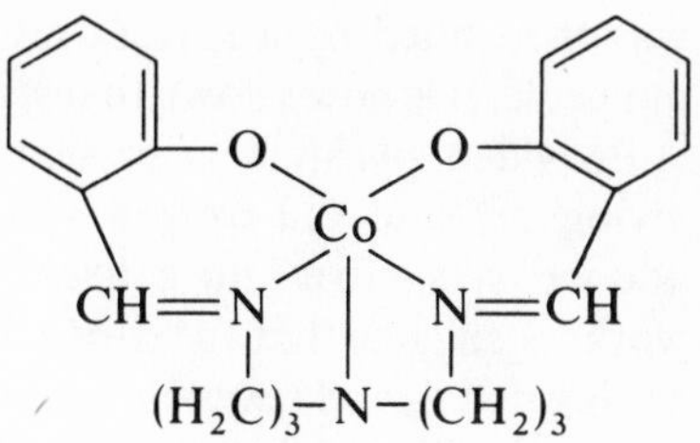

Both types of complex exist in a number of crystalline modifications, only some of which are active. Complexes of salen and related ligands absorb $\frac{1}{2}$ mole of oxygen per mole of complex. X-ray studies have shown the molecules to have a square planar structure and to be packed parallel in the crystal. The oxygenated form involves bridging oxygen. The crystal structure[17a] of the oxygen-inactive form of the bis(salicylaldehyde)-ethylenediimine cobalt(II) (Cobalt(II) salen) has been determined. The cobalt atoms are five-coordinate with square pyramidal stereochemistry. The complex exists as discrete dimers, each cobalt being bound to an oxygen atom of the other molecule in the dimer. It may well be that this is the basic reason for the difference in activity of the two forms. The monomer may absorb oxygen but steric (or electronic) effects prevent oxygen binding to the dimer. A feature of a number of the cobalt(II) oxygen carriers we shall consider is the presence of a set of four good co-planar ligand atoms. This provides supporting evidence for the suggestion that only the monomeric form can absorb oxygen. Previous explanations had involved differences in the crystal structure, the active carrier supposedly having holes in the crystal structure which were not present in the inactive forms. This viewpoint has been questioned.[17a] Type II chelates absorb 1 mole of oxygen per mole of complex. Their structure appears to be a tetragonal pyramidal one.

Magnetic measurements indicate that types I and II chelates have 1 and 3 unpaired electrons respectively. After oxygenation type I chelates were diamagnetic while type II chelates have one unpaired electron, suggesting the formation of a peroxide species in each case.

These chelates will undergo a number of cycles of oxygenation and deoxygenation, but their efficiency gradually deteriorates. It appears that this is at least partly due to the oxidation of the ligands, although it has been thought that the presence of a coordinated water molecule was necessary for activity and that this was gradually lost. The sixth position was then filled by a solvent molecule which is replaced by the oxygen molecule. It is now known that water is not necessary for carrier properties.

Provided pyridine is present, species I in solution absorbs oxygen giving 1:1 and 1:2 oxygen adducts, but in the solid state the pyridine adduct is inactive and a five-coordinate dimer. Calligaris and his co-workers suggest that the πMO formation between metal orbital and the conjugated ligand necessitates σ donation in the axial position to increase the availability of electrons on the metal atom. There is in fact a relationship between the pK_a of the base and the polarographic half wave potential of the Co(II)–Co(III) couple, so confirming this suggestion.

Evidence has been presented recently[17b] for the formation of a 1:1 adduct between a vanadium(III) salen complex and oxygen in pyridine. This is of great interest in view of the fact that vanadium(III) acts as an oxygen carrier in ascidians[17c].

Aminoacid and dipeptide complexes of cobalt(II)

A large number of cobalt(II) complexes with amino acids and dipeptides have oxygen-carrying properties[1,19]. Typical amino acids are histidine, histamine, lysine, arginine and serine. Studies have been carried out in solution, but some oxygenated solids have been isolated in which one mole of oxygen is added per two moles of complex. It is valuable to compare the behaviour of a range of metal complexes of histidine. Thus while iron(II) complexes are irreversibly oxidized to iron(III), the copper(II) and nickel(II) species are not affected by oxygen. The redox potential of the cobalt(II)–cobalt(III) couple lies in an intermediate state in terms of the equilibrium between M(II), O_2 and M(III): i.e. Fe < Co < Ni < Cu. However, it should be noted that a recent report[19a] has shown that the nickel complex of the tetrapeptide tetraglycine is readily oxidized by molecular oxygen under mild conditions. It appears that in this case nickel(II) is activated by peptide coordination so that it reacts with molecular oxygen and catalyses the oxidation of the peptide to a number of products (i.e. probably through the formation of nickel(III)).

Gillard and Spencer[19] have measured the uptake of oxygen by alkaline solutions of cobalt(II) with nineteen dipeptides and have observed three different types of stoicheiometric behaviour: (1) O_2:4Co:8 dipeptide:(2) O_2:2Co:4 dipeptide and (3) continuous oxygen uptake with catalytic oxygenation of the dipeptide. Oxygen uptake is very pH dependent. The 2:1 complex of glycylglycine with cobalt(II) at pH 10 reacts rapidly with oxygen giving yellow (or at higher concentrations, brown) solutions of binuclear, diamagnetic oxygenated species. Depending upon the ligand, these complexes decompose at a variety of rates to yield red mononuclear cobalt(III) chelates. The authors conclude that it appears that a minimum of three N donors is necessary for formation of an oxygenated complex.

Vaska's iridium complex[20,20a]

The complex $Ir(PPh_3)_2COCl$ behaves as an oxygen carrier. The oxygenated form is diamagnetic and has a trigonal bipyramidal structure, containing π-bonded oxygen (Figure 7.4). This compound has proved a useful

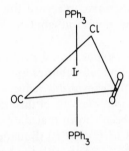

Figure 7.4 Vaska's iridium complex.

model for more complex systems. The parent complex only reacts with oxygen in benzene solution, taking up one mole of oxygen per mole of complex, with a colour change from yellow to red. Molecular weight studies show it to be monomeric in benzene. It yields hydrogen peroxide on treatment with acid, and is diamagnetic. The infrared spectrum of the carrier shows a band at 860 cm^{-1} characteristic of a coordinated peroxy group. The compound will also reversibly absorb other molecules such as hydrogen, chlorine and the hydrochloric acid molecule. It appears that the oxygen molecule is π-bonded only and it has been suggested that the carrier properties of the complex may be associated with this. Certainly Vaska's complex will reversibly bind a number of π-bonding ligands. The reversible addition of molecular oxygen may then be represented by Equation (8).

$$\underset{\underset{PPh_3}{|}}{\overset{\overset{PPh_3}{|}}{OC}} Ir \overset{Cl}{\diagup} \ + O_2 \ \underset{k_{-1}}{\overset{k_2}{\rightleftharpoons}} \ \underset{\underset{PPh_3}{|}}{\overset{\overset{PPh_3}{|}}{OC}} Ir \overset{Cl}{\underset{O}{\diagup}} \overset{O}{\diagdown} \qquad K = \frac{k_2}{k_{-1}} \qquad (8)$$

X-ray structural studies[21] show that both oxygen atoms are equidistant from the iridium atom and it should be noted that the structure could be 5- or 6-coordinate depending upon whether the coordination positions are towards the oxygen atoms or only to the double bond of the oxygen molecule. The oxygen–oxygen bond length (0·130 nm) is close to that of a superoxide species but this is not a true representation of the complex as this implies iridium in an oxidation state of II. Iridium(II) species (d^7) are paramagnetic. However, a peroxide structure implies iridium in the III oxidation state (d^6), consistent with a diamagnetic complex. It is probably best, at this stage, to regard this complex as a dioxygen complex of iridium(I) and not to attempt to interpret the O–O bond length.

The function of the phosphine ligand may be readily understood in broad terms. This is a group that stabilizes low oxidation states, and will therefore help to prevent the irreversible reduction of oxygen by electron transfer from the iridium ion. The role of this ligand has been examined in more detail.[20a] It has been shown that the rate constants for oxygen addition (k_2) to trans-[IrCl(CO)(R$_3$P)$_2$] (Equation (8)) and the stability of the dioxygen adduct (K) increase with the basicity of the group R in the phosphine ligand, provided the groups being considered are similar in geometry. The geometry of group R can have a considerable effect. Thus, oxygen addition does not occur when R = ortho-CH$_3$C$_6$H$_4$, due to the steric effect of the methyl group, although the electronic properties of the group are favourable. No reaction occurs, in addition, when R = C$_6$F$_5$ due to the lack of electron density on the metal. Vaska and Chen also noted a correlation between constants k_2 or K and the carbonyl stretching frequency (v_{CO}) in the complexes when steric effects are absent. Thus for R = C$_6$F$_5$, v_{CO} = 1994 cm^{-1} while values lie between 1966 and 1934 cm^{-1} for other phosphines which do absorb oxygen. These results are important in helping to define the conditions under which a transition metal ion will bind oxygen reversibly.

Complexes with dimethylglyoxime and related ligands

It has been seen that a number of oxygen-carrying cobalt(II) complexes involve ligands that provide four in-plane donor atoms, such as the 'salen' complexes discussed earlier. Thus corrin, porphyrin and dimethylglyoxime[22] complexes carry oxygen, while the use of EPR techniques has shown[22a] that phthalocyanine complexes also form 1:1 complexes with oxygen. Both the cobalt(II) corrinoids and phthalocyanins do this in the absence of a nitrogen donor ligand in the axial position, as is usually required. We will consider dimethylglyoxime complexes of a number of metal ions in detail and compare them with vitamin B_{12} oxygen carriers.

Bis(dimethylglyoximato)iron(II) readily reacts with oxygen in the presence of ligands[22] such as pyridine, ammonia, histidine or imidazole. The reaction may be reversed by bubbling nitrogen through the solutions. Strongly alkaline solutions of bis(dimethylglyoximato)nickel(II) also reversibly absorb oxygen, while the corresponding cobalt(II) complexes have been studied[24] as models for the oxygen carrier vitamin B_{12r}. Bis(dimethylglyoximato)cobalt(II), CoD_2H_2 (where $DH_2 = $ dimethyl-glyoxime) is a paramagnetic solid ($\mu = 1.70$) which will react with certain bases to give 1:1 and 1:2 adducts which are soluble in solvents such as acetone, benzene and dichloromethane. Such solutions absorb 0.5 moles of oxygen per mole of cobalt, giving intensely coloured products in which the oxygen is held only weakly, being readily removed by a stream of nitrogen or by the action of heat. These species are easily oxidized to cobalt(III), particularly in the presence of traces of water. The initial oxygen adducts are diamagnetic, μ-peroxo complexes which may be isolated as solids. The solutions gradually turn brown with the evolution of oxygen, and the formation of μ-superoxo species. (Equation 9).

$$2BCo(D_2H_2)-O_2-Co(D_2H_2)B \rightarrow BCo(D_2H_2)-O_2-Co(D_2H_2)B^+$$

$$+ 2Co(D_2H)B + \tfrac{1}{2}O_2 + H_2O \qquad (B = Base) \qquad (9)$$

On standing, the solution of the superoxo species (B = pyridine) undergoes further change, as is evidenced by the change in the EPR spectrum, apparently to the mononuclear peroxo complex py-$Co(D_2H_2)O_2^-$. Vitamin B_{12r}, on reaction with oxygen, gives a peroxo radical[25] having an EPR spectrum very similar to that of py-$Co(D_2H_2)O_2^-$, suggesting that the electronic environment of the Co–O_2^- group must be very similar in both cases. It is probable that the cobalt(II) chelates react with oxygen in two stages (Equations 10 and 11) to give the μ-peroxo complexes.

$$Co^{II} + O_2 \rightleftharpoons Co.O_2^- \qquad (10)$$

$$Co.O_2^- + Co^{II} \rightarrow Co-O_2-Co \qquad (11)$$

It is possible that steric reasons prevent the formation of the bridging peroxide in the case of vitamin B_{12r}. However, other cases have been reported[26] in which the oxygenation reaction proceeds no further than reaction (10), that is with the formation of mononuclear peroxo radicals.

SOME COMMENTS ON DIOXYGEN COMPLEXES

In the preceding sections it has been noted that a number of stoicheiometries for oxygen addition have been observed and that the oxygen in such complexes may be formulated as a peroxide or a superoxide on the basis of magnetic measurements and oxygen–oxygen bond lengths. Such formulations may, of course, be oversimplified, as has been demonstrated for Vaska's iridium complex. Again, in cobalt-containing systems it has been suggested that cobalt is present as cobalt(III). X-ray studies[27] on the adduct of oxygen with the bis(salicylaldehyde)ethylenediiminecobalt complex, containing dimethylformamide, show that the Co–O–O–Co group is twisted, as observed for peroxides, but the oxygen–oxygen bond length is much shorter than that expected for a peroxide, suggesting that complete transfer of electrons to oxygen from the cobalt has not occurred. In Table 7.3 are listed oxygen–oxygen bond lengths of oxygen in some dioxygen complexes, together with details of some other species for comparison purposes.

TABLE 7.3 Form of oxygen in carrier

Compound	O—O structure	O—O (nm)
H_2O_2	Peroxide O_2^{2-}	0·148
BaO_2	Peroxide O_2^{2-}	0·149
KO_2	Superoxide O_2^-	0·128
$Ir(PPh_3)_2COClO_2$	Superoxide O_2^-	0·130
$CoL_2{}^aO_2$	Peroxide	0·135
$[(NH_3)_5CoO_2Co(NH_3)_5]^{4+}(SCN)_4$	—	0·165
$[(NH_3)_5CoO_2Co(NH_3)_5]^{5+}(SO_4)(HSO_4)_3$	—	0.131

^{a} L = bis(salicylaldehyde)ethylenediimine.

It is clear, particularly for complexes involving bridging oxygen, that it is difficult to assign formal oxidation states to the metal and the oxygen. It is probably better to consider the grouping $M–O_2–M$ as one unit and not to attempt such a distinction.

The bonding of oxygen to the metal ions may be readily understood. Figure 7.5 illustrates the possible bonding of oxygen to Vaska's complex and in bridging in cobalt complexes.

In the first example, the oxygen molecule donates π electron density as in ethylene complexes, with back-bonding resulting from the transfer of electron density from an appropriate filled metal orbital to the antibonding orbitals of the oxygen molecule. According to the extent of electron transfer 'reduction' of the oxygen molecule will occur.

In the second ('end on') case, oxygen acts as a σ donor, with back-bonding as before. It is usually suggested that carrier properties result from π interactions only, i.e. similar to those in Vaska's complex.

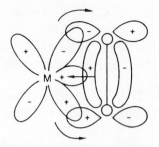

(a) π-bonding of oxygen to the metal ion.

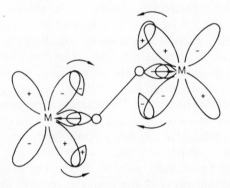

(b) σ-bonding of oxygen in the cation $[(NH_3)_5Co\!-\!O\!-\!O\!-\!Co(NH_3)_5]^{4+}$.

Figure 7.5 Oxygen-binding by a transition-metal ion.

Taube[5] has summarized criteria for oxygen-carrying properties. These will be considered for the reaction (12)

$$2M^{II}L_6 + O_2 \rightleftharpoons L_5M^{III}\!-\!O\!-\!O\!-\!M^{III}L_5 + 2L \qquad (12)$$

(1) Obviously for reversible oxygen absorption the equilibrium constant for this reaction must be close to unity. Taube has shown that for the half reaction (13), this must imply a value of E° of about -0.5 V.

$$L + M^{II}L_5H_2O \rightleftharpoons M^{III}L_6 + H_2O + \varepsilon \qquad (13)$$

(2) The side reaction of $M^{II}L_6$ with L_5MOOML_5 must not be too rapid. The rate of this reaction may be reduced by the presence of bulky L groups or by incorporating the complex into a high polymer or by reaction in the solid state.

(3) The $L_5M^{III}\!-\!\bar{O}\!-\!\bar{O}\!-\!M^{III}\!-\!L_5$ system will not be thermodynamically stable with respect to $2MOH^{2+} + H_2O_2$. The rate of this dissociation

must be low. This is accomplished for Co^{II}–Co^{III} systems by the fact that cobalt(III) is inert.

(4) Any oxidation state of the metal higher than $+3$ should not be readily accessible, for then complete reduction of the O–O bond could occur, and

(5) The equilibrium (12) should be achieved rapidly. This occurs if the complex in the II oxidation state is five-coordinate (preferably a square pyramid).

The varying activities of the cobalt complexes discussed in this chapter as oxygen carriers may readily be related to these criteria, particularly for example with the complexes of cobalt(II) with the Schiff base ligands salicylaldehydeethylenediimines. However, it should be noted that variation in ligand may have considerable effect on the relevant properties of these complexes. There is a delicate balance between partial electron transfer to oxygen, irreversible reduction of oxygen and failure to show oxygen-carrying properties. Thus the replacement of the chloro group in Vaska's salt by an iodo group (i.e. $Ir(PPh_3)_2Co$) gives a complex which is irreversibly oxidized by oxygen. The greater charge-releasing properties of the iodide result in the metal having greater electron density and hence a more facile reduction of oxygen. Appropriately, the oxygen–oxygen bond length in this compound is increased to 0·151 nm.

Before turning to consider natural oxygen carriers in some detail, it may be noted that many of the points covered in this discussion of model compounds lend themselves to the discussion of these natural carriers. Here it is the function of the protein to stabilize the lower oxidation state of the transition metal ion in addition to providing the open structure for receiving the oxygen molecule and generally adjusting the redox potential of the metal ion.

NATURAL OXYGEN CARRIERS

The hemoproteins

As previously noted, heme is a square planar iron(II) complex of proto-porphyrin (Figure 5.6). The four-coordinate iron may accept two additional ligands. In hemoproteins these are usually histidine groups, but the sixth position is often filled by weakly bound ligands, so explaining why they may be readily replaced by π-bonding ligands, such as cyanide and carbon monoxide. In such cases the high ligand field strength of these ligands results in an electronic rearrangement from the weak field, high spin complex to a low spin one. Iron(III) protoporphyrins, the hemins, may be formed by irreversible oxidation of the hemoproteins, particularly in acid solutions or in media of high dielectric constant. The question of the ease of oxidation of iron(II) porphyrins to iron(III) porphyrins is obviously very important. The redox potential of the couple depends upon a number of factors, such as the nature of the axial ligands, the substituents of the porphyrin ring and the configuration of the protein.

The structure of hemoglobin and myoglobin

A vast amount of work has been published on these two species and a number of reviews are available (References 28–30 and references cited by these authors). Much is known about the amino acid sequence in the globin, while hemoglobin and myoglobin are well known as the first macromolecules of this type to be intensively studied by X-ray diffraction. Perutz and Kendrew and their coworkers[31–41] have published data of high resolution on a number of hemoglobins and myoglobins respectively and have laid the foundation for the investigation of the molecular basis of the physiological function of these proteins.

There is, of course, as always, the problem of relating their results to the nature of the protein in solution, but it appears that there is no obvious reason for supposing that this extension is not valid.

Hemoglobin is a tetramer of molecular weight 64,450. It is made up of two identical pairs of units (the α and β units) roughly arranged in a tetrahedron. This explains why the molecule readily splits into halves and quarters. Each unit contains a heme group together with protein, and each unit may reversibly bind oxygen. However, uptake of oxygen by one unit enhances the ability of the other groups to take up oxygen, so that the ratio of successive oxygen-binding constants is $1:4:24:9$. This means that the more oxygen there is bound to hemoglobin, so the more likely it is that further oxygen will be bound. Conversely the less oxygen bound, the more easily will it be released. This phenomenon serves to emphasize the efficiency of hemoglobin as an oxygen carrier, and ensures that all the oxygen is readily released when required. An explanation of this cooperative effect is clearly essential to the full understanding of the molecule as a whole. There is some considerable distance between neighbouring heme groups and so the effect must be transmitted through the protein structure.

Of great significance are certain differences between the oxy and deoxy forms of hemoglobin. The deoxy form is high spin, paramagnetic, while the iron atom lies above the plane of the nitrogen donors towards the imidazole axial ligand with an Fe–N bond length of 0·29 nm. The oxy form is low spin, diamagnetic and has the iron atom in the plane of the porphyrin ring with a Fe–N bond length of 0·20 nm, due to the smaller size of the low spin ion. The high-spin d^6 form is too large to fit in the 'hole' in the porphyrin. This change in the size of the iron atom on absorbing oxygen is then magnified through the interactions in hemoglobin so that substantial conformational changes occur. This explains why crystals of oxyhemoglobin crumble on loss of oxygen. The papers of Perutz should be consulted for fuller details, but it may be noted that the heme changes its angle of tilt by a few degrees, the helical regions shift relative to each other, while the subunits rotate relative to each other. An overall effect is that the distance between heme groups is decreased on oxygenation. Clearly, the interactions between subunits must be considered. These were assumed to be peptide salt linkages, as dissociation of the subunits is enhanced by the presence of neutral electrolytes in high concentration, presumably by weakening these polar interactions. The phenomenon of conformational change is clearly associated with the heme:heme cooperative effect. Myoglobin shows no conformational changes on deoxygenation. Here there is only one heme group.

Hoard[41a] has shown by crystallographic studies that the iron porphyrins show a similar high spin–low spin size effect in terms of the position of the iron atom relative to the porphyrin ring (as noted in Chapter 5) and has suggested that this might be the trigger for the conformational changes that occur when hemoglobin picks up oxygen.

Myoglobin is a simpler molecule, rather similar to the monomeric units of hemoglobin, having a molecular weight of 17,500. The structure of deoxymyoglobin has been determined at 0·14 nm resolution. The protein contains 153 amino acid residues, with no disulphide or SH groups. Like hemoglobin, a large fraction has the α-helix structure. Another common feature revealed by the X-ray studies, is that all the polar groups of the protein are on the outside of the molecule, with a resulting hydrophobic interior. This is almost certainly a very important aspect of the structure of these two proteins and may be correlated with the known fact that iron(II) porphyrins are much more stable with respect to irreversible oxidation to iron(III) porphyrins in media of low dielectric constants, such as could be provided by the non-polar aromatic amino acid residues found inside the hemoglobin and myoglobin.

The binding of the heme group to the protein is clearly through the binding of the imidazole of a protein histidine group to the metal ion, but considerable extra stabilization occurs through a number of other heme:globin interactions. Thus recent high resolution X-ray studies on oxyhemoglobin have indicated the complexity of these interactions. Each heme group of oxyhemoglobin is in van der Waals contact with about sixty atoms of the protein. Almost all these atoms belong to amino acid residues common to the globins of all species of mammal hemoglobin examined so far, suggesting that they are all necessary for function.

Further specific stabilization of the heme : globin interaction may result from the interaction of the carboxyl groups of the porphyrin with basic groups of the protein. Thus, the X-ray model of myoglobin suggests that one such carboxylic acid grouping interacts with an arginine residue of the protein. Again the X-ray model shows that the vinyl groups of the heme are directed towards the hydrophobic interior of the peptide chains and hence a stabilizing interaction with these aromatic amino acids residues will result.

The Bohr effect

The oxygen affinity of hemoglobin varies with the pH of the medium. This is the Bohr effect and it may be attributed, in general terms, to the effect of pH on the interaction between the heme group and ionizable groups of the protein. The effect is complex, and the study of this phenomenon in hemoglobin is very difficult, due to the interactions between the four heme groups.

Recent evidence has suggested that an imidazole group of histidine is associated with oxygen uptake. Thus, the pH range in which the Bohr effect is operative is in the range in which imidazole dissociates, while pK_a measurements over a temperature range show that the heat of ionization of this 'oxygen-linked' group is characteristic of the heat of ionization of imidazole. It should be noted that this is not the iron-binding imidazole group.

Models for hemoglobin

A number of iron(II) complexes have been reported which show oxygen-carrying properties. Of special interest, however, are the results of Wang,[43,44] in which he synthesized a material having hemoglobin-like properties involving an environment of low dielectric constant for the oxygen molecule. Wang had previously suggested that the stability of oxyhemoglobin could be attributed to the hydrophobic environment produced by the protein, the presence of which has now been confirmed by X-ray studies.

Wang treated a benzene solution of the diethyl ester of hemin and an excess of 1-(2 phenylethyl)imidazole with aqueous alkaline sodium dithionite, under carbon monoxide. After centrifugation, the bright red benzene solution was mixed with a 10 % solution of polystyrene in benzene and dried in a warm stream of carbon monoxide. A transparent film was obtained having the 1-(2-phenylethyl)-imidazolecarbonmonoxyheme diethyl ester embedded in a matrix of polystyrene and the substituted imidazole. The bound carbon monoxide could be removed by treatment with nitrogen gas for several hours (as indicated by the absorption spectra). On exposure to air, the resulting material rapidly combined with oxygen giving a product having an oxyhemoglobin type spectrum. The oxygenation is reversible and the cycle may be repeated a number of times. The film may also be kept in air saturated water for days without oxidation to iron(III). No previous model showed such stability. In contrast, however, a film made with free heme instead of the diethylester was oxidized immediately to iron(III) on exposure to air. This suggests that here the free hemes are not embedded in the hydrophobic matrix but rather are exposed.

A number of other similar studies have also been reported.[45,46]

The reversible addition of oxygen to hemoglobin

A great deal of work has been carried out on the kinetic and thermodynamic aspects of the interaction of hemoglobin and myoglobin with oxygen. This is discussed in a number of recent reviews. Comparatively little has been done in this connection on model compounds. We shall not attempt to cover this material. However, much has yet to be done on the details of this reaction, particularly on the interactions between heme groups. Thus, it has been pointed out that the change in spin state, from high to low, produced in the ferrous ion by addition of oxygen would result in a change in metal ligand bond length of about 0·01 nm. This small change is amplified through the ligand structure in such a way that the reactivity of the second heme group to oxygen some 2·5 nm away, is enhanced considerably. Such changes will best be observed by the use of high resolution NMR coupled with X-ray structural data. At present little has been done in this connection.

We are concerned particularly with the reversible binding of the oxygen molecule to an iron(II) species. The importance of the five-coordinate open structure of this species and the hydrophobic nature of the neighbouring protein groups has already been emphasized. Griffith[47] had predicted that the oxygen molecule would be bound in a structure in which both oxygen atoms were equidistant from the metal. This is similar to that observed in Vaska's complex, and may be understood in terms of π-bonding alone.

Griffith has also suggested that the ligand field of the iron heme system causes a splitting of the normally degenerate antibonding molecular orbitals of the oxygen molecule. The separation between the two orbitals is great enough to force spin pairing. This, with the existence of low spin Fe(II), will account for the diamagnetism of the oxyhemoglobin system and may possibly contribute to the stability of the oxygen molecule with respect to reduction.

The role of the fifth ligand to iron, the imidazole, is important in two respects. It provides direct heme:globin interaction, and it will influence electron distribution in the heme-oxygen adduct. Imidazole is both a strong σ donor and a π acceptor ligand. The σ donation to the iron atom will in turn help to increase the donor strength of the metal t_{2g} orbital, with the acceptor properties being available to prevent too excessive a build-up of charge. This scheme is represented in Figure 7.6.

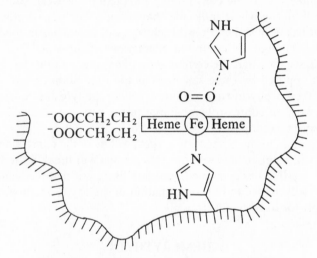

Figure 7.6   The binding of oxygen in hemoglobin.

The second imidazole group is included in this scheme, but the mode of interaction is not defined. It may be that the importance of this factor is not great. Myoglobin has been isolated[30] from *Aplysia*, a Mediterranean mollusc, which, while containing one heme group per molecule of molecular weight 20,000, has a very different amino acid composition and other properties from those of mammalian myoglobins. This species shows no Bohr effect and apparently no oxygen linked imidazole group, but its oxygen affinity is much lower than that of other myoglobins.

Perutz[41b] has recently suggested an explanation for the cooperative effect in hemoglobin, based upon a comparison of subunit interactions in the oxy- and deoxy forms of the molecule. The last amino acid residue in the α chain is arginine: the last but one is tyrosine. The corresponding residues in the β unit are histidine and tyrosine. In the deoxy form these residues interact with neighbouring subunits so that the molecule is constrained. In the oxy form these interactions do not occur and the residues may take up a number of positions. These interacting groups are clearly some distance from the iron atoms, but it was noted that the rearrangements resulting from the reaction with oxygen cause a narrowing of the pocket containing the tyrosine residue at the end of the chain and its expulsion from the pocket. Thus one constraint is removed. But this in turn causes the cleavage of the salt links between the terminal acid and other groups, so that the globin can now take up the conformation of the oxy form of hemoglobin. As each iron atom in the tetramer picks up an oxygen molecule so the constraining interactions between subunits are broken off till the whole molecule changes into the 'oxy' conformation. Indeed, it has been shown that this change will take place in the absence of O_2 when enough of the subunit interactions are broken.[41b]

We can see then that the combination of oxygen with the deoxy hemoglobin is opposed by these salt links as the iron cannot move into the plane of the porphyrin ring without expelling the tyrosine residues from their pockets. In contrast, hemoglobin in the non-constrained form *will* accept oxygen readily.

Hemoglobin, in the absence of oxygen, will be in the constrained form. Oxygen will combine and subunit interactions will break. However, the more oxygen that has combined, the more likely is it that the hemoglobin molecule will change to the conformation of the oxy form, in which case the affinity for oxygen will rise.

HEMERYTHRIN

This non-heme iron protein is found in the following four phyla: branchiopods, polychaetes, priapulids and sipinculids. The hemerythrin of *Golfingia gouldii* has a molecular weight of 107,000, and consists of eight identical subunits containing two iron atoms each. Each pair of iron atoms binds one molecule of oxygen, presumably in a similar fashion to that already noted for certain of the cobalt(II) synthetic oxygen carriers, that is with bridging oxygen. The protein occurs as the octameric species, but the existence of an equilibrium between this and the monomers was shown by the use of succinyl hemerythrin in recombination experiments with the native protein, and by observing the effect of added anions on the reaction of hemerythrin with mercaptan reagents. In the latter case it was suggested that the presence of these potential ligands speeded up this reaction, as the ligand would shift the octamer:monomer equilibrium towards the monomer with presumed increased exposure of SH groups.

Deoxyhemerythrin is a paramagnetic iron(II) species, but oxyhemerythrin may be formulated in a number of ways depending upon the extent to which electron transfer to the oxygen has occurred and the resulting oxidation state of the metal ion. As was the case for hemoglobin and myoglobin, the oxy- and deoxyhemerythrins may be oxidized to the iron(III) (met) species. Oxyhemerythrin is slowly converted to the deoxygenated met form unless oxygen is rigorously excluded. In fact it is not possible to prepare oxyhemerythrin which is free from iron(III) species.

Each subunit molecule contains some 113 amino acid residues and the amino acid sequence has been partly determined.[50] The nature of the metal-binding groups in the protein has been studied.[50a] Four of the seven histidine residues per subunit are ligands to iron, while studies of the nitration of hemerythrin monomer by tetranitromethane show that all five tyrosines in the apoprotein are nitrated but only three residues react in oxy- and methemerythrin suggesting that two are protected *via* coordination to iron (tyrosines 8, 109). There is no differential reactivity towards N-bromosuccinimide oxidation suggesting that tryptophan is not a ligand. Cysteine has also been rejected as a ligand. Assuming that the sixth position is occupied by oxygen or water and that the iron is 6-coordinate this means that two ligands are still unaccounted for. The cysteine group may be involved in the association of the monomers, as the octamer dissociates into monomers if this group is modified.

Oxygen binding

The nature of the oxygen binding in hemerythrin has been the subject of controversy. It has been suggested that oxyhemerythrin contains iron(III) and peroxide and alternatively iron(II) and oxygen. These are represented in Figure 7.7, together with other possibilities.

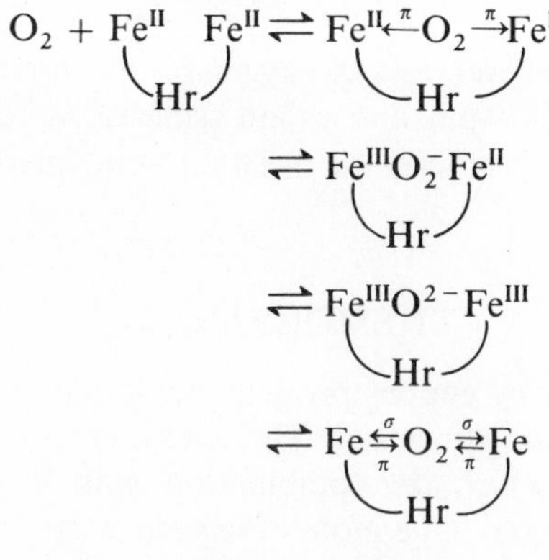

$$O_2 + Fe^{II} \underset{Hr}{\overset{}{\bigcup}} Fe^{II} \rightleftharpoons Fe^{II} \overset{\pi}{\leftarrow} O_2 \overset{\pi}{\rightarrow} Fe^{II} \underset{Hr}{\overset{}{\bigcup}}$$

$$\rightleftharpoons Fe^{III} O_2^- Fe^{II} \underset{Hr}{\overset{}{\bigcup}}$$

$$\rightleftharpoons Fe^{III} O^{2-} Fe^{III} \underset{Hr}{\overset{}{\bigcup}}$$

$$\rightleftharpoons Fe \overset{\sigma}{\underset{\pi}{\leftrightarrows}} O_2 \overset{\sigma}{\underset{\pi}{\rightleftharpoons}} Fe \underset{Hr}{\overset{}{\bigcup}}$$

Figure 7.7 Formulations of oxyhemerythrin.

A recent Mossbauer study[51] has examined the state of iron in hemerythrin. As a necessary preliminary the authors prepared a number of binuclear iron complexes, formally in the III oxidation state, with iron: iron interactions, and measured their Mossbauer spectra in the presence and absence of a magnetic field. The results for these systems were then compared with those obtained for oxy and deoxy hemerythrin and some methemerythrin species. Differences in magnetic susceptibilities were also measured by the NMR method for oxy and deoxyhemerythrin, the metaquo and various liganded met derivatives. This data showed that oxyhemerythrin and the various met species have lower effective magnetic moments than does deoxyhemerythrin. The low moment of the met species could either result from the formation of a low spin complex (i.e. in the presence of a strong ligand field) or to some interaction between iron atoms in each subunit, through the bridging ligand. The difference in susceptibility between deoxyhemerythrin and oxyhemerythrin was slightly lower than that observed for high spin iron(II) compounds, i.e. it is fairly consistent with the conversion of high spin iron(II) to low spin iron(II).

The Mossbauer spectra support the suggestion that there is iron-iron interaction in the methemerythrin species while a magnetic field had no effect upon the spectrum of the oxycompound, suggesting the species is diamagnetic, in keeping with susceptibility and EPR studies. This is consistent with either a peroxy bridged species $Fe^{III}-O_2^{2-}-Fe^{III}$ or a system containing two iron(II) atoms and an oxygen molecule. As we have noted before it is often difficult to assign formal oxidation states to such systems. However, these results have clearly indicated that both iron atoms have identical magnetic environments, that they are interacting and that the bridging oxygen active site hypothesis for hemerythrin is correct.

Some insight into the interaction between hemerythrin and oxygen is given by ORD studies. Oxyhemerythrin has absorption bands with maximum absorption at 330, 380 and 500 nm. The latter two bands are optically active. However, loss of oxygen results in practically the loss of the Cotton effect at 500 nm, and an intensification of the effect at 380 nm, suggesting that oxygen affects the orientation of the iron bonding groups of the protein.

HEMOCYANINS

These oxygen-carrying copper proteins are found in the hemolymph of various moluscs and arthropods. The deoxyhemocyanins are colourless copper(I) species, which on combination with oxygen became blue, suggestive of copper(II). One mole of oxygen is carried per two moles of copper.

A feature of the hemocyanins is their high molecular weight. They may, however, be broken down into subunits, depending upon the conditions of ionic strength and pH. The hemocyanin of *Loligo pealei* has a molecular weight of 3,750,000, while the subunits have values between 25,000 and 75,000. Some aspects of the physiology and biochemistry of the hemocyanins have recently been reviewed.[53] Many of their properties are unusual. Thus, certain hemocyanins show an extreme Bohr effect. Their sensitivity to pH change is such that a slight increase in the acidity of about 0·13 pH units in the tissue capillaries due to the influx of carbon dioxide results in the release of nearly one third of the oxygen carried by the hemocyanin of *Loligo pealei*. A number of general factors are not yet resolved, not surprisingly. Thus it is not known what is the smallest subunit which will reversibly bind oxygen.

Metal-binding groups

On treatment with cyanide, the copper may be removed from hemocyanins giving the apoprotein. Studies[54] on the —SH content (by titration with Ag^+ or Hg^{2+}) on native hemocyanins and the apoprotein from various species have led to conflicting reports as to whether copper is bound by —SH groups. In some cases the number of titratable groups is increased by the removal of the metal ion, while in other cases there is no change. It may well be that copper is bound in this way in certain hemocyanins.

Oxyhemocyanin has an absorption spectrum containing four bands, at 347, 440, 570 and 700 nm of molar extinction coefficients 8900, <500, 500 and <500 respectively. The low wavelength band is of interest as it has an extremely high molar extinction coefficient. Copper blue proteins usually show three absorption bands, at around 450, 600 and 750 nm. Those with more than one mole of copper per mole of protein show an additional band around 330–340 nm. Those species which do not have this band do not contain any copper(I) component or bind oxygen and a possible explanation[54,55] of the source of this band is that it results from charge-transfer occurring between a pair of copper atoms or between copper and oxygen. As has been noted earlier, in general the spectra of these blue complexes are unlike those of model complexes.

As with hemerythrin, ORD measurements on hemocyanins indicate changes on deoxygenation. Studies on hemocyanin and the apoprotein of *Loligo pealei* indicated[56] that the intrinsic Cotton effects of the two proteins did not change in the presence of oxygen. The configuration of the protein did not change therefore when copper or oxygen were absent. Negative Cotton effects in oxyhemocyanin only were observed at 347 and 580 nm. These were not seen in the deoxy form or in the apoprotein. This shows that the extrinsic Cotton effect is associated with the copper–oxygen chromophore. It has been suggested by Van Holde that the full ORD/CD study strongly suggests a partial copper(II) state, that oxygen binding results in a lower symmetry, and that imidazole groups of histidine are copper-binding groups.

EPR spectra

EPR data for copper proteins in general have been utilized as a tool for studying the extent of conversion of copper(I) to copper(II), the covalence of the metal–ligand bond, and possibly the nature of the donor atoms from the study of ligand hyperfine structure. However, little has been done on single crystals of copper proteins. Usually spectra are measured on frozen solutions. The EPR spectra of hemocyanins have not, therefore, been able to contribute much to the solution of these problems for these particular proteins.

The diamagnetic nature of deoxyhemocyanin has been amply confirmed by magnetic susceptibility measurements. However, oxyhemocyanin from *Cancer magister* also showed no EPR signal,[58] while a lobster hemo-cyanin has given an EPR signal[59] which accounts for one quarter of the total copper. Methemocyanins do have an EPR spectrum. These results are difficult to interpret. The absence of the signal may result from line broadening effects and magnetic susceptibility measurements are really necessary to confirm diamagnetism. In any case, a more fundamental underlying problem here is the familiar one of the nature of the copper–oxygen structure. As was noted in an earlier section, when discussing pairs of metal ions linked by oxygen ($Cu–O_2–Cu$ and also hemerythrin), it may be unrealistic to attempt to ascribe formal states to the metal and oxygen, and it is better to consider the group as a whole. Indeed, as has already been emphasized, the very existence of this problem may reflect a state of affairs that is essential for the function of these oxygen carriers.

Williams and Morpurgo[60] have compared the oxygen-carrying proper-ties of the hemocyanins with the oxygen-utilizing properties of other copper protein redox catalysts. They have suggested that the carrier properties of the hemocyanins may not necessarily be due to the presence of a hydrophobic environment for the oxygen molecule, but rather to the absence of oxidizable groups such as thiols in the environment of the metal. Water can be reduced and so this too must be excluded. (For this, of course, a hydrophobic environment *is* necessary.) Such a suggestion is consistent with the properties of many of the model oxygen carriers dis-cussed earlier. Williams and Morpurgo have been able to distinguish between oxidases, electron transferases and oxygen carriers for a number of iron- and copper-containing proteins on the basis of this suggestion.

Oxygen-binding

In the preceding discussion, it has been tacitly assumed that the oxygen molecule bridged the two copper atoms in hemocyanins (Cu–O=O–Cu,

or Cu–O=O–Cu). However, some recent studies[61] on the infrared spectrum of the carbon monoxide bound to hemocyanin suggest that the oxygen molecule may in fact be bound to only one copper in oxyhemocyanin. Hemocyanin binds CO if oxygen is excluded. The infrared spectrum showed v_{CO} at 2063 cm^{-1}, a value typical of mononuclear non-bridging carbonyl groups and higher by 100–200 cm^{-1} than would be expected for a bridging carbonyl. If it is assumed that dioxygen is similarly bound, then some other role must be suggested for the second metal atom, possibly involving a Cu–Cu bond or a Cu–L–Cu bridge.

REFERENCES

1. L. H. Vogt, H. M. Fargenbaum and S. E. Wiberley, *Chem. Rev.*, 1963, **63**, 269
2. J. A. Connor and E. A. V. Ebsworth, *Adv. in Inorganic and Radiochem.*, 1964, **6**, 279
3. F. Basolo and R. G. Pearson, *Mechanism of Inorganic Reactions*, (2nd Ed.), p. 641, Wiley, New York, 1968
4. E. Bayer and P. Schretzmann, *Structure and Bonding*, 1967, **2**, 181, Springer-Verlag, Berlin, 1967
5. H. Taube, *J. Gen. Physiol.*, 1965/66, **49** (No. 1, Pt. 2), 29
6. S. Fallab, *Angew. Chem. Int. Ed.*, 1967, **6**, 496
7. W. M. Latimer, *Oxidation Potentials*, (2nd Ed.), Prentice-Hall Inc., Englewood Cliffs, New Jersey, 1952
8. A. Werner and Mylius, *Z. anorg. Chem.*, 1898, **16**, 245
9. J. Bjerrum, *Metal Ammine Formation in Aqueous Solution*, P. Hasse and Son, Copenhagen, 1941; J. Simplicio and R. G. Wilkins, *J. Amer. Chem. Soc.*, 1967, **89**, 6092; J. Simplicio and R. G. Wilkins, *J. Amer. Chem. Soc.*, 1969, **91**, 1325; M. Mori, J. A. Weil and M. Ishigoro, *J. Amer. Chem. Soc.*, 1968, **90**, 615; G. G. Christoph, R. E. Marsh and W. P. Schaefer, *Inorg. Chem.*, 1969, **8**, 291; E. A. V. Ebsworth and J. A. Weil, *J. Phys. Chem.*, 1959, **63**, 1890; R. Davies, A. K. E. Hagopian and A. G. Sykes, *J. Chem. Soc. (A)*, 1969, 623; M. B. Stevenson and A. G. Sykes, *J. Chem. Soc. (A)*, 1969, 2293
10. W. P. Schaefer and R. E. Marsh, *J. Amer. Chem. Soc.*, 1966, **88**, 178; W. P. Schaefer, *Inorg. Chem.*, 1968, **7**, 725; W. P. Schaefer and R. E. Marsh, *Acta Cryst.*, 1966, **21**, 735

11. N. G. Vannerberg and C. Brosset, *Acta Cryst.*, 1963, **16**, 247
12. A. E. Martell and M. Calvin, *The Chemistry of the Metal Chelate Compounds*, Prentice Hall Inc., Englewood Cliffs, New Jersey, 1952
13. M. Calligaris, G. Nardin, L. Randaccio and A. Ripamonti, *J. Chem. Soc.* (*A*), 1970, 1069
14. M. Calligaris, D. Minichelli, G. Nardin and L. Randaccio, *J. Chem. Soc.* (*A*), 1970, 2411
15. S. Bruckner, M. Calligaris, G. Nardin and L. Randaccio, *Acta Cryst.*, 1969, **25**, 1671
16. C. Floriani and T. Calderazzo, *J. Chem. Soc.* (*A*), 1969, 945
17. G. Costa, A. Puxeddu and L. Nardin Stefani, *Inorg. Nucl. Chem. Letters*, 1970, 6, 191
    (a) R. De Iasi, S. L. Holt and B. Post, *Inorg. Chem.*, 1971, **10**, 1498
    (b) J. H. Swinehart, *Chem. Comm.*, 1971, 1443
    (c) D. B. Carlisle, *Proc. Roy. Soc.*, 1968, **B171**, 31
18. A. Earnshaw, P. C. Hewlett, A. King and L. F. Larkworthy, *J. Chem. Soc.* (*A*), 1968, 241
19. M. S. Michailidis and R. B. Martin, *J. Amer. Chem. Soc.*, 1969, **91**, 4683; R. D. Gillard and A. Spencer, *J. Chem. Soc.* (*A*), 1969, 2718 and references cited in these papers.
    (a) E. B. Paniago, D. C. Weatherburn and D. W. Margerum, *Chem. Comm.*, 1971, 1427.
20. L. Vaska, *Science*, 1963, **140**, 809
    (a) L. Vaska and L. S. Chen, *Chem. Comm.*, 1971, 1080
21. J. A. McGinnety, R. J. Doldens and J. A. Ibers, *Inorg. Chem.*, 1967, **6**, 2243
22. J. F. Drake and R. J. P. Williams, *Nature*, 1958, **182**, 1084
    (a) E. W. Abel, J. M. Pratt and R. Whelan, *Chem. Comm.*, 1971, 449
23. J. Selbin and J. H. Junkin, *J. Amer. Chem. Soc.*, 1970, **82**, 1057
24. G. N. Schrauzer and L. P. Lee, *J. Amer. Chem. Soc.*, 1970, **92**, 1551
25. J. H. Bayston, N. K. King, F. D. Looney and M. E. Winfield, *J. Amer. Chem. Soc.*, 1969, **91**, 2775
26. A. L. Crumbliss and F. Basolo, *Science*, 1969, **164**, 1168
27. M. Calligaris, G. Nardin and L. Randaccio, *Chem. Comm.*, 1969, 763
28. M. F. Perutz, *Europe. J. Biochem*, 1969, **8**, 455
29. G. Braunitzer, K. Hilse, V. Rudloff and N. Hilschmann, *Adv. in Protein Chem.*, 1964, 19, Academic Press, New York
30. A. Rossi Fanelli, E. Antonini and A. Capulo, ibid, 1964, 73, Academic Press, New York
31. J. C. Kendrew, H. C. Watson, B. E. Strandberg, R. E. Dickerson, D. C. Phillips and V. C. Shore, *Nature*, 1960, **185**, 422
32. J. C. Kendrew, R. E. Dickerson, B. E. Strandberg, R. C. Hart, D. R. Davies, D. C. Phillips and V. C. Shore, *Nature*, 1960, **185**, 422
33. C. L. Nobbs, H. C. Watson and J. C. Kendrew, *Nature*, 1966, **209**, 339
34. H. Muirhead, J. M. Cox, L. Mazzarella and M. F. Perutz, *J. Mol. Biol.*, 1967, **28**, 117
35. M. F. Perutz, H. Muirhead, J. M. Cox, L. G. Goaman, F. S. Matthews, E. L. McGandy and L. E. Webb, *Nature*, 1968, **219**, 29
36. M. F. Perutz, H. Muirhead, J. M. Cox and L. G. Goaman, *Nature*, 1968, **219**, 131
37. M. F. Perutz and H. Lehmann, *Nature*, 1968, **219**, 902

38. P. A. Bresscher, *Nature*, 1968, **219**, 606
39. W. Bolton, J. M. Cox and M. F. Perutz, *J. Mol. Biol.*, 1968, **33**, 283
40. W. Bolton and M. F. Perutz, *Nature*, 1970, **228**, 557
41. M. F. Perutz, J. F. Trotter, L. R. Howells and D. W. Green, *Acta Cryst.*, 1955, **8**, 241
    (a) J. L. Hoard, M. J. Hamour, T. A. Hamour and W. S. Caughey, *J. Amer. Chem. Soc.*, 1965, **87**, 2311
    (b) M. F. Perutz, *New Scientist and Science Journal*, 1971, 676
42. A. Nakahara and J. H. Wang, *J. Amer. Chem. Soc.*, 1958, **80**, 3168
43. J. H. Wang, A. Nakahara and E. B. Fleischer, *J. Amer. Chem. Soc.*, 1958, **80**, 1109
44. J. H. Wang, *J. Amer. Chem. Soc.*, 1958, **80**, 3168
45. M. Thojo and K. Shibata, *Arch. Biochem. Biophys.*, 1963, **103**, 401
46. M. Hatano, *Kagaku to Kogyo* (Tokyo), 1965, **18**, 926
47. J. S. Griffith, *Proc. Roy. Soc. A*, 1956, **235**, 23
48. S. Keresztes-Nagy, L. Lazer, M. H. Klapper and I. M. Klotz, *Science* 1965, **150**, 357
49. S. Keresztes-Nagy and I. M. Klotz, *Biochemistry*, 1965, **4**, 919
50. W. R. Groskopf, J. W. Holleman, E. Margoliash and I. M. Klotz, *Biochemistry*, 1966, **5**, 3779; G. L. Klippenstein, J. W. Holleman and I. M. Klotz, *Biochemistry*, 1968, 7, 3868
    (a) J. L. York and C. C. Fan, *Biochemistry*, 1971, **10**, 1659; J. L. York and C. C. Fan, *Fed. Proc. Fed. Amer. Soc. Exp. Biol.*, 1970, **29**, 463
51. M. Y. Okamura, I. M. Klotz, C. E. Johnson, M. R. C. Winter and R. J. P. Williams, *Biochemistry*, 1969, **8**, 1951
52. D. D. Ulmer and B. L. Vallee, *Biochemistry*, 1963, **2**, 1335
53. F. Ghiretti, Editor, *Physiology and Biochemistry of Hemocyanins*, Academic Press, New York, 1968
54. A. Ghiretti-Magaldi, C. Nazzolo and F. Ghiretti, *Biochemistry*, 1966, **5**, 1946
55. A. S. Brill, R. B. Martin and R. J. P. Williams, *The Electronic Aspects of Biochemistry*, Pullman (Ed), p. 529, Academic Press, New York, 1964
56. K. E. Van Holde, *Biochemistry*, 1967, **6**, 93; 1964, **3**, 1809
57. T. Nakamura and H. S. Mason, *Biochim. Biophys. Res. Comm.* 1960, **3**, 297
58. D. C. Gould and A. Ehrenberg, *Physiology and Biochemistry of Hemocyanins*, Ghiretti (Ed), p. 95, Academic Press, New York, 1968
59. W. E. Blumberg, *The Biochemistry of Copper*, J. Peisach, P. Aisen and W. Blumberg (Eds), p. 473, Academic Press, New York, 1968.
60. G. Morpurgo and R. J. P. Williams in *Physiology and Biochemistry of Hemocyanins*, Ghiretti (Ed), p. 113, Academic Press, New York, 1968
61. J. O. Alben, L. Yen and N. J. Farrier, *J. Amer. Chem. Soc.*, 1970, **92**, 4475

CHAPTER EIGHT

THE ALKALI METAL AND ALKALINE EARTH METAL CATIONS IN BIOLOGY

A considerable amount of information is available on the role of the Group IA and Group IIA metal ions in biological processes. Some of this can now be brought together in a fairly cohesive manner, while current developments in experimental approaches to studying the role of these ions suggest that much more will be achieved in this field in the coming years.

Some references have been made in Chapter I to the role of these elements, while the function of magnesium and calcium as activators of hydrolytic enzymes has been briefly discussed in Chapter 4. This latter topic will not be considered again. To set the scene for this chapter it is necessary to bring forward a number of facts concerning the occurrence of the four important metal ions, their role and the problems presented in the understanding of all this. We shall limit the discussion to that of sodium, potassium, magnesium and calcium cations, and in no sense will a comprehensive account be given. The aim of this chapter is to highlight some points of current interest.

THE DISTRIBUTION OF INORGANIC IONS

In general the concentration of inorganic ions varies in a systematic way inside and outside the cell. The cells are in contact with a fluid which supplies all essential materials. These substances must pass through the membrane that separates the inside of the cell from its external environment. In the other direction must pass the various waste products of the cell. It is clear, though, that the cell membrane does not in general allow molecules to pass through by simple diffusion. It is extremely selective and only certain molecules may pass through. In some cases molecules may only be allowed to travel in one direction, out but not in, for example. That this selectivity of the membrane must operate is shown by the fact that the ionic composition inside the cell is different from that outside it. Thus the interior of the cell contains large·amounts of K^+ but relatively little Na^+. ($[K^+]/[Na^+]$ lies in the range $\frac{3}{1} \rightarrow \frac{20}{1}$). Similarly there is more Mg^{2+} and phosphate inside than outside. The reverse situation[1] holds for Na^+, Ca^{2+} and Cl^-. Here the concentration is greater outside the cell than inside. These differences could not exist if the alkali metal cations freely diffuse in and out of the cells, and represent positive activity on the part of the cell membrane with due expenditure of energy. The way in which the cell membrane does this is an important area of research. It is particularly important in animal cells, more so than in bacterial cells where the membrane is much less permeable.

The distribution of the metal cations in this way inside and outside the cell allows a number of generalizations. Thus Mg^{2+} and K^+ are associated with intracellular activity and Ca^{2+} and Na^+ are involved in processes occurring outside the cell, and the two sets of ions may not replace each other. Mg^{2+} and K^+ are important activators of enzymes occurring inside the cell, while Ca^{2+} activates extra-cellular enzymes. The exclusion of Ca^{2+} from the cell under normal conditions explains why a sudden influx of Ca^{2+} acts as a trigger. Na^+ in general will only activate a few enzymes. These may not be activated by K^+ and Na^+ will not activate the K^+ dependent enzymes, though NH_4^+ will. Cell Mg^{2+} and K^+ are very important in the stabilization of RNA and the RNA/DNA synthetic systems. Thus, in resting bacterial cells (i.e. the effect produced by limiting the supply of certain requirements) the in-cell K^+, Mg^{2+} and phosphate is lower than normal. Once the limitation is removed, the cells grow but the first step prior to this is to increase the concentrations of Mg^{2+}, K^+, phosphate. This is an example of ion pumping. Once the metals are inside the cell, they will be bound by appropriate ligands in accordance with their known chemical reactivity. In general, we expect Na^+, K^+ to be bound to a single multidentate ligand, while Ca^{2+}, Mg^{2+} will act as a bridge between two ligands.

The fact that calcium is not utilized by the cell is associated with its extensive use as a structural factor, in bones and teeth for example. The concentration of calcium is controlled by two hormones, calcitonin and parathyroid hormone. The skeleton is the major reservoir of calcium, calcitonin inhibiting the release of calcium ions and parathyroid hormone mobilizing the element from the bones. Together they are probably responsible for the remodelling and maintenance of bone structure.

The existence of the sodium–potassium concentration gradient across the cell membrane has important implications in physiology. The difference in concentration results in the generation of a potential difference between the two sides of the membrane. Thus the inside of the cell is some 80 m volts negative with respect to the external surface of the membrane. The existence of this potential allows the nerve fibres to conduct impulses and hence muscles to contract.

The control of the permeability of the membrane is also dependent on the presence of divalent metal ions; probably for electrostatic interaction with two carboxylate groups.

Some details of the occurrence of metal ions are given in Table 8.1. Before considering detailed examples of the role of these metal ions in biology, we shall consider aspects of the chemistry of the Group IA and IIA cations, paying particular attention to complex formation and stability constants. It seems reasonable that the first stage in the transport of a cation through the cell membrane is the binding of that cation by components of the cell wall. It is worthwhile looking at the factors which determine the binding of these pairs of cations by model ligands. In particular we will discuss certain macrocyclic ligands, which will, for example, reverse the usual binding order for Na^+ and K^+ and bind K^+ exclusively.

TABLE 8.1 The distribution of the group IA and IIA cations in some biological systems[a]

	Na^+	K^+	Mg^{2+}	Ca^{2+}
Bacteria (dried)(mmoles/1000 g)				
E. coli	—	310	247	2·5
Aerobacter aerogenes	—	410	124	5·0
B. subtilis	—	1260	124	2·5
B. subtilis spores	—	231	206	400
B. cereus	—	1180	450	7·5
B. cereus spores	—	51	124	475
E. coli (wet)	80	250	20	5
Yeast cells	10	110	13	1·0
Skeletal muscle	27	92	22	3
Human red cell	11	92	2·5	0·1
Whole blood	85	44	1·57	—
Squid nerve (outside) (m molar)	440	22	—	—
Squid nerve (inside) (m molar)	49	410	—	—
Crab leg nerve (outside) (m molar)	510	12	—	—
Crab leg nerve (inside) (m molar)	52	410	—	—
Frog sartorius muscle (outside) (m molar)	110	2·6	—	—
Frog sartorius muscle (inside) (m molar)	15	125	—	—
Giant algal cells (m molar)				
Nitella translucens				
cytoplasm	55	150	—	—
vacuole	65	75	—	—
outside solution	1·0	0·1	—	—
Tolypella intricata				
chloroplasts	36	340	—	—
vacuole	7	100	—	—
outside	1·0	0·4	—	—

[a] For a full account of alkali metal distribution see Reference 2.

These ligands may well have considerable significance in the context of membrane selectivity.

One other problem which must be considered is the way in which the role of these metal ions may be studied. Their electronic properties, in contrast to those of the transition elements, do not lend themselves readily to this, although ^{23}Na and ^{43}Ca NMR have been used.

A number of review articles may be cited. A great deal of information is available[1,2] in 'The Alkali Metal Ions in Biology'. Other reviews have covered general aspects[3,4] of these four metal ions and also more specific topics, such as the dependence of enzymes on univalent cations[5] and K^+,[6] and cell potassium[7] and cell calcium.[8]

THE CHEMISTRY OF THE s BLOCK ELEMENTS

These elements are listed in Table 8.2, together with some physical data. Their chemistry is dominated by the tendency to attain the noble gas configuration, with the formation of ions M^+ and M^{2+} in Groups I and II respectively. Their chemistry is therefore essentially ionic chemistry, with

TABLE 8.2 Some properties of the group IA and group IIA cations

		Ionic radii		First ionization potential	Common coordination no.
		(Å)	(nm)	(e.V.)	
Lithium	Li^+	0·60	0·06	5·39	4, 6
Sodium	Na^+	0·95	0·095	5·14	6
Potassium	K^+	1·33	0·133	4·34	6, 8
Rubidium	Rb^+	1·48	0·148	4·18	8
Cesium	Cs^+	1·69	0·169	3·81	8
Ammonium	NH_4^+	1·45	0·145	—	—
Beryllium	Be^{2+}	—	—	9·32	2, 4
Magnesium	Mg^{2+}	0·65	0·065	7·64	6
Calcium	Ca^{2+}	0·99	0·099	6·11	6, 8
Strontium	Sr^{2+}	1·10	0·110	5·69	6, 8
Barium	Ba^{2+}	1·29	0·129	5·21	6, 8

the exception of lithium and particularly beryllium as indicated by the ionization potential data. The small size of the 'cations' of these two elements makes them good polarizers. Hence the chemistry of beryllium is covalent, while that of lithium shows some covalent properties. We are not, of course, particularly concerned with these, but magnesium is similar to lithium in many ways. This is an example of the diagonal relationship.

The smaller size of the Group II cations and their higher charge makes them better polarizers than the Group I cations. Univalent Group II metal cations are not obtained because the lattice energy of the MX system is very much lower than the lattice energy of the MX_2 system.

We are particularly concerned with the factors that result in ligands binding Na^+ or K^+, and Mg^{2+} or Ca^{2+} selectively. For small, simple ligands, particularly the anions of weak acids, the binding is determined by the polarizing power of the cation and therefore decreases with increase in size of the cation. This order may be changed by increasing the complexity of the ligand, as is shown for example by comparing data in Table 8.3 for acetate and nitrilotriacetate. In the latter case Ca^{2+} is bound more strongly than Mg^{2+}, while in the case of acetate formation constants follow the 'normal' sequence of $Mg^{2+} > Ca^{2+} > Sr^{2+} > Ba^{2+}$. Deviation from this order may then be due to a size effect. In general, for strong acid anions this order is also reversed, i.e. $K^+ > Na^+$ and $Ca^{2+} > Mg^{2+}$.

The importance of the size of the cation cannot be overemphasized. It is reflected in solution and structural aspects of the chemistry of the elements of Groups IA and IIA. Ionic radii are listed in Table 8.2. These are based upon the hard sphere model for ionic structures, with $r^+ + r^-$ equal to the internuclear equilibrium distance. The size of a cation is a limiting factor in determining its coordination number (the radius ratio rule). It is also bound up with the solubilities of Group IA and IIA metal salts. An ionic compound will dissolve if the energy required to break down the lattice (i.e. lattice energy) is more than compensated for by the stabilization of these ions by hydration. If the hydration energy is greater than the lattice energy, the compound will dissolve giving hydrated ions. Small cations will be good polarizers; they will therefore be heavily solvated and the hydration energy will be high. Thus lithium and magnesium salts are very soluble, and in general solubility will decrease as the groups are descended. Solubility is usually enhanced by the presence of large anions as this results in a lower value for the lattice energy of the crystal as the small cation cannot stabilize the lattice of the large anion. The same property (size) that confers high solubility upon magnesium and lithium salts also results in them being solvated. Thus magnesium sulphate which is soluble is a hydrate, $MgSO_4.7H_2O$; while insoluble barium sulphate is not solvated.

Complexes of the group IA and group IIA metal ions

These complexes are all labile. For Group IA cations, complexes are weakly formed, usually as solvates or with oxygen donors, although a number of species such as $Na(NH_3)_4I$ are known. In general, complexes

TABLE 8.3 Formation constant data[a] (log K) for group IA and IIA cations and possible probes at 25°

Ligand	Li^+	Na^+	K^+	Rb^+	Cs^+	Tl^+	Mg^{2+}	Ca^{2+}	Sr^{2+}	Ba^{2+}	Eu^{2+}	Nd^{3+}
EDTA	2.8	1.7	1.0	—	—	5.84	8.9	10.7	8.8	7.9	7.7	6.7
SO_4^{2-}	0.64	0.71	0.96	—	—	—	2.23	2.28	—	—	—	3.64
NO_3^-	-1.5	-0.5	-0.1	—	0.1	-0.3	—	0.28	0.82	0.92	—	—
$P_2O_7^{4-}$	2.4	2.3	1.5	—	2.3	1.69	5.79	5.0	4.66	—	—	—
Glycine	—	—	—	—	—	—	3.4	1.4	0.9	0.8	—	—
Acetate	—	—	—	—	—	—	0.82	0.77	0.44	0.41	—	—
Nitrilo-triacetate	—	—	1.0	—	—	4.4	5.3	6.4	5.0	4.8	—	—
Ethylenediamine	—	—	—	—	—	0.4	0.37	—	—	—	—	—
ATP^{4-}	—	—	1.0	—	—	2.0	—	—	—	—	—	—
Nonactin[21]	—	4.84	4.85	—	4.0	—	—	—	—	—	—	—
Nonactin (wet acetone)	—	2.32	4.30	—	2.60	—	—	—	—	—	—	—
Cryptate (a)[b]	4.30	2.80	<2.0	<2.0	<2.0	—	—	2.8	<2.0	<2.0	—	—
(b)	2.0	5.40	3.95	2.55	<2.0	—	<2.0	6.95	7.35	6.30	—	—
(c)	<2.0	3.9	5.4	4.35	<2.0	—	<2.0	4.4	8.00	9.5	—	—
(d)	<2.0	<2.0	2.2	2.05	2.2	—	<2.0	~2	3.40	6.0	—	—
(e)	<2.0	<2.0	<2.0	≤0.7	<2.0	—	<2.0	<2.0	~2	3.65	—	—
(f)	<2.0	<2.0	<2.0	≤0.5	<2.0	—	<2.0	<2.0	<2.0	—	—	—
Dicyclohexyl-18-crown-6[c]	—	—	2.02	1.52	0.96	—	—	—	3.24	3.51	—	—

[a] Taken from 'Stability Constants', Special publication 17, The Chemical Society, London, 1964.

[b] J. M. Lehn and J. P. Sauvage, Chem. Comm., 1971, 440.

[c] R. M. Izatt, D. P. Nelson, J. H. Rytting, B. L. Haymore and J. J. Christensen, J. Amer. Chem. Soc., 1971, 93, 1619.

The size of the cryptate cavity is estimated by the radius of the largest sphere that may fit in without strain. For (a), (b), (c), (d), (e) and (f) these values are 0.8, 1.15, 1.4, 1.8, 2.1 and 2.4 Å respectively.

either involve neutral ligands with a simple anion, or chelating anionic ligands such as the β-diketones together with neutral oxygen donor ligands such as the phosphine oxides. There has been some interest[13] in complexes of the alkali metals and a number of K^+ complexes have been examined by X-ray diffraction. Thus the complex with isonitrosoaceto-phenone involves 7-coordinate K^+, while that formed between potassium nitrophenolate and isonitrosoacetophenone has an irregular structure in which the potassium ion is surrounded by seven oxygen and one nitrogen donor. It appears that high coordination numbers are a feature of the heavier members of Group IA.

Nonactin (R = H);   *Monactin* (R = CH₃)

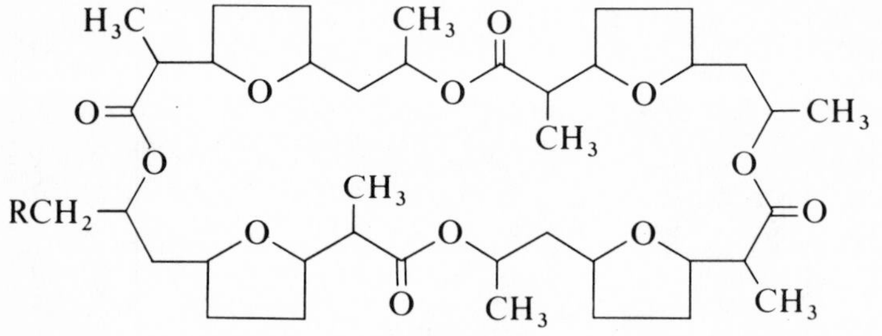

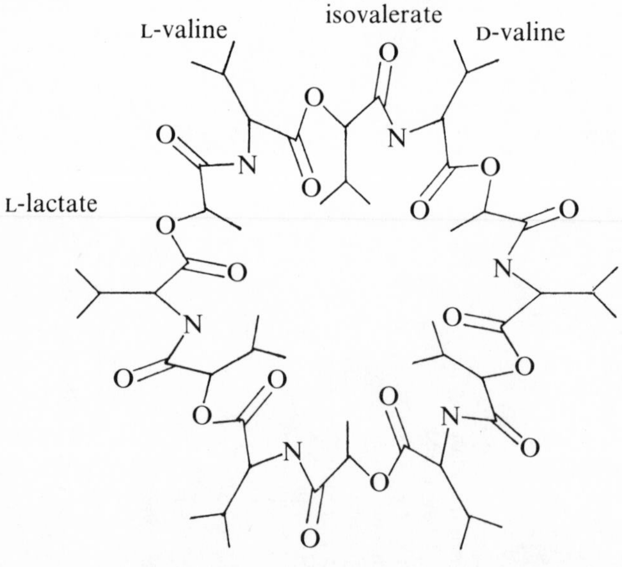

Figure 8.1   Some macrocyclic antibiotics.

As a result of their higher charge and smaller size, the alkaline earth elements form a wider range of complexes. Mg^{2+} tends to be bound by nitrogen donors, but the other Group IIA metal ions prefer oxygen donors although ammine and amine complexes of Ca^{2+} are known. EDTA complexes are also well known.

The formation constant data of Table 8.3 show, as noted above, that for many ligands, cations form complexes in the series $Li^+ > Na^+ > K^+ > Rb^+ > Cs^+$ and $Mg^{2+} > Ca^{2+} > Ca^{2+} > Sr^{2+} > Ba^{2+}$, but it is clear that simple ligands are known which will preferentially complex with either cation of the pairs Na^+, K^+; and Mg^{2+}, Ca^{2+}. Thus EDTA forms complexes with Na^+ but not with K^+, while the reverse is true for some β-diketones. However, a new class of ligand has been prepared recently, which has the property of selectively binding particular cations such as K^+ to a marked degree, largely due to a size effect. These are the cyclic ethers. A number of natural and synthetic cyclic peptides including some antibiotics such as nonactin, monactin, monamycin and valinomycin (Figures 8.1, 8.2) also have the property of selectively binding K^+

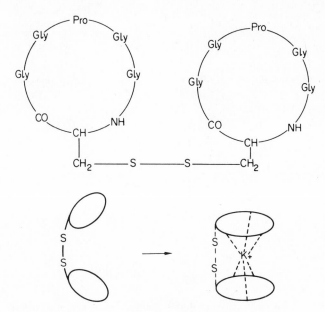

Figure 8.2 A synthetic peptide and its K^+ complex.

rather than Na^+ under certain conditions. Thus, nonactin[14] and valinomycin[15] have been used in the development of metal selective membrane electrodes which can be used to estimate K^+ in the presence of Na^+.

However, actinomycin[16] binds Na^+ rather than K^+. Here is the real point of interest; that different ligands may be designed which offer the prospect of selectively binding either Na^+ or K^+. Their potential as models for cell membranes is obvious. Similarly it has been postulated[17] that macrocyclic ligands offer the possibility of differentiating between divalent metal ions.

The cyclic ethers as ligands

Some examples are schematically represented in Figure 8.3. Many of these ethers have now been synthesized, with the number of oxygen atoms varying from 3 to 20. Many of them with between five and ten oxygen

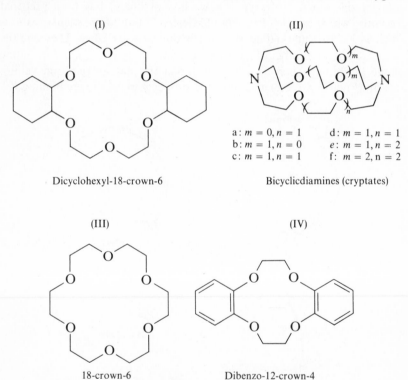

(I)

Dicyclohexyl-18-crown-6

(II)

a: $m = 0, n = 1$ d: $m = 1, n = 1$
b: $m = 1, n = 0$ e: $m = 1, n = 2$
c: $m = 1, n = 1$ f: $m = 2, n = 2$

Bicyclicdiamines (cryptates)

(III)

18-crown-6

(IV)

Dibenzo-12-crown-4

Figure 8.3   Some oxygen donor ligands. The macrocyclic ethers.

atoms form stable complexes with a range of cations, for example Li^+, Na^+, K^+, NH_4^+, Cs^+, Rb^+, Ag^+, Ca^{2+}, Sr^{2+}, Ba^{2+}, Cd^{2+}, Hg_2^{2+}, Hg^{2+}, La^{3+}, Tl^+, Ce^{3+}, Pb^{2+}. The binding strength[18] varies considerably from metal to metal. Suitable design of ligand can produce 'hole' sizes suitable for different cations. In most cases a 1:1 stoicheiometry holds.

The cyclic ethers have been termed 'crown compounds'. Individual examples are named by a prefix giving the number and kind of hydrocarbon rings, followed by a number giving the total number of atoms in the polyether ring, the name crown and then the number of oxygen atoms in the polyether ring. The nomenclature is easily understood by considering the examples in Figure 8.3.

The dicyclohexyl-18-crown-6 ligand will selectively complex with K^+ rather than Na^+, while of those studied so far, dicyclohexyl-15-crown-5 is the best complexing agent for Na^+, although dicyclohexyl-16-crown-5 is the most specific one for Na^+ in competition with the other alkali metal cations. The interactions between metal ion and ligand are electrostatic and the stability of the complex depends upon the relative positions of the oxygen atoms and the ratio of cation and hole diameters.[18] Similar considerations apply for Mg^{2+} and Ca^{2+}.

The polyethers (and antibiotics) also confer solubility upon alkali metal salts and alkaline earth metal salts. The substituents on the macrocyclic ring will make the complexes soluble in hydrophobic solvents. Thus dicyclohexyl-18-crown-6 will complex K^+ in KOH and hence allow its solution in benzene. The species (II) in Figure 8.3 is an example of a cryptate,[12] which will cause $BaSO_4$ to dissolve in water.

The cryptates are bicyclic in contrast to the cyclic ethers. This means that they can exert a 'three dimensional' selectivity compared with the 'two dimensional' selectivity of the cyclic ethers in terms of providing an appropriately sized central molecular cavity. The formation constants for the interaction of a range of cryptates with the alkali and alkaline earth metal cations have been measured[12a] and are included in Table 8.3. All are 1:1 complexes. Clearly there is a comparison between cavity size, ionic radius of the metal ion and formation constant. The stability of the complexes is increased on changing to an organic solvent. This, presumably, is a solvation effect.

Crystal structures have been reported[12b] for the complexes of sodium iodide, potassium iodide and cesium thiocyanate with cryptate (c) (m=n=1). In all cases the alkali metal ion is enclosed by the ligand and has been dehydrated. In contrast, the complexes of barium thiocyanate with certain cryptates have structures[12c] in which the barium ion has not been completely dehydrated. The Ba^{2+} ion is still enclosed by the hydrophobic shell of the ligand. All the hetero atoms of the cryptates are directed towards the inside of the cage.

The interaction of cations with the antibiotic nonactin and other cyclic peptides

Studies have been carried out on the interaction of cations with nonactin, an antibiotic that shows a high degree of cation specificity[19] for K^+ and

. Rb$^+$. The interaction of Na$^+$, K$^+$, Cs$^+$ with nonactin has been studied by 220 MHz proton magnetic resonance spectroscopy[20,21] in hexadeutero-acetone and hexadeuteroacetone-water mixtures. Formation constants have been calculated and are listed in Table 8.3. The main point of interest is that although all three ions bind to nonactin to similar extents in dry acetone, the binding of Na$^+$ and Cs$^+$ is dramatically reduced when water is present. Thus the formation constants in wet acetone for Na$^+$, K$^+$ and Cs$^+$ are 210, 2 × 10^4 and 400 respectively. The binding of K$^+$ to nonactin is obviously highly favoured in the more aqueous medium.

The results in dry acetone indicate that the free (i.e. non-hydrated) ions are all bound to the same extent by nonactin. In wet acetone the alkali metal must be stripped of its hydration shell before it may be bound in the nonactin cavity. Only in the case of K$^+$ is this favoured. This serves to emphasize the importance of the consideration of hydration energy in this context and in turn the question of ionic size and coordination number.

The NMR study has also shown the presence of conformational changes in nonactin on incorporation of the alkali iron; while a consideration of shifts in the NMR spectrum has suggested that the hydrogens whose resonances are shifted most are those near the four carbonyl groups, implying that tetrahedral coordination of K$^+$ by four carbonyl oxygens has occurred.

Eigen and his coworkers[22] have shown that the formation of the K$^+$-monactin complex in methanol occurs very rapidly. They have suggested that the fast replacement of the solvent shell of the alkali metal cation is a stepwise process, in which the twisted ring system opens up and then closes to the normal compact form. These conformational changes are represented in Figure 8.4.

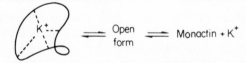

Figure 8.4   Schematic representation of the conformation of the K$^+$-monactin complex.

The cation binding properties of valinomycin (Figure 8.1) have also been studied spectroscopically,[23,24] by X-ray diffraction[25] and high resolution NMR.[26] The structure shown in Figure 8.1 is schematic, hydrogen-bonding effects will modify this. There are six ester carbonyl groups coordinated to the single potassium ion in K$^+$-valinomycin.

A synthetic peptide has been prepared[27] which will bind K^+ specifically. This is shown in Figure 8.2. Reduction of the dimer gives a monomer which will not form a complex. The dimer has been used as a model for the transport of K^+ through biological membranes. Ion transport is therefore linked to the redox state in this scheme. It has been suggested that the conformational change depicted in Figure 8.2 occurs when K^+ is bound by the dimer.

It is clearly possible then to conceive of membrane selectivity occurring by the provision of a suitable hole for metal binding in a macrocyclic ligand. In particular, the provision of an hydrophobic environment, which will require the loss of the hydration shell of the metal ion, will favour K^+ rather than Na^+, while the size of the hole will prevent the binding and transfer of metal ions of larger size.

The study of the alkali metal and alkaline earth metal cations

While certain of these ions have been followed in physiological studies by the use of radioactive tracers, there are few properties that may be utilized in the study of the environment of a bound cation. Sodium systems may be studied directly[28,29] by ^{23}Na NMR, while it has been suggested[30] recently that ^{43}Ca NMR may be utilized in the study of calcium-containing systems. The extension[29] of the technique to ^{36}K NMR appears unlikely.

Before moving to consider the use of other metal ions as probes, some examples will be given of the application of ^{23}Na NMR studies. Thus, interactions between $^{23}Na^+$ and soluble RNA in aqueous solution have been studied by this technique. At low concentrations of NaCl, the interactions obey a simple equilibrium model and the formation constant has been measured. Studies with chelating ligands, particularly the amine carboxylic acids, have also been carried out. The NMR parameter used was the ^{23}Na NMR relaxation rate. This was pH dependent due to the protonation of basic sites and the resulting effect on Na^+-ligand interaction. In particular, for Na^+ EDTA interaction a large change in the ^{23}Na relaxation rate occurred over the pH range 7·5–11 where the last proton is lost by a nitrogen of EDTA. This clearly illustrates the importance of Na–nitrogen binding. However, the ^{23}Na relaxation rate is unaffected by the presence of equimolar amounts of trimethylamine at pH 11, so indicating that the Na–nitrogen interaction in EDTA is due largely to a chelate effect. A comparison with histidine shows only a small increase in Na^+ interaction as the protons on the carboxyl, imidazole and amino groups are successively titrated. This serves to demonstrate the importance of additional carboxylate groups for the formation of stronger chelates. Again formation constants were measured. This type of approach is of particular potential in the study of biological problems and in addition it may be possible to carry out *in vivo* studies of small systems.

Probes

In order to examine the environment of alkali and alkaline earth cations when bound in biological systems, it is important that metal for metal substitution be carried out to allow the use of modern physical techniques. The presence of a paramagnetic cation will obviously be particularly useful, both for its own properties and as a perturber of the resonances of other nuclei such as hydrogen, as used by Cohn. Alternatively a main group cation must have properties such as nuclear spin so that nuclear resonance studies are possible.

There have been a number of developments in the area of metal for metal substitution recently. These have been discussed by Williams.[4] As noted in Chapter 3 any replacement must involve isomorphous replacement. The new metal ion must be of similar size and prefer the same stereochemistry as the replaced ion. Sometimes a metal ion may satisfy these requirements but is not effective in generating a biologically active species because it prefers to bind to different groups. Thus Mn(II) might be expected to be a suitable probe for Ca(II), but the preferred binding groups are dissimilar and so in fact Mn(II) has been used as a probe for Mg(II). A better replacement for Mg(II) in terms of size would be Ni(II), but the same problem exists in that Ni(II) prefers different binding groups. Some possible probes were given in Table 3.6 together with details of ionic radii. There is no probe available for sodium.

Potassium

Potassium may be replaced by a series of cations and so it is possible to compare the activity of these systems and arrive at mechanistic conclusions. Tl^+ has recently attracted[31,32] attention as a probe for K^+. Thallium binds[32] ligands more strongly than K^+ and is some ten times more effective[32] in activating the enzymes pyruvate kinase and diol dehydratase due to the increase in binding. In addition it is known[33] that erythrocytes accumulate thallium in place of potassium, so it appears that thallium is a most suitable probe. The main relevant probe property is that of NMR (Tl205, spin $= \frac{1}{2}$), but the shift in the UV band at 215 nm and the quenching of fluorescence on complex formation offer alternative probe properties. An example of thallium-205 NMR lies in the work of Kayne and Reuben.[31] These workers studied rabbit muscle pyruvate kinase which shows an absolute requirement for a divalent and a monovalent cation. The function of the divalent cation[34] has been investigated by the use of Mn^{2+}, but Kayne and Reuben carried out a double probe experiment using Tl^+ and Mn^{2+}. They have shown that the monovalent and divalent binding sites are very close together inasmuch as there is good evidence for interaction between bound Tl^+ and Mn^{2+}. It is suggested that this result implies a more direct role for the monovalent cation than the usually accepted one of stabilization of conformational changes.

An alternative probe for K^+ is Cs^+ and hence the use of cesium NMR.

Magnesium

A range of cations have been used to replace Mg^{2+} on a number of occasions. Reference has already been made to the work of Cohn[35] on the use of Mn^{2+} and the resulting perturbation of proton magnetic resonance. The presence of this paramagnetic species[36] also allows the use of EPR. A number of chemical similarities between Mg^+ and Mn^{2+} are in accord with its good probe properties, e.g. the sulphates $MgSO_4.7H_2O$, $MnSO_4.7H_2O$. Nickel has also been successfully used on one occasion, despite normally preferring to bond to different ligands.

Calcium

Mn^{2+} and Eu^{2+} are bivalent ions of appropriate ionic radii for the replacement of Ca^{2+}. Mn^{2+} is not a good example for reasons discussed earlier. Eu^{2+} appears to be a potentially useful probe[4] however, as in addition to being paramagnetic it is a Mossbauer nuclide. If the requirement for a bivalent metal ion is lowered, then the lanthanide elements in the III oxidation state may also be considered as probes. Again early work suggests[38] much promise here. Eu(III), Gd(III) and Nd(III) in particular[39] have been considered. The rare earth metal ions form complexes which are similar to those formed by the calcium ion. In the example of Nd(III) it has been suggested on the basis of a careful study of the spectra of model compounds that neodymium is bound to the protein bovine serum albumin by carboxylate groups. Hence it is suggested that calcium ion is also bound by carboxylate in this protein. This, of course, is quite in accord with what would have been expected from the known behaviour of calcium. The triply charged rare earth ions will naturally bind more strongly than Ca^{2+}, on the basis of polarizing power criteria.

The particular advantage of the lanthanide ions is that they offer a range of ions having slightly varying properties such as size. Replacement of Ca^{2+} by a series of these ions should then allow much insight into the correlation of biological activity, physical properties of the probe and the environment of the metal ion and so in turn allow greater insight into the role of Ca^{2+}.

There have been a number of studies of the effect of replacing calcium by lanthanum, particularly on the transport of Ca^{2+} and the properties of membranes. Thus, it is thought that Ca^{2+} affects the conductance of cations through nerve membranes by binding to the membrane, replacing Mg^{2+} and hence activating certain enzymes. These changes are blocked in lobster giant axons by the action[40] of La^{3+}, La^{3+} acting 'as if it were equivalent to an extraordinarily high Ca^{2+} concentration'. This is an illustration of the greater binding properties of the triply charged La^{3+}. Again[41] the efflux of ^{45}Ca from the squid giant axon is inhibited by added $LaCl_3$, the lanthanum ions competing with external calcium for binding sites on the cell membrane. Other examples are known in which La^{3+} competes successfully with Ca^{2+} for calcium binding sites.

SOME EXAMPLES OF THE ROLE OF GROUP IA AND GROUP IIA CATIONS IN BIOLOGICAL PROCESSES. METAL IONS IN BACTERIAL CELLS

Magnesium and potassium ions are present in high concentrations in bacterial cells, while Ca^{2+} is present to a lesser extent. Some interesting studies have been reported[42] in which bacteria were grown in chemostat conditions in which different cations were made the growth limiting factor.

When magnesium was made the growth limiting factor the growth of bacteria was directly proportional to $[Mg^{2+}]$. The plot of bacterial concentration against $[Mg^{2+}]$ may be extrapolated back through the origin. The RNA content varied in a similar fashion, due to changes in the cellular ribosome content. It appears therefore that Mg^{2+} is an integral part of ribosome structure. It is known[43] to affect the stability of this structure both *in vivo* and *in vitro*.

Magnesium is also found in the outer membrane of the cell. Phospholipid-containing extracts[44] from *P. aeruginosa* contain phosphate and Mg^{2+} in the ratio $7:1$.

The growth rate of bacteria is similarly dependent upon $[K^+]$. Again, it has been suggested that K^+ is associated largely with ribosome content. On the basis of *in vitro* studies it appears that the K^+/Mg^{2+} ratio of the suspending medium is important in stabilizing these particles. A high ratio tends to result in dissociation of the ribosomes into subunits and the unfolding of the RNA chain. A low ratio results in tighter cross links in the RNA and ribosomal aggregation.

There is evidence for calcium[45] in cell walls, but spores appear to contain much larger amounts of Ca^{2+} (Table 8.1). It is thought to be intracellular Ca^{2+} in these cases and to be associated with dipicolinic acid (2,6-pyridinedicarboxylic acid). Bacterial cells will also require other cations as activators for enzyme processes, but it is difficult to control the concentration of these species and to demonstrate a need for them.

Some workers have attempted to demonstrate the role of multivalent cations in bacterial cell wall structure by adding metal sequestering agents such as EDTA. It is known that Ca^{2+}, Mg^{2+} and Zn^{2+} are all components of the cell wall of *Pseudomonas aeruginosa* and it is assumed that the divalent species is necessary to bridge neighbouring carboxylate groups. Incubation of *P. aeruginosa* with EDTA causes loss of the structural integrity of the cell wall and eventually death. It is possible, by this treatment, to produce osmotically fragile species (osmoplasts) due, it is suggested, to chelation of the metal ion by EDTA and its removal from the cell wall. On addition of multivalent cations osmotically stable forms were restored. Monovalent cations could not do this, probably because they could not act as bridges. However, only those cells treated with divalent cations were capable of multiplication, those restored with trivalent cations were capable of normal respiration but not of multiplication.

When lysozyme and EDTA were used together, osmoplasts were formed which could not be restored with Ca^{2+}. This suggests that the impaired cell walls were more susceptible to lysis. It has been suggested that a lipopolysaccharide component is liberated on attack by EDTA. This is presumed to be a subunit cross linked to other subunits and to other compounds *via* cations. On removal of the cations the subunit is liberated.

ION TRANSPORT THROUGH MEMBRANES

The permeability properties of cell membranes are dependent upon the structure of the membrane and the presence of enzyme systems that operate the ion pump. The carrier must fulfil the appropriate requirements for carrying one metal and not another, as already outlined. In discussing membranes we are not only concerned with cell outer membranes[47,48]. There are the membranes that separate the internal components of the cell. The mitochondrial membranes have been much discussed. There are many types of tissue, for example the skin and urinary bladder of amphibians, which control ionic balance.

The control of cell permeability is a most important aspect of medicine. This will be discussed in more detail in Chapter 9. Drugs may, for example, act by conferring liposolubility upon other species or by preventing the transport of ions. In the latter case they may operate either by inhibiting the transport enzymes (e.g. by the cardiac glycosides) or, as in the case of tetra-alkyl ammonium salts, by blocking the access of K^+ to its binding site in the membrane. Obviously it is necessary to have an increased appreciation of the mechanisms controlling the permeability of membranes before the action of many drugs, such as the diuretics, may be understood. In this way the design of new, more effective drugs is then possible.

Reference has already been made to the sodium pump, associated with the active transport of sodium and potassium across the cell membrane (i.e. against a concentration gradient). It appears, however, that more than one mechanism exists for the transfer of Na^+, and that there are ion pumps for Cl^- transport and for the transport of other cations. Some will transport cation and anion together. An enormous amount of work has been carried out on the influx and afflux of ions. Frog skin in particular has received much attention. Its properties are quite remarkable. Thus it can transfer NaCl against a concentration gradient of around 100 to 1. Frog skin was the membrane for which certain experimental techniques (such as the use of microelectrodes) and certain basic concepts were developed. Thus it was suggested[49] in 1958 that the outer surface of the skin (the cells of stratum germinativium) could be regarded as possessing passively Na^+-permeable outer membranes and K^+-permeable inner membranes with a Na^+/K^+ exchange pump located at the inner membrane.

With the possible exception of certain plant cells, ion transport is associated with phosphorylation. Thus, if cells in general are cooled to 0°C active transport of ions ceases and the membrane behaves as a normal Donnan membrane. The same result is caused by inhibitors that affect these enzymes. Active transport is replaced by a flow of ions towards an equilization of ion content inside and outside the cell.

The sodium pump[50]

It appears that ion pumping occurs at a limited number of positions in the membrane of the cell. These are sites that may absorb an appropriate ion. The mode of transport of the ion through the membrane has been the subject of much controversy. The simplest concept is that of a carrier in the membrane which can combine with the ion to be transported, carry it to the other side and then release it. This could equally well involve a series of carriers.

The experimental approach to the problem involved a search for a cell component that was phosphorylated only when the ion pump was operating. The carrier is now regarded as a phosphoprotein (in which the phosphate is bound to a serine group) which binds K^+ but not Na^+. On phosphorylation it will bind Na^+ but not K^+. Phosphorylation and Na^+ binding occur at the inner surface of the membrane (with ATP being converted to ADP). The system is then transported to the outer cell where it is dephosphorylated giving the original phosphoprotein which releases Na^+ and binds K^+ prior to its return to the inner surface. It is then phosphorylated. K^+ is released into the cell and Na^+ is bound prior to efflux from the cell. This is schematically represented in Figure 8.5. A feature of ion transport is the presence of Na^+, K^+ activated transport ATPase.

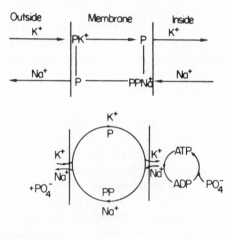

P = phosphoprotein

PP = phosphorylated protein

Figure 8.5 Schematic representations of the sodium pump.

There is still the question of the binding of Na^+ and K^+ by phosphorylated and dephosporylated protein respectively. Our earlier discussion of ligand preferences is relevant here. It is likely that carboxylate and phosphate groups are available on the phosphoprotein. Selectivity could be conferred by the hydrophobic nature of the protein side chains which would exclude the strongly hydrated species (Na^+, Ca^{2+}, Mg^{2+}). Such an environment, with negatively charged oxygen donors, could readily account for selective K^+ binding. The Na^+ selective site, one less hydrophobic, could then reject K^+ by the provision of a limited size hole for the metal binding site. The argument could be extended to cover M^{2+} ions. Thus Mg^{2+} would be favoured over Ca^{2+} by the presence of a nitrogen base.

The enzyme system involved in ion transport (Na^+, K^+ activated ATPase)
'Transport' ATPase was first noted by Skou,[51] bound to a membrane in the leg nerves of the shore crab *Carcinus maenas*. Attempts to remove it

from the membrane result in its inactivation. It is now generally accepted that this membrane enzyme system plays an essential role in the transport of Na^+ and K^+ across cellular membranes. It is extremely widespread. For maximum activity there is a requirement for Mg^{2+} in concentrations equivalent to that of the substrate, in addition to Na^+ and K^+. The actions of the monovalent cations are dependent upon each other. It is known that the active transport of Na^+ and K^+ is inhibited by the cardiac glycosides. These also affect the respiration of whole cells but not the mitochondrial enzyme systems. It is very significant that the cardiac glycosides (particularly the water soluble ouabain) also specifically inhibit the membrane bound transport ATPase system.

The original studies on ^{14}C labelled ATP indicated that an intermediate might exist during the hydrolysis of ATP by the Na^+, K^+ ATPase, which could have the role of ion carrier. This is the phosphoprotein earlier discussed. Much work has been carried out on this system, and it is now believed that Na^+ is necessary for the formation of the intermediate phosphorylated complex and K^+ is necessary for its subsequent dephosphorylation. There is evidence that ouabain inhibits the K^+-dependent step but it is not clear if it also inhibits the Na^+-dependent step.

There have been a number of attempts to postulate detailed models for the overall phenomenon of ion transport, concentrating particularly on the act of transport. The complexity of the cell membrane is a limiting factor on the validity of such models.

One scheme has been suggested by Opit and Charnock.[52] They have assumed that ion pairs are formed between the monovalent cations of the cell cytoplasm and the anionic side chains of the inner facing proteins of the membrane. To initiate the cycle a critical number of such sites absorb Na^+, the number normally absorbed being dependent on the intracellular sodium content. The attainment of this critical number of bound ions induces a redistribution of electron density along the polarizable protein backbone of the membrane so that the active centres of the Na^+/K^+ ATPase become reactive to the substrate MgATP. This results in the incorporation of the phosphoryl group into the enzyme protein (i.e. the phosphoprotein discussed earlier). The presence of this phosphorylated group results in a further redistribution of charge and a change in molecular configuration so that the anionic side chains become strongly K^+ preferring. In linked $Na^+ - K^+$ transport this must also result in a rotation of the protein molecule to face the opposite surface so allowing K^+ from outside the cell to replace the Na^+ originally there, the site now being K^+ specific. The result of this twist also renders the phosphorylated active site susceptible to hydrolysis, so that the phosphate is liberated and the membrane reverts to the original configuration carrying the K^+ with it, into the cell. As the Na^+ concentration in the cell increases again, once more sufficient Na^+ binding occurs to reinitiate the process. Supporting evidence for this scheme is suggested by Charnock and Opit in their review article.[50]

THE ROLE OF Ca^{2+} IONS IN CONTROL MECHANISMS

The exclusion of calcium ions from the cell under normal conditions means that a controlled influx of Ca^{2+} may be used in trigger or control mechanisms, for example in muscle contraction. Inside the cell there will be sites having a higher affinity for Ca^{2+} than for Mg^{2+}. An influx of Ca^{2+} will result in calcium ions competing effectively for these sites and hence initiating a new process, whether it be by specific enzyme activation or by some less well defined mechanism. Control of the permeability of cell walls is therefore the key feature of many trigger phenomena. Once the phenomenon has occurred the Ca^{2+} pump must reject Ca^{2+} so deactivating the relevant sites. A similar control mechanism will not occur through the influx of Na^+ as K^+ will bind more strongly than Na^+. In this section we will examine a number of phenomena resulting from the sudden appearance of Ca^{2+} in a cell.

Contraction in muscle[53]

Before considering the control of muscle contraction it is necessary to outline the sliding filament model[54] for muscle contraction. The major constituents of muscle fibres are the proteins myosin and actin, which constitute some 54% and 27% respectively of the structural protein of muscle. The remainder comprises tropomyosin (12%) and troponin and actinin. It is beyond the scope of this book to consider the function of these latter proteins in detail. We will limit it to that of myosin and actin.

Each muscle fibre is made up of some 1000 fibrils. Myosin and actin are organized in the fibrils in the form of two types of filament, one thick and the other thin. Myosin is a long, thin molecule of overall length 150 nm and includes a thicker portion or 'head' 20 nm long. The head is 4 nm in breadth, that is it is twice the thickness of the thin portion of the molecule. The thick filaments are made up of either 180 or 360 myosin molecules. The actin molecule is a globular protein, some 5·5 nm in diameter. The thin filaments are made up of double stranded chains of these actin molecules. The two proteins differ very much in their primary structure. Muscle contraction and relaxation result from the sliding of one filament over the other, as schematically depicted in Figure 8.6.

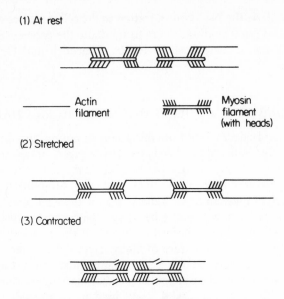

Figure 8.6 The sliding filament model for muscle contraction.

The key to understanding muscle contraction is that of the interaction between the two types of filaments, i.e. between the myosin heads and the actin monomers. Substantial interaction may occur between the filaments, the cross linkages so formed resulting in the locking of the muscle.

Associated with the myosin heads is the presence of a site for ATP hydrolysis which is activated by Ca^{2+}. This is the reaction that provides the energy for contraction. The addition of ATP and Mg^{2+} to filament systems in the absence of Ca^{2+} ions does not lead to the splitting of ATP. Mg^{2+} ions cannot activate the ATPase site. However, the MgATP complex prevents the formation of cross links, so allowing the actin filaments to slide past the myosin filaments without interaction. If calcium ions are then introduced into the system the ATPases on the heads of the myosin molecules are activated, ATP is hydrolysed and interactions can then occur between the filaments, resulting in the sliding of one filament against another and the contraction of the muscle. X-ray studies[55] on muscle have shown that there is a movement of the myosin heads which could result in a pushing or pulling effect. Provided ATP is present, the removal of Ca^{2+} ions will then reverse this effect.

Much effort has gone into the study of the biochemistry of muscle contraction, in particular with reference to the ATPase site on myosin. However, myosin and actin are very dependent upon each other and it is doubtful whether information obtained from the study of myosin in isolation may be extrapolated to muscle. It appears that myosin readily binds Ca^{2+} and ATP, and it has been suggested that they are bound at separate sites on myosin. The muscle system is also Mg^{2+} dependent. Thus, when the fibrils in muscle are examined for ATPase activity, it was found that, in the presence of Mg^{2+}, ATP is split very slowly, with no contraction. ATP is split readily in the presence of Ca^{2+} but again with no contraction, while Ca^{2+} and Mg^{2+} must both be present for ATP-splitting with contraction to occur.

The control mechanisms for the appearance and disappearance of Ca^{2+} ions in the muscle cells will be discussed in the following section.

Transmission of nerve impulses[56, 57]

The nervous system of higher organisms is usually discussed in terms of the central nervous system and the peripheral nervous system. The latter consists of the connecting nerve fibres, or axons, and is responsible for the transmission of signals between the central nervous system and the functional parts of the organism, that is 'input' from receptor organs and 'output' to muscles and glands. Each nerve cell is in contact with other nerve cells through synapses, at which point reception of signals occurs. This causes a response and a conduction of a signal *via* the axon. The actual site of excitation in the axon is the axon membrane. It has long been known that nerve activity is associated with the production of electric currents. Figure 8.7 shows a cross section of a nerve fibre. In the squid giant nerve fibres occur having a diameter of up to 1 mm, so allowing the study of nerve transmission by physical techniques.

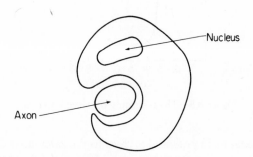

Figure 8.7   Cross section through a nerve fibre (schematic).

The nerve impulse can be regarded as an electrical impulse conducted along the membrane due to the potential difference between the inside and outside of the cell. In particular, it is regarded as a depolarization wave associated with a temporary reversal of the selectivity of the membrane, allowing an influx of Na^+. Normally there is a large excess of K^+ inside, and an excess of Na^+ and Cl^- outside. The resting cell membrane may be treated as a Cl^- or a K^+ electrode. In the latter case the potential is given by $E = (RT/F) \log_e[K^+]_{out}/[K^+]_{in}$. Some K^+ concentrations inside and outside nerve cells are included in Table 8.1. Using the values for the squid nerve cells, it may be shown that the resting potential should be -74 mV, and that on depolarization the value should be $+55$ mV. In Figure 8.8, the method[56] of recording these potentials is shown and a typical result given. If a nerve is stimulated some distance away from the recording electrodes, a delay occurs while the impulse is conducted to the electrodes. There is then a rapid change in the potential difference which dies away again as the impulse moves on past the electrodes. This is the action potential, while the shape depicted in Figure 8.8 is termed a spike. It may be seen

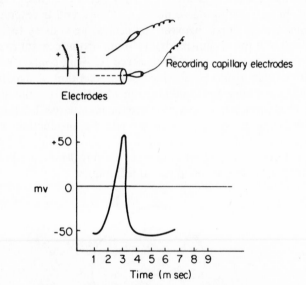

Figure 8.8   The recording of nerve impulses.

that the total change in potential is close to the calculated value (74 + 55 mV). The phenomena may be understood in terms of the change in potential difference, produced when an electrical impulse is used to depolarize the membrane, causing a locally depolarized region that affects the regions on either side of it. While the influx of Na^+ into the cell is rapid, the amount transferred is very small. The Na^+ will then be pumped out by the sodium pump, causing the fall of the spike.

The hypothesis that the action potential results from a reversal of the ion selectivity of the membrane was proved by showing that the excitability of isolated nerve fibres was lost when sodium was absent from the external medium. The change in intracellular cation distribution has also been directly observed by the use of sodium and potassium selective microelectrodes.

For the spike to be propagated there has to be a mechanism for the amplification of a partial depolarization of the membrane potential. However, this may be understood. As the nerve membrane is depolarized, so sodium ions will enter the axon and by virtue of their positive charge will reinforce the lowering of the initial resting potential. This will cause an increase in the sodium permeability and a further reinforcement of the lowering of the potential.

Transmission across synaptic gaps between nerve fibres

When the action potential reaches the end of a nerve fibre there is a different, non-electrical, mechanism for transmission. The impulse has to be transmitted across a gap of about 15–20 nm between the two cells. The use of the electron microscope has shown the presence of presynaptic vesicles which contain chemical neurotransmitters, such as acetylcholine. The impulse causes the ejection of the acetylcholine which diffuses across to the end plate membrane of the next axon and initiates a fresh action potential. However, an acetylcholine esterase is present in the end plate membrane of the nerve so that the acetylcholine is rapidly destroyed, so preventing too large a depolarizing effect. It has been shown that the liberation of acetylcholine is greatly reduced and the neuro-transmission blocked when the concentrations of Ca^{2+} ions and Mg^{2+} ions are lowered and raised respectively. This may be related to the membrane structure of the vesicle. Normally this is stabilized by Mg^{2+}. Hence a raised concentration of $[Mg^{2+}]$ will adversely affect the release of acetylcholine. Calcium ions will bind preferentially over magnesium ions, but normally are kept outside the cell. However, the depolarization effect that alters the selectivity of the membrane will also allow the entry of Ca^{2+} ions. These will bind strongly, displacing Mg^{2+} ions, and may cause[4] the membranes of the vesicle and the outer membrane to come together and collapse, so ejecting the contents of the vesicle. The acetylcholine then initiates a new impulse in the next axon.

Transmission across neuromuscular synaptic gaps

Here the intracellular space may be between 50 to 100 nm. Again, there is evidence for the presence of the neurotransmitter acetylcholine which diffuses across this gap and initiates muscle contraction. Experiments have shown that acetylcholine will depolarize the muscle fibre and set up impulses if applied to the synaptic region. Certain drugs that cause paralysis such as the curare alkaloids compete effectively with acetylcholine for the receptor site and produce a blockage of the neuromuscular impulse.

Muscle contraction will then be initiated by the release of Ca^{2+} ions from the longitudinal vesicles of the fibrils, followed by diffusion to the active site and the activation of the ATPase on the myosin. The fact that Ca^{2+} is released by the arrival of the action potential has been demonstrated[58] by the use of murexide. Toads were fed on murexide and injected with dimethyl sulphoxide to increase the permeability of membranes to large molecules. Hence the murexide passes *via* the blood stream into the sarcoplasm of the toad's muscle. Subsequent experiments show that an appropriate stimulus causes the release of Ca^{2+} into the sarcoplasm as demonstrated spectrophotometrically by the colour change resulting from murexide–Ca^{2+} interaction. Relaxation of the muscle then results from the active accumulation of Ca^{2+} by the membranes of the endoplasmic reticulum in striated muscle.

REFERENCES

1. H. H. Ussing, 'The Alkali Metal Ions in isolated systems and Tissues'. *Handbuch der Experimentellen Pharmakologie*, p. 1, Springer-Verlag, Berlin, 1960
2. P. Kruhoffer, J. H. Thaysen and N. A. Thorn, 'The Alkali Metal Ions in the Organism,' *Handbuch der Experimentellen Pharmakologie*, p. 196, Springer-Verlag, Berlin, 1960
3. *Regulation of the Inorganic Ion content of cells*, CIBA Foundation Study Group No. 5, 1960
4. R. J. P. Williams, *Quart. Rev. Chem. Soc.*, 1970, **24**, 331
5. C. H. Suelter, *Science*, 1970, **168**, 789
6. H. J. Evans and G. J. Sorger, *Ann. Rev. Plant Physiol.*, 1966, **17**, 47
7. P. R. Kernan, *Cell-Potassium*, Butterworths, London, 1965
8. C. P. Bianchi, *Cell-Calcium*, Butterworths, London, 1968
9. C. J. Pederson, *J. Amer. Chem. Soc.*, 1967, **89**, 7017; 1970, **92**, 386, 391
10. J. L. Dye, M. G. De Backer, V. A. Nicely, *J. Amer. Chem. Soc.*, 1970, **92**, 5226
11. D. St. C. Black and I. A. McLean, *Tet. Letters*, 1969, 3961
12. B. Dietrich, J. M. Lehn and J. P. Sauvage, *Tet. Letters*, 1969, 2885; 2889
    (a) J. M. Lehn and J. P. Sauvage, *Chem. Comm.*, 1971, 440
    (b) B. Metz, D. Moras and R. Weiss, *Chem. Comm.*, 1971, 444
    (c) B. Metz, D. Moras and R. Weiss, *J. Amer. Chem. Soc.*, 1971, **93**, 1806
13. M. A. Bush and M. R. Truter, *J. Chem. Soc. (A)*, 1971, 745; M. A. Bush, H. Luth and M. R. Truter, *J. Chem. Soc. (A)*, 1971, 740; M. R. Truter, *Chemistry in Britain*, 1971, **7**, 203
14. L. A. R. Pioda and W. Simon, *Chimia*, 1969, **23**, 72
15. M. S. Frant and J. W. Ross, *Science*, 1970, **167**, 987
16. C. Moore and B. C. Pressman, *Biochem. Biophys. Res. Comm.*, 1964, **15**, 562
17. R. J. P. Williams, *Analyst*, 1953, **78**, 586

18. R. M. Izatt, J. H. Rytting, D. P. Nelson, B. L. Haymore and J. J. Christensen, *Science*, 1969, **164**, 443
19. P. Mueller and D. V. Rudin, *Biochem. Biophys. Res. Comm.*, 1967, **26**, 398
20. J. H. Prestegard and S. I. Chan, *Biochemistry*, 1969, **8**, 3921
21. J. H. Prestegard and S. I. Chan, *J. Amer. Chem. Soc.*, 1970, **92**, 4440
22. H. Diebler, M. Eigen, G. Igenfutz, G. Mass and R. Winkler, *Pure Appl. Chem.*, 1969, **20**, 93
23. D. H. Haynes, A. Kowalsky and B. C. Pressman, *J. Biol. Chem.*, 1969, **244**, 502
24. V. T. Ivanov, J. A. Laine, N. D. Abdulaev, L. B. Senyavina, E. M. Popov, Yu. A. Orchinnikov and M. M. Shemyakin, *Biochem. Biophys. Res. Comm.*, 1969, **34**, 803
25. M. Pinkerton, L. K. Streinauf and P. Dawkins, *Biochem. Biophys. Res. Comm.*, 1969, **35**, 512
26. M. Ohnishi and D. W. Urry, *Ibid*, 1969, **36**, 194
27. R. Schwyzer, Aung Tun-Kyi, M. Caviezel and P. Moser, *Helv. Chim. Acta*, 1970, **53**, 15
28. F. W. Cope, *Proc. Natl. Acad. Sci. U.S.*, 1966, **54**, 225
29. T. L. James and J. H. Noggle, *J. Amer. Chem. Soc.*, 1969, **91**, 3424; *Proc. Natl. Acad. Sci. U.S.*, 1969, 644
30. R. G. Bryant, *J. Amer. Chem. Soc.*, 1969, **91**, 1870
31. F. J. Kayne and J. Reuben, *J. Amer. Chem. Soc.*, 1970, **92**, 220
32. J. P. Manners, K. G. Morallee and R. J. P. Williams, *Chem. Commun.*, 1970, 965
33. P. J. Gehring and P. B. Hammond, *J. Pharmacol.*, 1964, **145**, 215
34. A. S. Mildvan and M. Cohn, *J. Biol. Chem.*, 1965, **240**, 238; 1966, **241**, 1178
35. M. Cohn, *Q. Rev. Biophysics*, 1970, **3**, 61
36. G. L. Cottam and R. L. Ward, *Arch. Biochem. Biophys.*, 1969, **132**, 308
37. E. J. Peck and W. J. Ray, *J. Biol. Chem.* 1969, **244**, 3748
38. K. G. Morallee, E. Nieboer, F. J. C. Rossotti, R. J. P. Williams, A. V. Xavier, and R. A. Dwek, *Chem. Commun.*, 1970, 1132
39. E. R. Birnbaum, J. E. Gomez and D. W. Darnall, *J. Amer. Chem. Soc.*, 1970, **92**, 5288
40. M. Takata, W. F. Pickard, J. Y. Lettvin and J. W. Moore, *J. Gen. Physiol.*, 1966, **50**, 461
41. C. Van Breemen and P. De Weer, *Nature*, 1970, **226**, 760
42. D. W. Tempest, *Symp. Soc. Gen. Microbiol.*, 1969, **19**, 87
43. J. S. Edelman, P. O. P. Tso, J. Vinograd, *Biochim. Biophys. Acta*, 1960, **43**, 393
44. R. C. Gordon and R. A. McLeod, *Biochem. Biophys. Res. Comm.*, 1966, **24**, 684
45. R. J. P. Williams and W. E. C. Wacker, *J. Amer. Med. Assoc.*, 1967, **20**, 18
46. M. A. Asbell and R. G. Eagon, *J. Bact.*, 1966, **92**, 380
47. R. D. Keynes, *Q. Rev. Biophys.*, 1969, **2**, 177 and references cited
48. E. A. C. MacRobbie, *Q. Rev. Biophys.*, 1970, **3**, 251
49. V. Koefoed and H. H. Ussing, *Acta Physiol. Scand.*, 1958, **42**, 298
50. J. S. Charnock and L. J. Opit, in *The Biological Basis of Medicine*, Bittar and Bittar (Eds), Academic Press, 1968, 69
51. J. C. Skou, *Biochim. Biophys. Acta*, 1957, **23**, 394; J. C. Skou, *Physiol. Rev.*, 1965, **45**, 596
52. L. J. Opit and J. S. Charnock, *Nature*, 1965, **208**, 471
53. J. R. Bendall, *Muscles, Molecules and Movement*, Heinemann, London, 1969
54. J. Hansen and H. E. Huxley, *Symp. Soc. Exp. Biol.*, 1955, **9**, 228
55. H. E. Huxley and W. Brown, *J. Mol. Biol.*, 1967, **30**, 383

56. A. L. Hodgkin, *Conduction of the nervous impulse*, Liverpool University Press, 1967
57. B. Katz, *Nerve, Muscle and Synapse*, McGraw Hill, New York, 1966
58. F. F. Jobsis and M. J. O'Connor, *Biochem. Biophys. Res. Comm.*, 1966, **25**, 246

CHAPTER NINE

METAL IONS AND CHELATING AGENTS IN MEDICINE

The concentrations of the naturally occurring metal ions in living systems are carefully controlled within fine limits. This control in general is exercised by certain proteins and hormones. Clearly, disorders must arise if this balance is upset. Conversely, the interrelationships between metal ions and binding substances in the body are so complex that disorders or diseases involving the metal-binding substances may result in the presence of high or low concentrations of metal ion, compared with that normally present. Accordingly, analysis of body fluids and tissues for trace metal ions could become an important diagnostic procedure. It does appear that a number of major diseases are associated with changes in concentration of the trace metal ions in certain tissues and body fluids, while it has long been appreciated that the balance of alkali metals in and out of the cell is affected in many diseased conditions.

A trace metal ion that is essential for the activity of enzyme systems may well become toxic if the concentration is raised above certain limits. The level beyond which a metal ion becomes toxic is dependent upon its location. Thus a doubling of the extracellular concentration of K^+ leads to heart disorder and possibly death, but this concentration is still well below the normal concentration of intracellular K^+.

A deficiency of any metal ion, for example zinc or the transition elements, will result, in the appropriate cases, in reduced enzyme activity with a breakdown of the normal metabolic processes.

It appears that the role of metal ions in medicine can be considered under three headings: (a) metal poisoning, where excess metal is ingested, (b) abnormalities resulting from a breakdown of the body's own control system and (c) the diagnostic analysis of metal ions in body tissues and fluids.

(a) Metal poisoning

The presence of excess metal ions or foreign metal ions may be dangerous. Thus lead, mercury and arsenic poisoning are well-known occupational hazards. Iron poisoning is also common and leads to the condition known as siderosis. This is a common affliction among the Bantu tribe in South Africa where it results from drinking beer brewed in iron pots. When taken orally only a small fraction of the metal ion will be absorbed from the gastrointestinal tract and some may be filtered off by the kidneys, but enough may be absorbed to displace the normal catalytic metal ions from their binding site on a protein and so disrupt enzyme action. Alternatively, it may disrupt membrane and transport properties. Thus, in the case of poisoning by lead, mercury or arsenic, there is a tremendous increase in the excretion of copper and zinc, which have been displaced from their normal binding site by the foreign ion. Metal poisoning of this type is therefore treated by the use of chelating agents which will bind more strongly to the metal or metalloid than do the normal metal-binding groups of the apoprotein.

Poisoning involving the alkali and alkaline earth cations has a rather different underlying reason for its effect, that is the effect of the metal ion upon the osmotic balance of the system, and also the binding of the metal ion by negative centres of the protein or other macromolecules. The behaviour of the metal ion is not unusual: it is a concentration effect.

(b) Failure of the metal ion control system
It is not always possible to decide whether an abnormal metal ion concentration is the cause or the result of a diseased condition. Thus Wilson's disease (hepatolenticular degeneration) results from an accumulation of copper in the tissues, particularly in the liver, brain and kidneys. The extent of the symptoms is dependent upon the amount of copper present and may be reversed by the use of chelating agents. The actual mechanism for the *deposition* of copper in the tissues is not certain, however, although the disease itself is genetically induced. The symptoms may be controlled by the use of metal chelating agents, in particular penicillamine (Table 9.1) or $Na_2CaEDTA$. The calcium salt of EDTA is used to prevent EDTA labilizing calcium from the skeleton by combining with calcium ions and hence disturbing the equilibrium distribution of these ions at the ultimate expense of the calcium of the skeleton. There are problems in the use of EDTA. It has to be injected. In addition, it cannot act directly by chelation of the undesirable metal ion on the enzyme site, but has to operately indirectly by removing the metal ion in solution so causing the release of the metal ion on the enzyme. Treatment with EDTA is therefore a long drawn-out process.

A further example involves the treatment of calcium deposition. The blood is rich in calcium salts, as they are rejected by the cell, but these salts are not very soluble (N.B. a size effect of Ca^{2+}) and so they are precipitated, resulting in the formation of stones, the hardening of arteries and the accumulation of calcium in cataracts in the eye. Cholesterol is also involved in the formation of gall stones and the development of atherosclerosis. These conditions have all been treated with EDTA, with some measure of success, in order to chelate out the calcium ions.

A number of cases involving metal ion deficiency will be considered later. One example will be considered at this stage. Copper deficiency produces defects in the structural stabilization of the fibrous protein of connective tissues in a number of animals. A number of abnormalities are associated with this. Of particular note is the fact that animals of a number of species fed on copper-deficient diets died as a result of structural lesions of major arteries. This structural weakness arises from the absence of cross-linking between polypeptide chains. These cross-linkages result from condensation reactions following the oxidative deamination of amino groups of lysyl side chains. It is thought that this latter reaction is catalysed by a copper-containing enzyme and so is inhibited by copper deficiency.

(c) Metal ion concentration as a diagnostic test

Of the metal ions that have been examined in detail in this context, copper appears to be the most valuable pointer to diseased conditions. Thus for infectious hepatitis, serum copper levels up to three times the normal value of 350 $\mu g/100$ ml have been observed. This is due to an accumulation of ceruloplasmin. Other diseases associated with high copper concentrations in the blood are leukemia, lymphomas, rheumatoid arthritis, psoriasis, cirrhosis, nephritis and Hodgkin's disease. High copper levels are associated with other phenomena, but clearly the presence of a high serum copper concentration is of diagnostic value when used in conjunction with other tests. Copper analysis can also be used to help monitor the effectiveness of treatment for certain of these conditions, as the copper level is directly proportional to the severity of the disease. This situation holds for hepatitis and for malignancies. In the latter case it iş also noteworthy that the increase in serum copper level appears before any clinical changes are apparent.

DRUG ACTIVITY

In this chapter we are specifically concerned with the medicinal aspects of metal ions and drugs which act *via* chelation of metal ions. It should be emphasized that texts on pharmacology should be consulted for a detailed discussion of the basis of drug activity. In general, drugs act by controlling the action of cells (particularly nerve and muscle cells) by acting on subcellular structures or by disturbing enzyme behaviour.

There are many drugs which are suspected of acting *via* chelation. A study of Table 9.1 indicates that many of the drugs listed there will act

TABLE 9.1 The structure of some drugs

Phenacetin

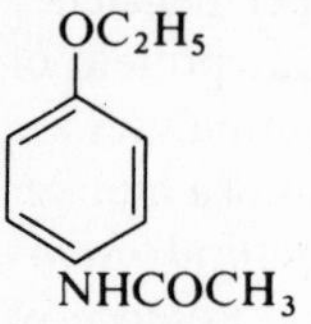

OC$_2$H$_5$

NHCOCH$_3$

(analgesic)

Aspirin

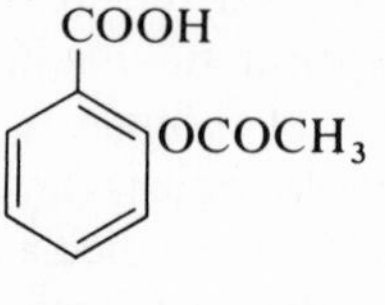

COOH

OCOCH$_3$

(analgesic)

Amphetamine

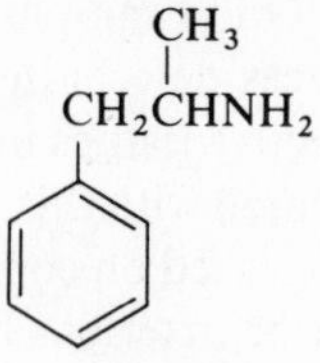

CH$_3$

CH$_2$CHNH$_2$

(stimulant)

Isoproniazid

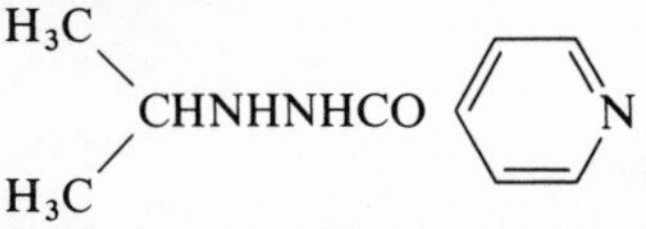

H$_3$C

    CHNHNHCO⟨N⟩

H$_3$C

(T.B. drug, antidepressant)

Isonicotinic acid hydrazide

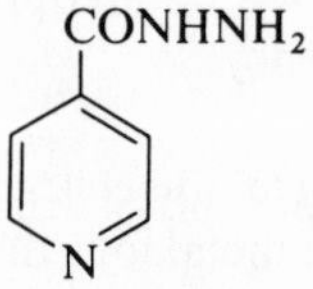

CONHNH$_2$

⟨N⟩

(T.B. drug)

Nialamide

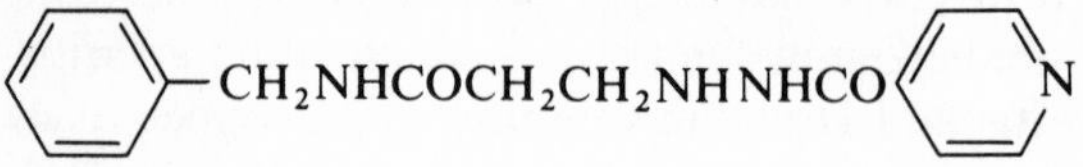

⟨⟩—CH$_2$NHCOCH$_2$CH$_2$NHNHCO⟨N⟩

(antidepressant)

Sulphanilamide

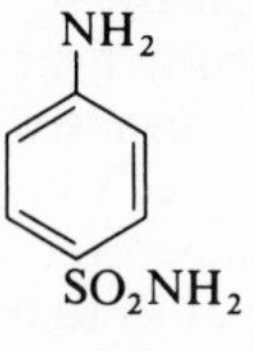

NH$_2$

SO$_2$NH$_2$

(bactericide)

Acetazolamide

(carbonic anhydrase inhibitor)

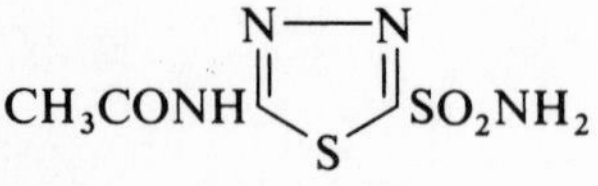

CH$_3$CONH⟨N——N / S⟩SO$_2$NH$_2$

(anticonvulsant, diuretic)

Dimercaprol

(British Anti Lewisite)

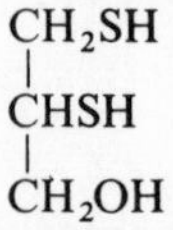

CH$_2$SH

CHSH

CH$_2$OH

(in metal poisoning)

Ferrioxamine B

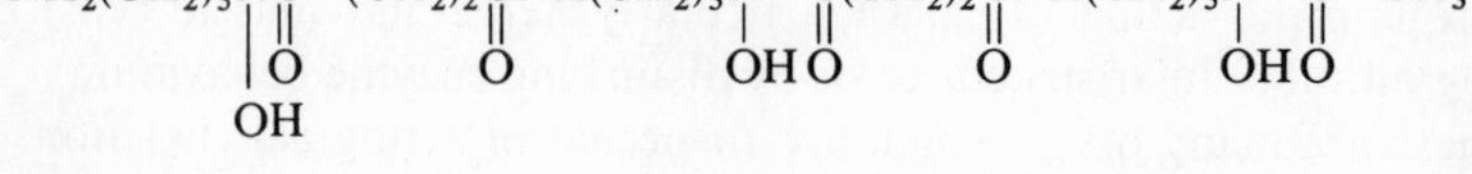

NH$_2$(CH$_2$)$_5$NC—(CH$_2$)$_2$CNH(CH$_2$)$_5$N—C(CH$_2$)$_2$CNH(CH$_2$)$_5$N—C—CH$_3$

$\quad\quad$ | $\parallel$ $\quad\quad\quad$ $\parallel$ $\quad\quad\quad\quad$ | $\parallel$ $\quad\quad\quad$ $\parallel$ $\quad\quad\quad\quad$ | $\parallel$

$\quad\quad$ O $\quad\quad\quad$ O $\quad\quad\quad\quad$ OH O $\quad\quad$ O $\quad\quad\quad\quad$ OH O

$\quad$ OH

(iron chelator)

TABLE 9.1 *continued*

Kojic acid

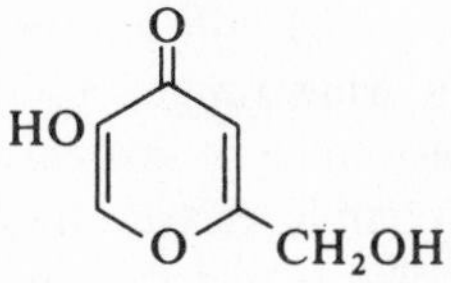

HO — (structure) — O — CH$_2$OH

(antibiotic)

D.M.D.C.

$$(CH_3)_2N-\overset{\overset{\textstyle S}{\|}}{C}-SH$$

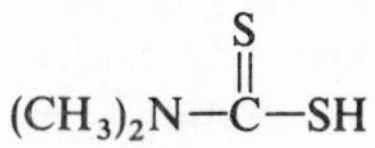

(antibiotic)

Ethambutol

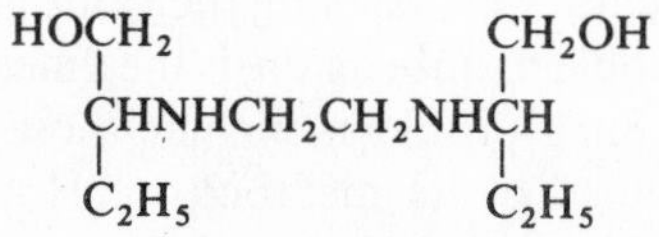

$$\underset{C_2H_5}{\underset{|}{\overset{HOCH_2}{\overset{|}{CHNHCH_2CH_2NHCH}}}} \quad \underset{C_2H_5}{\underset{|}{\overset{CH_2OH}{\overset{|}{}}}}$$

(T.B. drug)

as good ligands *in vitro*. However, it is not really possible to relate the activity of a drug to its *in vitro* properties. Thus *in vivo* a potentially chelating drug has to compete for a metal ion with a wide range of other ligands and there is no guarantee that it will do this effectively. On the other hand, there are well established cases where the medicinal behaviour of a compound may be attributed to the control or inhibition of enzyme activity *via* metal ion chelation. An example of this is disulfiram (tetra-ethylthiuram disulphide) which is used in the treatment of chronic alcoholism in cases where the patient wants to be cured. The drug inhibits the metalloenzyme aldehyde oxidase and so the metabolism of ethanol stops with the formation of acetaldehyde, so producing unpleasant symptoms and discouraging further indulgence.

disulfiram

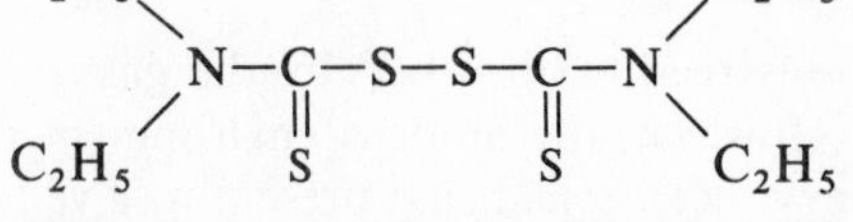

$$\underset{C_2H_5}{\overset{C_2H_5}{N}}-\overset{\overset{\textstyle}{}}{C}-S-S-\overset{}{C}-N\overset{C_2H_5}{\underset{C_2H_5}{}}$$

Normally the properties of a drug are associated with the shape of the molecule (or part of the molecule) and its ability to combine with a receptor site. Thus the role of the sulphonamides in controlling infections is associated with the presence of a *para*-NH$_2$ group which will fit in the bacterial enzyme receptor and so prevent the growth and reproduction of the bacteria. The activity of a drug may be inhibited by substances which compete for the receptor with the drug, or with the substrate receptor site for the drug site.

Certain metal complexes, usually with substituted 1,10-phenanthroline ligands, have a considerable activity against a wide range of micro-organisms, including some which have a high resistance to normal anti-biotics. The pharmacological properties of these complexes are a function of the complex cation as a whole, its shape and charge distribution. The presence of organic ligands and additional substituents helps to confer liposolubility on such a complex and hence favour its dissolution in and passage through the biological membrane into the cell.

There are many difficulties in assessing the value and mode of action of new drugs. The basic problem, once again, is the difference between *in vitro* and *in vivo* behaviour. Such problems include those of the side effects of the drug, or the products of its metabolism. How easily is the drug absorbed? How stable is it *in vivo*? How and where is the drug bound in its cycles around the body? The pH of the body solution and the pK_a of the drug will determine to what extent it is ionized *in vivo*. This is very important as in many cases biologically active species are weak acids or weak bases and only one form is active. Again, only unionized species are liposoluble. Sometimes the drug itself is not the active species, the pharmacologically active form is one that is metabolized from it. One drug may also interact with another. They may reinforce each other's activity, or they might have dangerous properties when together. All these are questions which must be answered. Finally there is the question of the readiness with which the body will develop a tolerance to the drug.

In the following sections we will enlarge upon each of the areas that have been touched upon in the introduction to this chapter.

THE CONTROL OF METAL ION CONCENTRATION
IN VIVO

The removal of metal ions

The only metal ions readily absorbed by the gastrointestinal tract are Na^+, K^+, Ca^{2+}. However, the modern environment may well result in enough of the heavy metal ions being present to give toxic effects. These in general will be 'soft acids' and so will prefer to bind to 'soft bases', in particular the SH groups of enzymes, displacing the active metal. Lead, arsenic, mercury, iron and beryllium poisoning have all been treated with a range of chelating agents. The requirement is that of a multidentate chelate to coordinatively saturate the metal ion to be removed, one that will give a soluble, easily excreted complex, while both chelating agent and metal complex must be non-toxic.

Reference has already been made to problems associated with the use of EDTA. Another important chelating agent in medicine is BAL, British Anti Lewisite or Dimercaprol. This was developed as a protection against the poison gas Lewisite, $Cl-CH=CH-AsCl_2$, which acted by binding to the SH groups of certain enzymes with resulting inactivation. However BAL is an efficient antidote as it binds more strongly to the arsenic than the enzyme and so binds it preferentially, the resulting complex being excreted.

British Anti Lewisite   *Penicillamine*

CH_2SH $HS-C(CH_3)_2$
| |
$CHSH$ $H_2N-CHCOOH$
|
CH_2OH

$$\begin{matrix} -S \\ \diagdown As \\ -S \diagup \end{matrix} + \begin{matrix} HS-CH_2 \\ | \\ HS-CH \\ | \\ HO-CH_2 \end{matrix} \rightarrow \begin{matrix} -SH \\ \\ -SH \end{matrix} + As\begin{matrix} S-CH_2 \\ | \\ S-CH \\ | \\ HO-CH_2 \end{matrix}$$

BAL has been used for treating As, Hg, Au, Te, Tl and Bi poisoning but is ineffective against lead. All these chelating agents will tend to be toxic in high quantities due to the removal of other metal ions. The big drawback in the use of such chelating agents is their lack of selectivity.

Iron is a well known 'tonic' and is used in treating anaemia. However, excess iron has a number of toxic effects, including irritation of the mucosa, giddiness, diarrhoea and may cause cardiac collapse. Desferrioxamine (Table 9.1) has been used to remove excess iron. This has a high affinity for it and so prevents its absorption. Desferrioxamine is a polyhydroxamic acid that is found as a complex with iron in certain microorganisms. This in itself indicates some specificity.

Penicillamine is used for removing toxic excesses of copper, usually in cases of Wilson's disease. Normally the body copper content is between 100–150 mg, the copper being found in highest concentration in the locus caeruleus in the brain stem. It is of interest that copper is found in particularly large amounts in the liver of newborn children, where it is stored in a protein having 2% copper content. This has dropped to the normal level at three months and the liver has developed the ability to synthesize the copper protein ceruloplasmin.

Copper concentrations up to one hundred times greater than normal have been observed in patients with Wilson's disease. The copper is deposited in a number of tissues but in particular is found in the liver, brain and kidneys. It may be seen as brown or green rings in the cornea. The presence of this high copper concentration leads to liver and nervous system disorders, usually in that sequence. The neurological disorder is associated with a severe tremor.

Penicillamine (as the D-isomer) is the most effective treatment. A dose of 1–2 g daily will lead to the excretion of 8–9 mg of copper in new patients, although this will fall as the treatment progresses and the excess copper is used up. Fortunately, the treatment does not appear to deplete normal stores of copper. As the symptoms are reversed by this treatment it appears that the brain damage is a biochemical rather than a structural one. One recent suggestion is that the effect of the Cu is to inhibit a cell-membrane ATPase

The origin of the high copper concentrations is not clear at present. It appears that the accumulation of copper is associated with a deficiency of the copper protein ceruloplasmin. Copper is stored in certain copper proteins in the liver and it has been suggested that transfer to apoceruloplasmin does not occur due to genetically-controlled deficiency. Eventually free copper ions penetrate the cell walls.

EDTA has been successfully used in the treatment of lead and vanadium poisoning, while beryllium poisoning has been treated by aurinetricarboxylic acid. Sodium salicylate has also been used.

Aurinetricarboxylic acid

Deficiency of metal ions

The best known example is probably that of anaemia in which there is a deficiency of iron, the red cells of the blood containing less hemoglobin than is normal. The deficiency may be associated with a poor diet, or *via* loss through bleeding. Pernicious anaemia results from a deficiency of vitamin B_{12} and may be treated with cyanocobalamin. Sickle-cell anaemia is restricted to equatorial Africa and is associated with a genetically controlled defect of hemoglobin synthesis. Patients suffering from sickle cell anaemia have abnormal hemoglobin, differing in a single amino acid residue from normal hemoglobin. Sickle cell anaemia is also associated with a resistance to malaria.

Iron is absorbed as Fe(II) in the duodenum, oxidized to Fe(III) and bound by apoferritin giving ferritin. The availability of apoferritin is a limiting factor on iron absorption. The iron is released from the ferritin store into the blood as Fe(II), where it combines with a globulin to give transferrin, again as Fe(III). As such it is transported to the bone marrow where it is eventually incorporated into hemoglobin. It is stored in the marrow as ferritin and hemosiderin. It is also stored in these forms in the liver, spleen and kidney. Iron released during erythrocyte destruction is also redistributed as transferrin. The iron stores may become overloaded if treatment with iron is unnecessarily prolonged, leading to problems of toxicity.

In treatment, ferrous sulphate is used, with appropriate coatings to protect from moisture and oxidation. Ascorbic acid is sometimes given with it to aid absorption. Chelate complexes have been used, such as ferric ammonium citrate, in the hope that the iron would be released slowly avoiding a build-up of concentration while the slow conversion to the ferrous form will prevent too rapid an absorption.

DRUGS THAT APPEAR TO INVOLVE INTERACTION WITH METAL IONS

The cardiac glycosides

Digitalis and a number of closely related drugs have a specific and powerful action on the heart. Digitalis itself is the dried leaf of the foxglove plant, while its active constituents are the glycosides. They are found in a number of plants having digitalis action. Digitalis affects heart muscle, which is much more sensitive to it than are other muscles, causing an increase in the force of contraction, with a resulting slowing of the cardiac rate from the rapid, irregular and ineffective beat of the failing heart. Toxic doses of digatalis result in an excessive slowing of the heart rate.

It will be recalled that reference to the cardiac glycosides has been made in the context of the distribution of the alkali and alkaline earth metal cations in the cell. In particular it was noted that ouabain is an effective inhibitor of the active transport of ions across cell membranes by interfering with the transport ATPase. Studies on heart muscle preparations show that the activity of the cardiac glycosides is modified by altering the K^+ concentration, while toxic doses cause a loss of K^+ from the heart and a gain in Na^+. The symptoms resulting from an overdose of cardiac glycosides can be treated by increasing the plasma K^+ concentration. Similarly, depletion of plasma K^+ concentration in patients treated with digitalis may induce toxicity. There are a number of other parallels that indicate that the action of cardiac glycosides is due to inhibition of transport ATPase. Thus, magnesium catalyses transport ATPase and decreases the toxic action of the cardiac glycosides. It is not possible at present to fully understand the action of the cardiac glycosides purely in terms of the inhibition of the cardiac transport ATPase system.

The contraction of muscle is dependent upon the presence of calcium ions in the muscle cells. It is known that ouabain produces an increase of calcium uptake into beating heart muscle; again the mechanism is uncertain. However, a decrease in calcium concentration diminishes the toxicity of the cardiac glycosides while an increase in calcium concentration has the reverse effect. Calcium and digitalis provide an example of the synergic effect. The symptoms produced by over-digitalization have been reduced by the injection of the sodium salt of EDTA due to the resulting decrease in plasma calcium concentration.

The urinary system—mercurial diuretics

The diuretics are drugs that promote the formation of urine. The mercurial diuretics are particularly well known and are very effective. All are derivatives of mercuripropanol $R-CH_2-CH-CH_2-HgX$, where R is
$$\underset{OH}{|}$$
a polar, hydrophilic group. The mercurials are given by intramuscular injection, as they are too toxic for intravenous injection and too poorly absorbed to be given orally. The organogroups confer liposolubility on the mercury. The drug is taken up selectively in the tissues. All the organomercurial diuretics are broken down in the kidneys, giving mercuric ions, which may be the active agent. It is quite clear, however, that the mercurials operate by inhibiting the SH-containing enzymes. They are known, for example, to inhibit ATPase and succinic dehydrogenase in kidney slices. This same property of mercurials has resulted in their being used to control bacterial infections, where again they combine with SH groups in the bacteria.

Carbonic anhydrase inhibitors

Acetazolamide (Table 9.1) and related drugs have a diuretic action probably by specifically inhibiting the action of carbonic anhydrase by binding to the zinc of that enzyme.

The coagulation of blood

Blood is composed of plasma (water, organic solutes, inorganic ions and proteins, including albumin, globulin and fibrinogen) and cellular elements (platelets, erythrocytes and leucocytes). Na^+ and K^+ levels in blood are regulated by the cells of the renal tubules, under the influence of the adrenal cortical hormones. Kidney failure results in a rise in K^+ concentration, with resulting toxic effects. Depletion of Na^+ occurs through perspiration and leads to cramp and fatigue. Fibrinogen is the protein concerned with the coagulation of the blood, which occurs when tissues are damaged. A clot is formed by the thrombin catalysed polymerization of fibrinogen into fibrin. Platelets have a specific role in causing bleeding to stop. The processes involved in coagulation are as follows:

$$\text{prothrombin} \xrightarrow[\text{Ca}^{2+}]{\text{thromboplastin}} \text{thrombin}$$

$$\text{fibrinogen} \xrightarrow{\text{thrombin}} \text{fibrin}$$

The formation of thromboplastin also involves Ca^{2+}.

It is obviously of importance under certain conditions to have drugs which will prevent the coagulation of blood, for example the prevention of thrombosis. One approach would be to remove the calcium. This is done *in vitro* by precipitating the calcium as an insoluble salt, but clearly this cannot be done *in vivo*. It is really a question of controlling the calcium ion concentration. This has been done with EDTA and sodium citrate. Unusually high concentrations of the antibiotics, the tetracyclines also affect blood coagulation, because of their high affinity for calcium.

ANTIMICROBIAL DRUGS

Many of these are obviously good ligands. The above mentioned effect of the tetracyclines on calcium dependent processes illustrates this point. It appears therefore, that the activity of these antimicrobial drugs, or at least some of them, is dependent upon metal ion complexing. It is known in some cases that their efficiency is increased when coordinated. It may be that the metal complex is more liposoluble than the drug, and so the metal ion helps to transport it across the cell membrane. Alternatively, in other cases, the metal ion itself may be the toxic agent and the role of the coordinated antibiotic is one of carrier across the membrane.

Oxine (8-hydroxyquinoline) has antifungal and antibacterial properties when chelated to Fe(III). Albert has shown that the ligand and Fe(III) are both inactive when separated from each other, but are highly effective together, particularly when present in the 1:1 molar ratio. Albert has suggested that the 3:1 complex alone is capable of penetrating the cell membrane. Presumably the coordinatively unsaturated 2:1 and 1:1 complexes bind too strongly to the membrane. Once the 3:1 complex is inside the cell it must break down to the 2:1 and 1:1 species. It appears that the metal is necessary for the entry of the oxine molecule into the cell. In support of the suggestion that liposolubility is an important effect in determining the toxicity of this complex, it should be noted that the substitution of lipophobic groups into the oxine results in a decrease in antibacterial action.

The antitubercular drug isoniazid (Table 9.1) acts by interfering with the metabolism of the tubercle organism. However, the chelated drug is more effective, again presumably due to the fact that the complex is more liposoluble than the drug itself. A similar reason probably accounts for the effectiveness of copper ions in enhancing the activity of the antitubercular drug thiacetazone (*p*-acetamidobenzaldehyde thiosemicarbazone).

Thiacetazone

$$CH_3CONH \quad \text{—} \quad CH{=}NNH \cdot \overset{\overset{\displaystyle S}{\|}}{C}NH_2$$

Other examples where the antibacterial action of drugs is enhanced by metal ions are kojic acid, bacitracin and sodium dimethyldithiocarbamate (D.M.D.C.). The last named drug requires copper. Structures are given in Table 9.1.

The possibility of metal ion involvement in the tetracyclines has attracted some attention. It has been shown that there is a correlation between the possession of antibacterial properties and the ability to form stable chelates with calcium ions. A similar correlation has been drawn between active tetracyclines and the ability to form 2:1 complexes with Cu^{2+}, Ni^{2+} and Zn^{2+}.

The action of tetracyclines is directed towards the ribosomes of the microorganism cells. These are dependent upon Mg^{2+} for their structural integrity. The TB drug ethambutol is also known to act by interference with a bacterial ribosome Mg^{2+}-spermidine complex. It appears likely that the chelation of Mg^{2+} is involved in the activity of these drugs. Thus the inhibition of the growth of *E. coli* by tetracyclines may be reversed by the addition of high concentrations of Mg^{2+}.

The antibacterial properties of hexachlorophane are dependent upon the presence of a bridge, such as methylene, linking the 2′, 2 positions. This will then allow the two phenol groups to coordinate.

Hexachlorophane

The monoamine oxidase inhibitors

These are used as stimulants. They are often derivatives of hydrazine. There is a correlation between *in vitro* monoamine oxidase inhibiting power and the clinical effectiveness of these compounds. That they are good chelating agents *in vivo* is demonstrated by one side effect, the decalcification of bone. Presumably therefore the inhibitors act by binding to the copper of monoamine oxidase.

THE USE OF SYNTHETIC METAL CHELATES

One example, that of the iron oxine complex, has been given. Others are discussed by Dwyer and Shulman. There are metal chelates that are active at low concentration against a range of bacteria, fungi and viruses. These complexes are coordinatively saturated and so operate by physical interactions. The size, charge distribution, shape and redox potentials of metal chelates can be varied regularly. They are, therefore, useful reagents with which to investigate pharmacological problems. It is clear that metal chelates can act in a number of ways. Thus they may inactivate a virus by occupying sites on its surface which would normally be utilized in the initiation of the infection of the host cell. The first step in the infection would be an adsorption reaction involving electrostatic interactions. Alternatively, the complex cations may penetrate the cell wall and prevent virus reproduction.

The importance of the receptor site principle is demonstrated by the fact that the toxicity of certain of these cationic complexes is matched by onium salts of the ligand. It is also demonstrated in the toxic action of certain chelates towards mice and rats. Thus *tris* (2,2′ bipyridyl) iron(II) sulphate and *tris* (1,10-phenanthroline) ruthenium(II) iodide cause paralysis with general symptoms similar to those of the curareform drugs.

The active component of curare is D-tubocurare which deactivates the membrane of the motor end plate of a nerve to the transmitter substance acetylcholine, so causing paralysis.

Acetylcholine has a charged head: $Me_3 \overset{+}{N} CH_2 CH_2 OCOMe$. The effective ionic radius of the head is about 0·320 nm. The receptor site for acetylcholine is assumed to include a depression to receive the cationic head, with maximum interaction, together with suitable groups to interact with the carbonyl oxygen and to hydrogen bond with the ether oxygen. D-tubocurare is a large molecule, having two positively charged centres. Its toxic action is probably due to a blocking of the acetylcholine receptor site.

The cationic complexes again probably act by competing with acetylcholine for the receptor site. Neutral complexes have no effect. The effectiveness of any cationic complex is associated with the overall peripheral charge of the complex. Thus $[Co(NH_3)_6]^{3+}$ is twice as potent as $[Co(NH_3)_5 NO_2]^{2+}$.

GENERAL REFERENCES

M. J. Seven and L. A. Johnson (Eds), *Metal binding in Medicine*, Lippincott, Philadelphia, 1960

M. J. Seven and L. A. Johnson (Eds), *Biological Aspects of Metal Binding*, Fed. Proc., 1961, Supp. 10

A. Shulman and F. P. Dwyer, in *Chelating Agents and Metal Chelates*, F. P. Dwyer and D. P. Mellor (Eds), Academic Press, New York, 1964, 383

J. Peisach, P. Aisen and W. E. Blumberg (Eds), *The Biochemistry of Copper*, Academic Press, 1966. This contains a number of relevant articles, p. 175, 475, 503

A. C. Cotzias and A. C. Foradoris, in *Biological Basis of Medicine*, Bittar and Bittar (Eds), Academic Press, New York, 1968, **1**, 105

A. Albert, *Selective Toxicity*, 4th Edition, Methuen, London, 1968

W. C. Bowman, M. J. Rand and G. B. West, *Textbook of Pharmacology*, Blackwell, Oxford, 1970

Symposium on Trace Elements in the Metabolism of Connective Tissue, *Fed. Proc.*, 1971, **30**, 983. This includes reports on the role of iron, copper, manganese and zinc, together with a general review on the implications of trace metals in human diseases by J. T. McCall, N. P. Goldstein and L. H. Smith

C. L. Comar and T. Bronner (Eds), *Mineral Metabolism: An Advanced Treatise*, Academic Press, New York, 1962, **2**, 371

INDEX